Lecture Notes in Physics

The Editorial Policy for Proceedings

The series Lecture Notes in Physics reports new developments in physical research and teaching – quickly, informally, and at a high level. The proceedings to be considered for publication in this series should be limited to only a few areas of research, and these should be closely related to each other. The contributions should be of a high standard and should avoid lengthy redraftings of papers already published or about to be published elsewhere. As a whole, the proceedings should aim for a balanced presentation of the theme of the conference including a description of the techniques used and enough motivation for a broad readership. It should not be assumed that the published proceedings must reflect the conference in its entirety. (A listing or abstracts of papers presented at the meeting but not included in the proceedings could be added as an appendix.)

When applying for publication in the series Lecture Notes in Physics the volume's editor(s) should submit sufficient material to enable the series editors and their referees to make a fairly accurate evaluation (e.g. a complete list of speakers and titles of papers to be presented and abstracts). If, based on this information, the proceedings are (tentatively) accepted, the volume's editor(s), whose name(s) will appear on the title pages, should select the papers suitable for publication and have them refereed (as for a journal) when appropriate. As a rule discussions will not be accepted. The series editors and Springer-Verlag will normally not interfere with the detailed editing except in fairly obvious cases or on technical matters.

Final acceptance is expressed by the series editor in charge, in consultation with Springer-Verlag only after receiving the complete manuscript. It might help to send a copy of the authors' manuscripts in advance to the editor in charge to discuss possible revisions with him. As a general rule, the series editor will confirm his tentative acceptance if the final manuscript corresponds to the original concept discussed, if the quality of the contribution meets the requirements of the series, and if the final size of the manuscript does not greatly exceed the number of pages originally agreed upon.

The manuscript should be forwarded to Springer-Verlag shortly after the meeting. In cases of extreme delay (more than six months after the conference) the series editors will check once more the timeliness of the papers. Therefore, the volume's editor(s) should establish strict deadlines, or collect the articles during the conference and have them revised on the spot. If a delay is unavoidable, one should encourage the authors to update their contributions if appropriate. The editors of proceedings are strongly advised to inform contributors about these points at an early stage.

The final manuscript should contain a table of contents and an informative introduction accessible also to readers not particularly familiar with the topic of the conference. The contributions should be in English. The volume's editor(s) should check the contributions for the correct use of language. At Springer-Verlag only the prefaces will be checked by a copy-editor for language and style. Grave linguistic or technical shortcomings may lead to the rejection of contributions by the series editors.

A conference report should not exceed a total of 500 pages. Keeping the size within this bound should be achieved by a stricter selection of articles and not by imposing an upper limit to the length of the individual papers.

Editors receive jointly 30 complimentary copies of their book. They are entitled to purchase further copies of their book at a reduced rate. As a rule no reprints of individual contributions can be supplied. No royalty is paid on Lecture Notes in Physics volumes. Commitment to publish is made by letter of interest rather than by signing a formal contract. Springer-Verlag secures the copyright for each volume.

The Production Process

The books are hardbound, and the publisher will select quality paper appropriate to the needs of the author(s). Publication time is about ten weeks. More than twenty years of experience guarantee authors the best possible service. To reach the goal of rapid publication at a low price the technique of photographic reproduction from a camera-ready manuscript was chosen. This process shifts the main responsibility for the technical quality considerably from the publisher to the authors. We therefore urge all authors and editors of proceedings to observe very carefully the essentials for the preparation of camera-ready manuscripts, which we will supply on request. This applies especially to the quality of figures and halftones submitted for publication. In addition, it might be useful to look at some of the volumes already published. As a special service, we offer free of charge LATEX and TEX macro packages to format the text according to Springer-Verlag's quality requirements. We strongly recommend that you make use of this offer, since the result will be a book of considerably improved technical quality. To avoid mistakes and time-consuming correspondence during the production period the conference editors should request special instructions from the publisher well before the beginning of the conference. Manuscripts not meeting the technical standard of the series will have to be returned for improvement.

For further information please contact Springer-Verlag, Physics Editorial Department V, Tiergartenstrasse 17, W-6900 Heidelberg, FRG

W. Dieter Heiss (Ed.)

Chaos and Quantum Chaos

Proceedings of the Eighth Chris Engelbrecht Summer School on Theoretical Physics Held at Blydepoort, Eastern Transvaal South Africa, 13-24 January 1992

Springer-Verlag Berlin Heidelberg GmbH

Editor

W. Dieter Heiss
Department of Physics
University of the Witwatersrand, Johannesburg
Private Bag 3, Wits 2050, South Africa

ISBN 978-3-662-13902-8 ISBN 978-3-540-47495-1 (eBook)
DOI 10.1007/978-3-540-47495-1

Originally published by Springer-Verlag Berlin Heidelberg New York in 1992
Softcover reprint of the hardcover 1st edition 1992

Typesetting: Camera ready by author/editor
58/3140-543210 - Printed on acid-free paper

Christian Albertus Engelbrecht

8 October 1935 - 30 July 1991

Chris Engelbrecht was the founder of the series of South African Summer Schools in Theoretical Physics. He negotiated its structure and its funding, determined its specific form and by applying his personal attention, he ensured that each school was relevant and of a high standard.

Born in Johannesburg where he received his school education, he studied at Pretoria University for a BSc and MSc degree before going to Caltech where he obtained a PhD in 1960. Back in South Africa he held appointments as theoretical physicist at the Atomic Energy Board (1961-1978) and at Stellenbosch University (1978-1991).

Apart from his research and excellence in teaching, he served physics and science on numerous bodies. He was elected President of the SA Institute of Physics for two terms - 1987 - 1991. It is a fitting memorial to him and a tribute to his selfless, excellent and dedicated service to the cause of physics and his fellow scientists, to henceforth name this series

The Chris Engelbrecht Summer Schools in Theoretical Physics.

Preface

Chaos and the quantum mechanical behaviour of classically chaotic systems have been attracting increasing attention. Initially, there was perhaps more emphasis on the theoretical side, but this is now being backed up by experimental work to an increasing extent. The words 'Quantum Chaos' are often used these days, usually with an undertone of unease, the reason being that, in contrast to classical chaos, quantum chaos is ill defined; some authors say it is non–existent. So, why is it that an increasing number of physicists are devoting their efforts to a subject so fuzzily defined?

Short pulse laser techniques make it possible nowadays to probe nature on the border line between classical and quantum mechanics. Such experimental back–up is direly needed, since, in the case of classically chaotic systems, the formal tools have so far turned out to be insufficient for an understanding of this border line.

The fact that the conceptual foundations of quantum mechanics are being challenged – or, at least, subjected to a search for deeper understanding – is of course ample explanation for this new field being so attractive.

We were fortunate that we could assemble seven leading experts who have made major contributions in the field. The emphasis of the school was on quantum chaos and random matrix theory. The material presented in this volume is a reflection of lucid and nicely coordinated presentations. What it cannot reflect is the friendly working atmosphere that prevailed throughout the course.

The Organizing Committee is indebted to the Foundation for Research Development for its financial support, without which such high–level courses would be impossible. We also wish to express our thanks to the Editors of *Lecture Notes in Physics* and Springer–Verlag who readily agreed to publish and assisted in the preparation of these proceedings.

Johannesburg
South Africa
September 1992

W D Heiss

Contents

The Problem of Quantum Chaos

Boris V. Chirikov

Budker Institute of Nuclear Physics
630090 Novosibirsk, RUSSIA

Abstract: The new phenomenon of quantum chaos has revealed the intrinsic complexity and richness of the dynamical motion with discrete spectrum which had been always considered as most simple and regular one. The mechanism of this complexity as well as the conditions for, and the statistical properties of, the quantum chaos are explained in detail using a number of simple models for illustration. Basic ideas of a new ergodic theory of the finite-time statistical properties for the motion with discrete spectrum are discussed.

1. Introduction: the theory of dynamical systems and statistical physics

The purpose of these lectures is to provide an introduction into the theory of the so-called *quantum chaos*, a rather new phenomenon in the old quantum mechanics of finite-dimensional systems with a given interaction and no quatized fields. The quantum chaos is a "white spot" far in the rear of the contemporary physics. Yet, in opinion of many physicists, including myself, this new phenomenon is, nevertheless, of a great importance for the fundamental science because it helps to elucidate one of the "eternal" questions in physics, the interrelation of dynamical and statistical laws in the Nature. Are they independently fundamental? It may seem to be the case judging by the striking difference between the two groups of laws. Indeed, most dynamical laws are time-reversible while all the statistical ones are apparently not with their notorious "time arrow". Yet, one of the most important achievements in the theory of the so-called *dynamical chaos*, whose part is the quantum chaos, was understanding that the statistical laws are but the specific case and, moreover, a typical one, of the nonlinear dynamics. Particularly, the former can be completely derived, at least in principle, from the latter. This is just one of the topics of the present lectures.

Another striking discovery in this field was that the opposite is also true! Namely, under certain conditions the dynamical laws may happen to be a specific case of the statistical laws. This interesting problem lies beyond the scope of my lectures, so I just mention a few examples. These are Jeans' gravitational instability, which is believed to have been responsible for the formation of stars and eventually of the celestial mechanics (the exemplary case of dynamical laws!); Prigogine's "dissipative structures" in chemical reactions; Haken's "synergetics"; and generally, all the so-called "collective instabilities" in fluid and plasma physics (see, e. g., Ref. [1–3]). Notice, however, that all the most fundamental laws in physics (those in quantum mechanics and quantum field theory) are, as yet, dynamical and, moreover, exact (within the boundaries of existing theories). To the contrary, all the *secondary laws*, both statistical ones derived from the fundamental dynamical laws and vice versa, are only approximate.

By now the two different, and even opposite in a sense, mechanisms of statistical laws in dynamical systems are known and studied in detail. They are outlined in Fig. 1 to which we will repeatedly come back in these lectures. The two mechanisms belong to the opposite limiting cases of the general theory of Hamiltonian dynamical systems. In what follows we will restrict ourselves to the Hamiltonian (nondissipative) systems only as more fundamental ones. I remind that the dissipation is introduced as either the approximate description of a many-dimensional system or the effect of external noise (see Ref.[103]). In the latter case the system is no longer a pure dynamical one which, by definition, has no random parameters.

The first mechanism, extensively used in the *traditional statistical mechanics* (TSM), both classical and quantal, relates the statistical behavior to a big number of freedoms $N \to \infty$. The latter is called *thermodynamic limit*, a typical situation in macroscopic molecular physics. This mechanism had been guessed already by Boltzmann, who termed it "molecular chaos", but was rigorously proved only recently (see, e. g., Ref. [4]). Remarkably, for any finite N the dynamical system remains *completely integrable* that is it possesses the complete set of N commuting integrals of motion which can be chosen as the action variables I. In the existing theory of dynamical systems this is the highest order in motion. Yet, the latter becomes chaotic in the thermodynamic limit. The mechanism of this drastic transformation of the motion is closely related to that of the quantum chaos as we shall see.

The second mechanism for statistical laws had been conjectured by Poincare at the very beginning of this century, not much later than Boltzmann's one. Again, it took half a century even to comprehend the mechanism, to say nothing about the rigorous mathematical theory (see, e.g., Refs.[4–6]). It is based on a strong local instability of motion which is characterized by the Lyapunov exponents for the linearized motion. The most important implication is that the number of freedoms N is irrelevant and can be as small as $N = 2$ for a conservative system, and even $N = 1$ in case of a driven motion

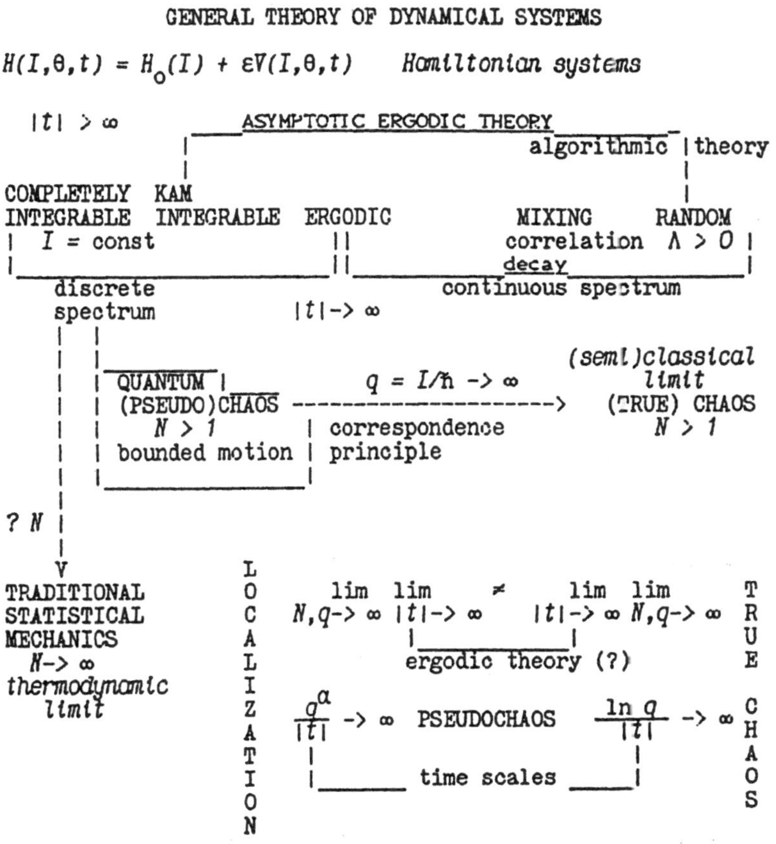

Figure 1: The place of quantum chaos in modern theories: action-angle variables I, θ; number of freedoms N; Lyapunov's exponent Λ; quasiclassical parameter q; Planck's constant $\hbar$. Two question marks indicate the problems in a new ergodic theory nonasymptotic in N and $|\,t\,|$.

that is one whose Hamiltonian explicitly depends on time. In the latter case the dependence is assumed to be regular, of course, for example periodic, and not a sort of noise.

This mechanism is called *dynamical chaos.* In the theory of dynamical systems it constitutes another limiting case as compared to the complete integrability. The transition between the two cases can be described as the effect of "perturbation" εV on the unperturbed Hamiltonian H_0, the full Hamiltonian being

$$H(I,\theta,t) = H_0(I) + \varepsilon V(I,\theta,t) \tag{1.1}$$

where I, θ are N-dimensional action-angle variables. At $\varepsilon = 0$ the system is completely integrable, and the motion is quasiperiodic with N basic frequencies

$$\omega(I) = \frac{\partial H_0}{\partial I} \tag{1.2}$$

Depending on initial conditions ($I(0)$) the frequencies may happen to be commensurable, or linearly dependent, that is the scalar product

$$m, \omega(I) = 0 \tag{1.3}$$

where m is integer vector.

This is called *nonlinear resonance*. The term nonlinear means the dependence $\omega(I)$. The interaction of nonlinear resonances (because of nonlinearity) is the most important phenomenon in nonlinear dynamics. The resonances are precisely the place where chaos is born under arbitrarily weak perturbation $\varepsilon > 0$. Hence the term *universal instability* (and chaos) of nonlinear oscillations [6]. The structure of motion is generally very complicated (fractal), containing an intricate mixture of both chaotic and regular motion components which is also called *divided phase space*. According to the Kolmogorov—Arnold—Moser (KAM) theory, for $\varepsilon \to 0$, most trajectories are regular (see, e. g., Ref. [7]). The measure of the complementary set of chaotic trajectories is exponentially small ($\sim \exp(-c/\sqrt{\varepsilon})$), hence the term *KAM integrability* [8]. Yet, it is everywhere dense as is the full set of resonances (1.3). A very intricate structure!

Even though the mathematical theory of dynamical systems looks very general and universal it actually has been built up on the basis of, but of course is not restricted to, the classical mechanics with its limiting case of the dynamical chaos. The quantum mechanics as described by some dynamical equations, for example, Schrödinger's one, for a specific dynamical variable ψ well fits the general theory of dynamical systems but turns out to belong to ... the limiting case of regular, completely integrable motion.

This is because the energy (frequency) spectrum of any quantum system *bounded in phase space* is always discrete and, hence, its time evolution is *almost periodic.* The ultimate origin of this quantum regularity is discreteness of the phase space itself inferred from the most fundamental uncertainty principle which is the very heart of the quantum mechanics. In modern mathematical language it is called *noncommutative geometry* of the phase space. Hence, the full number of quantum states within a finite domain of phase space is also finite. Then, what about chaos in quantum mechanics?

On the first glance, this is no surprise since the quantum mechanics is well known to be fundamentallly different as compared to the classical mechanics. However, the difficulty, and a very deep one, arises from the fact that the former is commonly accepted to be the universal theory, particularly, comprising the latter as the limiting case. Hence, the correspondence principle which requires the transition from quantum to classical mechanics in all cases including the dynamical chaos. Thus, there must exist a sort of quantum chaos!

Of course, one would not expect to find any similarity to classical behavior in essentially quantum region but only sufficiently far in the quasiclassical domain. Usually, it is characterized formally by the condition that Planck's constant $\hbar \to 0$. I prefer to put $\hbar = 1$ (which is the question of units), and to introduce some (big) quantum parameter q. Generally, it depends on a particular problem, and may be, for instance, the quantum (level) number. The quasiclassical region then corresponds to $q \gg 1$ while in the limit $q \to \infty$ the complete rebirth of the classical mechanics must occur somehow.

Notice that unlike other theories (of relativity, for example) the quasiclassical transition is rather intricate. Actually, this is the main topic of these lectures. Thus, the quantum chaos we are going to discuss is essentially a quasiclassical phenomenon in finite (essentially few-dimensional) systems with bounded motion. These restrictions are very important to properly understand the place of the new phenomenon - quantum chaos - in the general theory of dynamical systems, and to distinguish the former from the old mechanism for statistical laws in infinite systems $N \to \infty$. The latter nature is sometimes well hidden in a particular model as, for example, the nonlinear Schrödinger equation (Lecture 8).

The number of papers devoted to the studies of quantum chaos and related phenomena is rapidly increasing, and it is practically impossible to comprise everything in this field. In what follows I have to restrict myself to some selected topics which I know better or which I myself consider as more important. The same is true for references. I apologize beforehand for possible omissions and inaccuracies. Anyway, I refer in addition to a number of recent reviews [9–14], and to these proceedings.

My presentation below will be from a physicist's point of view even though the whole problem of quantum chaos, as a part of quantum dynamics, is essentially mathematical.

The main contribution of physicists to the studies of quantum chaos is in extensive numerical (computer) simulations of quantum dynamics, or numerical experiments as we use to say. But not only that. First of all, numerical experiments are impossible without a theory, if only semiqualitative, and without even rough estimates to guide the study. Mathematicians may consider such physical theories as a collection of hypotheses to prove or disprove them. What is even more important, in my opinion, that those theories require, and are based upon, a set of new notions and concepts which may be also useful in a future rigorous mathematical treatment.

I would like to mention that with all their obvious drawbacks and limitations the numerical experiments have very important advantage (as compared to the laboratory experiments), namely, they provide the complete information about the system under study. In quantum mechanics this advantage becomes crucial because in the laboratory one cannot observe (measure) the quantum system without a radical change of dynamics.

We call numerical experiments the *third way of cognition* in addition to traditional theoretical analysis, and to the main source of the knowledge and the Supreme Judge in science, the Experiment.

Laboratory experiments are vitally important for the progress in science not simply to prove or disprove some theories but to eventually discover, on a very rare occasion though, new fundamental laws of nature which are taken for granted in numerical experiments and theoretical analysis.

As an illustration of dynamical chaos, both classical and quantal, I will make use of the following "simple" model. In the classical limit it is described by the so-called *standard map*: $(n, \theta) \to (\bar{n}, \bar{\theta})$ where

$$\bar{n} = n + k \cdot \sin\theta; \qquad \bar{\theta} = \theta + T \cdot \bar{n} \tag{1.4}$$

Here n, θ are the action-angle dynamical variables; k, T stand for the strength and period of perturbation. Notice that in full dimensions parameter T is actually $\omega T/n_0$ where ω is the perturbation frequency, and n_0 stands for some characteristic action. The phase space of this model is an infinite cylinder which can be also "rolled up" into a torus of cirqumference

$$C = \frac{2\pi m}{T} \tag{1.5}$$

with an integer m to avoid discontinuities. Notice that map (1.4) is periodic not only in θ but also in n with period $2\pi/T$. The latter is a nongeneric symmetry of this model. In the studies of general chaotic properties it is a disadvantage. Nevertheless, the model is very popular, apparently because of its formal and technical symplicity combined with the actual richness of behavior. It can be interpreted as a mechanical system—the rotator driven by a series of short impulses, hence the nickname—"*kicked rotator*".

The quantized standard map was first introduced and studied in Ref. [15]. It is described also by a map: $\psi \to \bar{\psi}$ where

$$\bar{\psi} = \hat{R}_T \hat{F}_k \psi \tag{1.6}$$

and where

$$\hat{F}_k = \exp(-ik \cdot \cos\hat{\theta}), \qquad \hat{R}_T = \exp\left(-i\frac{T\hat{n}^2}{2}\right) \tag{1.7}$$

are the operators of a "kick" and of a free rotation, respectively. Momentum operator is given by the usual expression: $\hat{n} = -i\partial/\partial\theta$.

Sometime it is more convenient to use the symmetric map

$$\bar{\psi} = \hat{R}_{T/2}\hat{F}_k\hat{R}_{T/2}\psi \tag{1.8}$$

which differs from Eq. (1.6) by the time shift $T/2$, and which is, moreover, time reversible. In the most interesting case of a strong perturbation ($k \gg 1$) the operator $\hat{F}_k$ couples approximately $2k$ unperturbed states. Also, parameter T can be considered as an effective "Planck's constant" [103].

Notice that in classical limit the motion of model (1.4) depends on a single parameter $K = kT$ but after quantization the two parameters, k and T, can not be combined any longer.

Even though the standard map is primarily a simple mathematical model it can serve also to approximately describe some real physical systems or, better to say, some more realistic models of physical systems. One interesting example is the peculiar diffusive photoeffect in Rydberg (highly excited) atoms (see, e. g., Refs [14, 16, 104] for review).

The simplest 1D model is described by the Hamiltonian (in atomic units):

$$H = -\frac{1}{2n^2} + \varepsilon \cdot z(n,\theta)\cos\omega t \tag{1.9}$$

where z stands for the coordinate along the linearly polarized electric field of strength ε and frequency ω.

Another approach to this problem is constructing a map over a Kepler period of the electron [17]: $(N_\Phi, \phi) \to (\bar{N}_\Phi, \bar{\phi})$ where

$$\bar{N}_\Phi = N_\Phi + k \cdot \sin\phi; \qquad \bar{\phi} = \phi + \frac{\pi}{(2\omega)^{1/2}}(-\bar{N}_\Phi)^{-3/2} \tag{1.10}$$

Here, $N_\Phi = E/\omega = -1/2\omega n^2$, and perturbation parameter

$$k \approx 2.6\frac{\varepsilon}{\omega^{5/3}} \tag{1.11}$$

if the field frequency exceeds that of the electron: $\omega n^3 \gtrsim 1$.

Linearizing the second Eq. (1.10) in N_Φ reduces the Kepler map to the standard map with the same k, and parameter

$$T = 6\pi\omega^2 n^5 \tag{1.12}$$

Thus, the standard map describes the dynamics locally in momentum. In this particular model momentum N_Φ is proportional to energy as the conjugate phase $\phi = \omega t$ is proportional to time.

In quantum mechanics, instead of solving Schrödinger's equation with Hamiltonian (1.9) one can directly quantize a simple Kepler map (1.10) to arrive at a quantum map (1.6) with the same perturbation operator $\hat{F}_k$ (1.7) but with a different rotation operator

$$\hat{R}_\nu = \exp(-2i\pi\nu(-2\omega\hat{N}_\Phi)^{-1/2}) \tag{1.13}$$

Here parameter $\nu = 1$ (one Kepler's period) for quantum map (1.6), and $\nu = 1/2$ for symmetric map (1.8).

Notice that in Kepler map's description a new time (τ) is discrete (the number of map's iterations), and moreover, its relation to the continuous time t in Hamiltonian (1.9) depends on dynamical variable n or N_Φ:

$$\frac{dt}{d\tau} = 2\pi n^3 = 2\pi(-2\omega N_\Phi)^{-3/2} \tag{1.14}$$

In quantum mechanics such a change of time variable constitutes the serious problem: how to relate the two solutions, $\psi(t)$ and $\psi(\tau)$? For further discussion of this problem see Ref. [14]. Besides, map's solution $\psi(N, \tau)$ does not provide the complete quantum description but only some averaged one over the groups of unperturbed states [17].

These difficulties are of a general nature in attempts to make use of the Poincaré map for conservative quantum systems. The straightforward approach would be, first, to solve the Schrödinger equation, and then to construct the quantum map out of $\psi(t)$. Usually, this is a very difficult way. Much simpler one is, first, to derive the classical Poincaré map, and then to quantize it. However, generally the second way provides only an approximate solution for the original system. The question is how to reconcile the both approaches?

Another physical problem—the Rydberg atom in constant and uniform magnetic field, I will refer to below, is described by the Hamiltonian (for review see Ref. [18]):

$$H = \frac{p_z^2 + p_\rho^2}{2} - \frac{1}{r} + \frac{\omega L_z}{2} + \frac{\omega^2\rho^2}{8} \tag{1.15}$$

Here $r^2 = \rho^2 + z^2 = x^2 + y^2 + z^2$; ω is the Larmor frequency in the magnetic field along z axis, and L_z stands for the component of angular momentum (in atomic units). Unlike the previous model the latter one is conservative (energy preserving). It is simpler for theoretical studies and, hence, more popular among mathematicians. Physicists prefer time-dependent systems or, to be more precise, the models described by maps which greatly facilitate numerical experiments.

An important class of conservative models are billiards, both classical and quantal [19–21, 9, 105]. Especially popular is the billiard model called "stadium" [20]. Interestingly, instead of a quantum ψ wave one may consider classical linear waves, e. g., electromagnetic, sound, elastic etc. In the latter case the billiard is called "cavity". Of course, this problem has been studied since long ago, yet only recently it was related to the brand-new phenomenon of "quantum" chaos [22, 23] (see also Refs.[105, 106].

Quantum (wave) billiards are the limiting (and a simpler) case of the general dynamics of linear waves in dispersive media. It seems that the case of a spatially random medium does attract the most attention in this field. A striking example is the celebrated phenomenon of the *Anderson localization.* True, this is a statistical rather than dynamical problem. On the other hand, one may consider the random potential as a typical one, and the averaged solution as the representation of typical properties in such systems. Instead, in the spirit of the dynamical chaos, one can extend the problem in question onto a class of regular (but not periodic) potentials.

Recently, a deep analogy has been discovered between this rather old problem of wave dynamics in configurational space (in a medium) and of the dynamics in momentum space, particularly, the excitation of a quantum system by driving perturbation [24, 25]. Remarkably, that while the latter problem is described by a time-dependent Hamiltonian the former is a conservative system. This interesting and instructive similarity is discussed in Ref. [26].

2. Asymptotic statistical properties of classical dynamical chaos

To understand the phenomenon of quantum chaos it should be put into the proper perspective of recent developments in physics. The central focus of this perspective is the conception of classical dynamical chaos which has destroyed the deterministic image of the classical physics. What is the dynamical chaos? Which should be its meaningful definition?

This is one of the most controversial questions even in classical mechanics. There are two main approaches to the problem. The first one is essentially mathematical [4, 7]. The terms dynamical chaos and randomness are abandoned from rigorous statements, and left for informal explanations only,

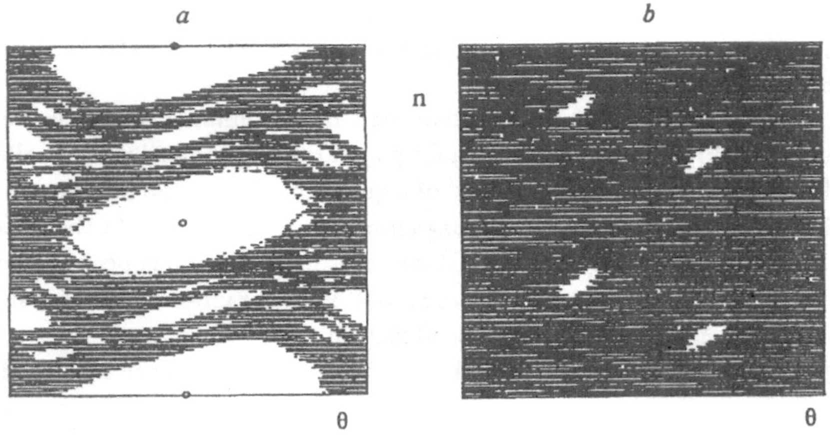

Figure 2: A fractal nonergodic motion component for the standard map, $K = 1.13$ (a); almost ergodic motion, $K = 5$ (b). Each hatched region is occupied by a single trajectory (after Ref.[14]).

usually in quotes, even in Ref. [27] where a version of the rigorous definition of dynamical randomness (chaos) was actually given. This is not the case in Chaitin's papers (see, e. g., Ref. [28]) but his approach is somewhat separated from the rest of ergodic theory, and is related to a new, *algorithmic theory* of dynamical systems started in the sixties by Kolmogorov (see Refs [27, 28]).

In the mathematical approach to the definition of dynamical chaos a hierarchy of statistical characteristics, such as ergodicity, mixing, K, Markov and Bernoulli properties etc, is introduced. In this hierarchy each property supposed to imply all the preceeding ones (see Fig. 1). However, the latter is not the case in the very important and fairly typical situation when the motion is restricted to a *chaotic component* usually of a very complicated (fractal) structure which occupies only a part of the energy surface in a conservative system or even a submanifold of lesser dimensions (see, e. g., Ref. [29]).

In Fig. 2a an example of the fractal chaotic component for the standard map is shown [14]. The motion is not ergodic as a chaotic trajectory covers about a half of the phase plane only (cf. Fig. 2b for a bigger perturbation K with only tiny islets of stability filled up by regular trajectories). For still bigger K the motion looks like completely ergodic. However, this has not been as yet rigorously proved. Numerical experiments are also not a reliable proof, at least not the direct one, because in computer representation any quantity is discrete. An indirect indication is the dependence of measured chaotic area μ_c on the spatial resolution (discreteness) Δ. Numerically [30]

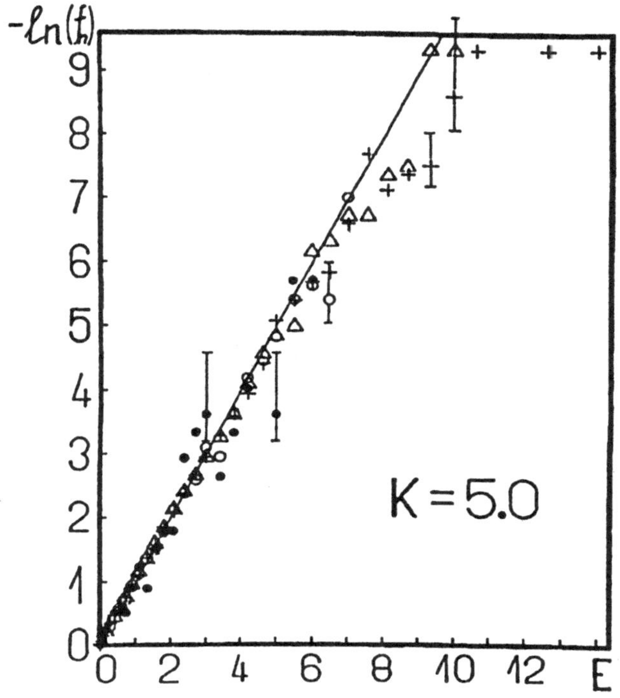

Figure 3: Normalized distribution function $f_n(E)$ in the standard map for various time intervals. The straight line is theoretical dependence $f_n = \exp(-E); E = (\Delta n)^2/\tau k^2$; statistical errors are shown in a few cases (after Ref.[6]).

$$\mu_c(\Delta) \approx \mu_c(0) + \alpha\Delta^\beta \tag{2.1}$$

with nonzero $\mu(0)$ and fractal exponent $\beta \approx 0.5$.

Being nonergodic the motion in the hatched domain in Fig. 2*a* is nonintegrable as the trajectory fills up a finite area of $\mu(0) \neq 0$. Hence, no motion integrals exist in this region. From the physical viewpoint there is a good reason to term such a motion chaotic. Anyway, the ergodicity, being the weakest statistical property, is neither necessary nor sufficient for the meaningful statistical description.

In this respect the most important property is mixing that is the correlation decay in time. It implies statistical independence of different parts of a trajectory as the separation in time between them becomes large enough. The statistical independence is the crucial property for the probability theory to be really applicable [31]. Particularly, the central limit theorem predicts Gaussian fluctuations which is, indeed, in a good agreement with the numerical data for the standard map (Fig. 3).

At average, the motion is described by the diffusion equation (also a

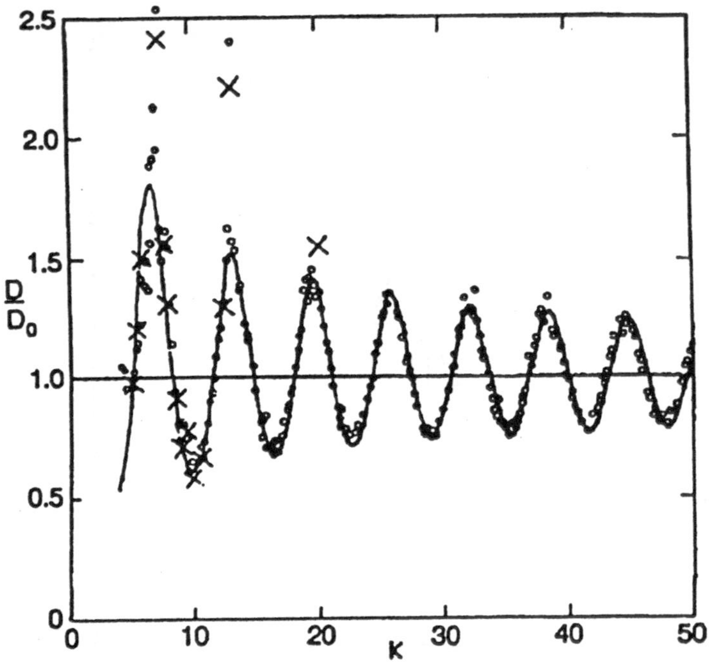

Figure 4: Classical (circles) and quantum (crosses) diffusion in the standard map; solid line is a simple theory; $D_0 = k^2/2$ (after Refs.[32, 33]).

typical statistical law) with the rate [32]

$$D \equiv \frac{\langle (\Delta n)^2 \rangle}{\tau} = \frac{k^2}{2}\kappa(K) \tag{2.2}$$

where function $\kappa(K)$ accounts for short-time correlations [33] (see Fig. 4).

The property of mixing is equivalent to continuous power spectrum of the motion which is the Fourier transform of the correlation function. This is just sufficient to provide the meaningful statistical description with its most important process of *relaxation* for an arbitrary initial distribution function $f(n,0) \to f_s(n)$ to some unique steady state. In case of the standard map on a torus, for example, the latter is ergodic

$$f_s(n) = f_e(n) = \frac{1}{C} \tag{2.3}$$

if $K \gg 1$ is big enough. The relaxation is asymptotically exponential [14]

$$f(n,\tau) - \frac{1}{C} \approx \exp\left(-\frac{2\pi^2|\tau|D}{C^2}\right) \cdot \cos\left(\frac{2\pi n}{C}\right), \qquad |\tau| \to \infty \tag{2.4}$$

with characteristic relaxation time

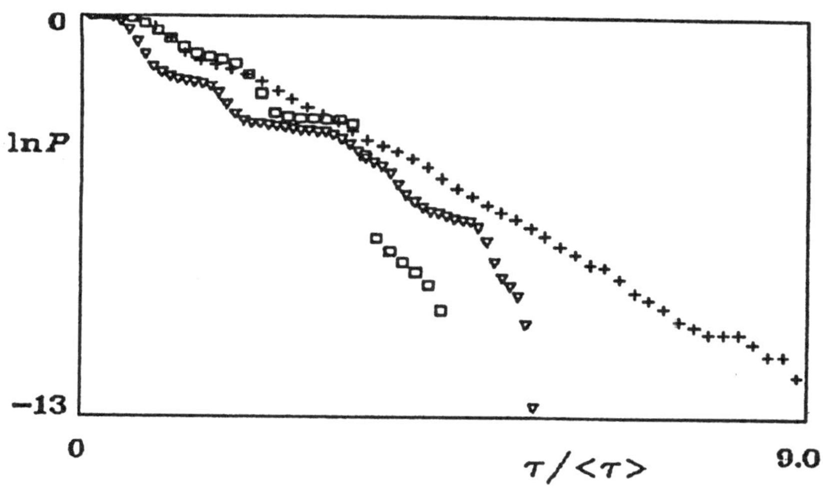

Figure 5: Statistics of Poincaré recurrences in discrete spectrum (regular motion): $N_\omega = 5, <\tau>= 6.3$ (squares); $N_\omega = 10, <\tau>= 10$ (triangles); $N_\omega = 100, <\tau>= 5.4$ (crosses); $\omega_1 = 1$.

$$\tau_e = \frac{C^2}{2\pi^2 D} \tag{2.5}$$

Notice that both diffusion and statistical relaxation proceed in two directions of time. The theory of dynamical chaos does not need the popular but superficial conception of "time arrow". True, the corresponding diffusion equation

$$\frac{\partial f(n,\tau)}{\partial \tau} = \frac{1}{2}\frac{\partial}{\partial n} D \frac{\partial f}{\partial n} \tag{2.6}$$

is irreversible in time. However, this is simply because the distribution function $f(n,\tau)$ is a *coarse-grained* phase density, averaged over phase θ. The *fine-grained* (exact) phase density $f(n,\theta,\tau)$ obeys the Liouville equation which is time-reversible as are the motion equations. Being time-reversible the statistical relaxation is *nonrecurrent* that is even the exact phase density $f(n,\theta,\tau)$ would never come back to the initial $f(n,\theta,0)$. Unlike this almost all trajectories are recurrent, according to the Poincaré theorem, independent of the type of motion (regular or chaotic). The difference is in the distribution of recurrence times: in discrete spectrum this time is strictly bounded from above while for chaotic motion an arbitrary long recurrence time can occur with some probability.

In Fig. 5 an example of the statistics for Poincare's recurrences is shown in regular motion with N_ω incommensurable frequencies randomly distributed within the interval $(0, \omega_1)$. Numerically [34], the upper bound is approximately

$$\frac{\tau_{max}}{\langle\tau\rangle} \approx 0.8N_\omega = 0.8\omega_1\rho_\omega \tag{2.7}$$

where $\langle\tau\rangle \approx 6/\omega_1$ is the mean recurrence time under given conditions (particularly, for a given set of frequencies), and $\rho_\omega = N_\omega/\omega_1$ is the density of frequencies. The latter quantity is going to play the central role in the problem of quantum chaos.

For $\tau < \tau_{max}$ the distribution function is close to exponential

$$P(\tau) = e^{-\tau/\langle\tau\rangle} \tag{2.8}$$

as predicted by the probability theory for a random process with continuous spectrum which is the limit for $N_\omega \to \infty$. Actuallly, in the above example, the spectrum is discrete but this apparently crucial property turns out to only restrict the random behavior to a finite time interval proportional to the frequency density.

Typically, chaotic motion possesses much stronger statistical properties than mixing. Here we come to the second approach to the definition of dynamical chaos which is essentially physical (see, e. g., Refs [5, 6]). In this approach the conception of random trajectories in a dynamical system is introduced from the beginning, and it is related to the strong (exponential) *local instability* of motion. This is characterized by a positive Lyapunov's exponent Λ or, more generally, by the Kolmogorov—Sinai (KS) dynamical entropy [4].

The main difficulty here is in that the instability itself is not sufficient for chaotic motion. One additional condition is boundedness of the motion to exclude, for example, the hyperbolic motion which hardly can be termed chaotic. Further, the separated unstable periodic trajectories must be also excluded, possibly, by the requirement of some minimal dimensions of a chaotic component. To the best of my knowledge, the complete set of conditions for an arbitrary motion component to be considered chaotic has not been found as yet, and it constitutes a difficult problem. Nevertheless, such a difinition of classical dynamical chaos is commonly accepted in the physical literature.

The crucial quantity Λ characterizes *linearized equantions of motion.* For example, in the standard map these are

$$\bar{\eta} = \eta + k \cdot \cos\theta(\tau) \cdot \xi; \qquad \bar{\xi} = \xi + T \cdot \bar{\eta} \tag{2.9}$$

where new dynamical variables, $\xi = d\theta$ and $\eta = dn$, form the additional tangent space. Lyapunov's exponent is defined by the limit

$$\Lambda = \lim_{|\tau|\to\infty} \frac{1}{|\tau|} \ln v(\tau) > 0 \tag{2.10}$$

where $v^2 = \xi^2 + \eta^2$, and $v(0) = 1$ is assumed. The last inequality in (2.10) means exponential instability of motion. The instability is time reversible as well as Λ. Actually, there are two (for a $2D$ map) Λ of opposite signs ($\Lambda_1 + \Lambda_2 \equiv 0$). The latter condition is equivalent to the area preservation in Hamiltonian systems.

In the standard map [6]

$$\Lambda \approx \begin{cases} 0.07K & K \ll 1 \\ \ln \frac{K}{2} & K \gg 1 \end{cases} \tag{2.11}$$

The first expression holds, of course, within the chaotic component of motion only which decomposes, for $K \ll 1$, into infinitely many exponentially narrow chaotic layers ($\Delta n \lesssim \exp(-\pi^2/\sqrt{K})$).

Remarkably, the main condition for chaos ($\Lambda > 0$) is related to *linear equations* (2.9) with time-dependent coefficients though. As this dependence ($\theta(\tau)$) is very complicated for a chaotic motion, the mathematical analysis of those linear equations is almost as difficult as that of the original nonlinear ones. However, numerically the criterion $\Lambda > 0$ is much simpler than, say, the spectrum or correlation function as the former requires much shorter computation time because the instability is fast. Actually, one needs to discern between the exponential and linear instabilities. The latter is always present in nonlinear oscillations due to the dependence of motion frequencies on initial conditions (see Eq. (1.2) and Refs [35, 36]).

According to the algorithmic theory of dynamical systems the information $J(t)$ associated with the chaotic trajectory of length t is asymptotically

$$\frac{J(t)}{|t|} \to \Lambda; \quad t \to \pm\infty \tag{2.12}$$

that is just proportional to the rate of exponential instability. This is the most important implication of the Alekseev—Brudno theorem (see Ref. [27]). It means that for each new time interval one needs a new information which cannot be extracted from the measurement, to arbitrarily high but finite accuracy $\nu > 0$, of any preceeding section of the trajectory (even the infinite one!).

Obviously, over some finite time interval the prediction of a chaotic trajectory is possible depending on the *randomness parameter* [37]

$$r = \frac{\Lambda |t|}{|\ln \nu|} \tag{2.13}$$

Prediction is restricted to a finite domain of *temporary determinism* ($r \lesssim 1$) which goes over, as r increases, to the infinite region of *asymptotic randomness* ($r \gg 1$). Notice that the average information per unit time (2.12) does not depend on the measurement accuracy $\nu > 0$.

For the regular motion with discrete spectrum the specific information decreases with time

$$\frac{J(t)}{|t|} \to \frac{\ln|t|}{|t|} \tag{2.14}$$

and the prediction is asymptotically possible contrary to the conclusion in Ref. [35].

Another way to understand the requirement of *exponential* instability for chaos is to consider the so-called *symbolic dynamics* (see, e.g., Ref.[27]) which is a mathematical description of the trajectory recording to a finite accuracy. For a map the total number of symbolic trajectories $M^\tau = \exp(\tau \ln M)$ grows exponentially in time where the total number of symbols $M \sim 1/\nu$ determines the measurement accuracy. If the motion instability is also exponential, then *all* the symbolic trajectories are realized for a sufficiently large map's period $T \gtrsim (\ln M)/\Lambda$.

The ultimate origin of the complexity (particularly, unpredictability) of a chaotic trajectory lies in the continuity of the phase space in classical mechanics. This is no longer true in quantum mechanics which leads to the most important peculiarity of the quantum chaos.

On the first glance the important condition for chaos $\Lambda > 0$ is not invariant with respect to the change of time. To avoid this difficulty the instability should be considered not in time but rather in the oscillation phase, e. g. θ for the standard map, or per map's iteration like in Eq. (2.11). In other words, the appropriate quantity is a dimensionless entropy, e.g., $\Lambda \to \Lambda/ < \omega >$ where $< \omega >$ is some average frequency of the motion.

To summarize, the physical definition reads: the *dynamical chaos is exponentially unstable motion* bounded, at least, in some variables.

Remarkably, the instability is determined from the linear equations, the role of nonlinearity being to bound the unstable motion. On the other hand, any motion can be described equivalently by the linear Liouville equation for the fine-grained distribution function or phase space density. Being a stronger statistical property the exponential instability implies the continuous spectrum and, hence, the correlation decay. Yet, the latter is not always expotential but may be instead a power-law one (see, e. g., Ref. [29]).

The role of exponential instability in the statistical description of dynamical systems is not completely clear, it seems to be only sufficient but not a necessary condition. Nevertheless, the conception of random trajectories of a purely dynamical system is of the fundamental importance as it destroys the mysterious image of the random and opens the way for quantitative studies in this large part of natural phenomena. Indeed, the theory of dynamical chaos shows that the random processes are not controlled by some qualitatively different laws, to account for by means of some additional statistical hypotheses, but constitute a very specific, even though typical, part of general dynamics. An interesting question if there are "more random", or "true random", processes remains, as yet, open.

3. The correspondence principle and quantum chaos

Absence of the claassical-like chaos in quantum mechanics apparently contradicts not only with the correspondence principle, as mentioned above, but also with the fudamental statistical nature of quantum mechanics. However, even though the random element in quantum mechanics ("quantum jumps") is inavoidable, indeed, it can be singled out and separated from the proper quantum processes. Namely, the fundamental randomness in quantum mechanics is related only to a very specific event – the *quantum measurement* – which, in a sense, is foreign to the proper quantum system itself.

This allows to divide the whole problem of quantum dynamics into two qualitatively different parts: (i) the proper quantum dynamics as described by the wave function $\psi(t)$, and (ii) the quantum measurement including the registration of the result, and hence the ψ collapse.

Below I am going to discuss the first part only, and to consider ψ as a specific dynamical variable ignoring the common term for ψ, the probability amplitude. Variable ψ obeys some purely dynamical equation of motion, e.g., the Schrödinger equation. This part of the problem is essentially mathematical, and it naturally belongs to the general theory of dynamical systems.

As to the second part of the problem – the quantum measurement – this is a hard nut for physicists. Currently, there is no common opinion even on the question whether this is a real physical problem or an ill-posed one so that the Copenhagen interpretation of (or convention in) quantum mechanics answers all the admissible questions. In any event, there exists as yet no dynamical description of the quantum measurement including the ψ collapse. An interesting recent discussion of this question in the light of quantum cosmology can be found in Ref.[38]. In my opinion, one could find more "earthy" problems as well. Below I comment about the quantum measurement on a few occasions, but I will not discuss it in any detail as this certainly goes beyond the frame of my lectures here.

Recent breakthrough in the understanding of quantum chaos has been achieved, particularly, due to the above philosophy of separating the dynamical part of quantum mechanics accepted, explicitly or more often implicitly, by most researchers in this field.

Currently, there are several approaches to the definition of quantum chaos. The first natural move was to extend onto the quantum mechanics the classical definition of dynamical chaos as exponentially unstable motion. One of a few physicists who still adheres to this philosophy is Ford [39]. He insists that *the quantum chaos is deterministic randomness in quantum mechanics over and above that contained in wavefunction or the expansion postulate.* The latter refers to the quantum measurement as mentioned above. Some mathematicians implicitly accepted the same definition, and "succesfully" constructed the quantum analogue to the classical KS-entropy (see, e.g., second Ref.[39]).

For bounded in phase space quantum systems the quantum KS-entropy is identically zero because of discrete spectrum, and the classical-like chaos is impossible. Is it possible for unbounded quantum motion ? The answer is *yes* as was found recently but the examples of such a chaos are rather exotic. The first one was briefly mentioned in Ref.[40]. We consider here another example following the second Ref.[41] (for a more physical example see Ref.[42] while some general consideration are presented in Ref.[43], and a rigorous mathematical treatment is given, e.g., in second Ref.[39]).

Consider the flow on an N-dimensional torus specified by the equation

$$\dot{\theta}_i = \nu_i(\theta) \tag{3.1}$$

If $N \geq 3$ the classical chaos is possible with positive Lyapunov exponents that is the solution of the linearized equations is exponentially unstable. Consider now the Hamiltonian system

$$H(n,\theta) = \sum_k n_k \nu_k(\theta) \tag{3.2}$$

linear in momenta n_k canonically conjugated to coordinates θ_k. Then, the equations for n_k coincide (in reverse time) with the linearized equations (3.1). Hence, as soon as system (3.1) is chaotic the momenta of system (3.2) grow exponentially fast.

It is easily verified that the density $f(\theta,t) = |\psi(\theta,t)|^2$ of quantized system (3.2) obeys exactly the same (continuity) equation

$$\frac{\partial f}{\partial t} + \sum_k \frac{\partial (f\nu_k)}{\partial \theta_k} = 0 \tag{3.3}$$

as for classical system (3.1) with the same (particularly, chaotic) solution. The peculiarity of this and similar examples is in that to achieve the true chaos not only the quantum motion must be unbounded and, hence, of a continuous spectrum but the momenta have to grow exponentially in time.

This is why most physicists reject the above definition of quantum chaos and adhere to another one which reads (see, e.g., Ref.[11]): *quantum chaos is the quantum dynamics of classically chaotic systems* whatever it could happen to be, I would add.

Logically, this is most simple and clear definition. Yet, it is completely inadequate and even helpless, in my opinion, just because that chaos my turn out to be a perfectly regular motion, much surpassing that in the classical limit. The point is that the discreteness of quantum spectrum supresses any transitions for a sufficiently weak perturbation, no matter what is the corresponding classical motion [44]. For example, in the standard map this occurs if the perturbation parameter $k \leq 1$ independent of classical parameter $K = kT$ which controls the transition to chaos. This specific quantum stability is also called *perturbative localization*, or *transition localization.*

For this reason Berry proposed [45] to use the term *"quantum chaology"* which essentially means studying the absence of chaos in quantum mechanics.

My position is somewhere in between. I would like to define the quantum chaos in such a way to include some essential part of the classical chaos. It would be natural to include the mixing property which provides the meaningful statistical description of quantum dynamics. The difficulty is in that the discrete spectrum prohibits even the mixing in the sense of the ergodic theory. Yet, it turns out that the *finite-time analogues* of all the asymptotic properties in the ergodic theory, mixing including, can be formulated as we shall see below (cf. Fig. 5 as an example). For this reason, I currently adhere to the following definition: *the quantum chaos is finite-time statistical relaxation in discrete spectrum.*

A drawback of this definition is that such a chaos occurs also in the classical systems of linear waves as already mentioned. The term quantum chaos (in this definition) is, nevertheless, meaningful, in my opinion. Unlike classical linear waves, which are no doubt a limiting approximation to generally nonlinear waves, the linear quantum mechanics is as yet the fundamental and universal theory.

In such interpretation the classical-like asymptotic (infinite-time) chaos remains as an important limiting pattern to compare with the true quantum dynamics.

4. The uncertainty principle and the time scales of quantum dynamics

The main difficulty in the problem of quantum chaos is in that one needs to reconcile the quantum discrete spectrum, which apparently prohibits any dynamical chaos, with the correspondence principle, which does require some chaos, at least, sufficiently far in the quasiclassical region. But this is also the principal importance of the phenomenon of quantum chaos which reveals the deep interrelation between the two opposites – order and chaos – in the theory of dynamical systems (see Fig.1). To put it another way, the quantum chaos, properly interpreted, unveils a very complicated and reach nature of what has been, and still is, considered as a dull order, the almost periodic motion of discrete spectrum.

The other side of this difficulty is discreteness of the phase space in quantum mechanics, the size of an elementary cell being $\sim \hbar = 1$.

We resolved the above difficulty by introducing the characteristic time scales of the quantum motion on which the latter is close to the classical chaotic dynamics [41]. Actually, the first of those time scales had been discovered and explained by Berman and Zaslavsky already in 1978 [46], and was subsequently confirmed in many numerical experiments (see, e.g., Refs.[47]). We call it *random time scale* for the reasons given below.This scale is characterized, generally, by the estimate

$$t_r \sim \frac{\ln q}{\Lambda} \tag{4.1}$$

where q is some big quantum (quasiclassical) parameter, and Λ stands for the Lyapunov exponent.

In the standard map $\Lambda \approx \ln(K/2)$ (see Eq.(2.11))and there are two quantum parameters: k and $1/T$. The transition to the classical limit corresponds to $k \to \infty$, $T \to 0$ while the classical parameter $K = kT = const$. It may seem strange that perturbation period $T \to 0$ in the classical limit. This is because in full dimensions $T \sim 1/n_0$ (see Eq.(1.4) and below), and characteristic action $n_0 \to \infty$. General estimate (4.1) takes now the form [41]

$$\tau_r \sim \frac{|\ln T|}{\ln(K/2)} \tag{4.2}$$

This corresponds to the optimal (least spreading) configuration of the initial $\psi(0)$, a coherent state.

The physical meaning of this time scale is in the fast (exponential) spreading of the initially narrow wave packet. Thus, the exponential instability is

present in quantum mechanics as well but only on a very short time interval (4.1,2).

This can be explained in two ways. On the one hand, the initial wave packet can not be less, in size, than a quantum phase-space cell. On the other hand, in Hamiltonian systems, the local instability leads not only to the expansion in a certain direction but also to the contraction in some other direction which rapidly brings the initial wave packet to the size of the quantum cell.

Accoding to the Ehrenfest theorem a wave packet follows the beam of classical trajectories but only as long as it remains narrow, that is only on time scale (4.1). Nevertheless, characteristic time interval τ_r grows indefinitely in quasiclassical region, as $T \to 0$, in accordance with the correspondence principle. However, the transition to the classical chaos is (conceptually) difficult as it includes two limits ($T \to 0\,(q \to \infty)$ and $t \to \infty$) which do not commute (see Fig.1) . This is a typical situation in the quasiclassical region as was stressed, particularly, by Berry [10].

Substituting t_r (4.1) for t into Eq.(2.13) we arrive at the quantum randomness parameter

$$r_q \sim \frac{\ln q}{|\ln \nu|} \gtrsim 1 \tag{4.3}$$

The latter inequality is the condition for the motion of a narrow wave packet to be random. It is equivalent to

$$q\nu \gtrsim 1 \tag{4.4}$$

Again, the transition to the classical chaos includes two noncommuting limlts: $q \to \infty$, $\nu \to 0$.

The first time scale (4.1) is rather short, and the important question is: what happens next ? Numerical experiments revealed [15,41] that some classical-like chaos persists on a much longer time scale t_R, generally, of the order

$$\ln t_R \sim \ln q \tag{4.5}$$

which means some power-law dependence $t_R \sim q^{\alpha}$ (see Fig.8 below).

For the quantized standard map

$$\tau_R \sim k^2 \tag{4.6}$$

On this time scale the diffusion in n proceeds and, moreover, closely follows classical diffusion in all details, again in agreement with the correspondence

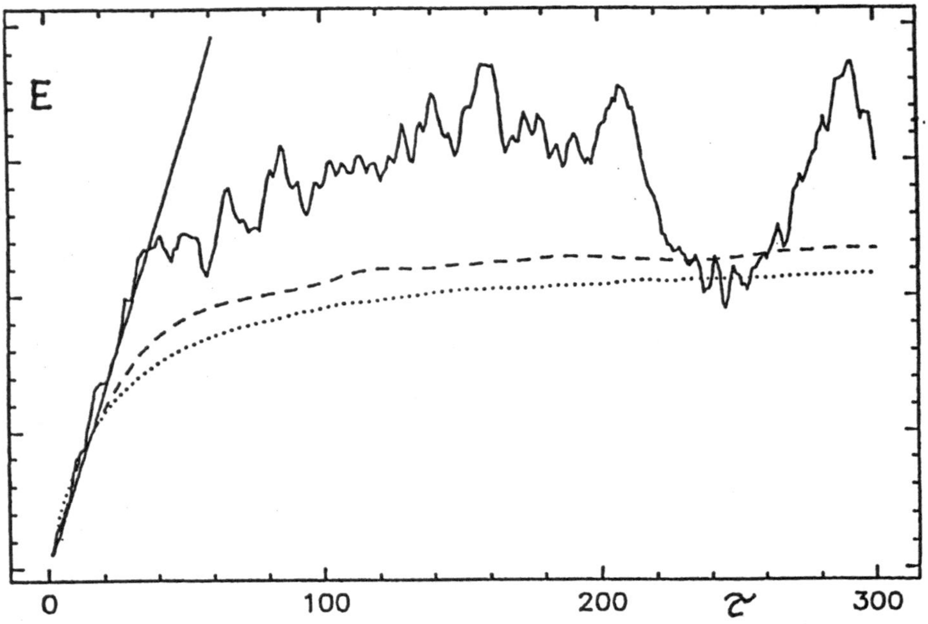

Figure 6: Quantum diffusion in the standard map: $K = 10; k = 6.56; T = 1.52; E =< n^2 > /2$ is the energy. Solid line - a single run; dashed and dotted lines - different averages over 10^4 runs; straight line - classical diffusion (after Ref.[48]).

principle. Subsequently, these numerical results were confirmed both numerically (see, e.g., Ref.[48]) as well as analitically [49]. In the Fig.6 the data from Ref.[48] are reproduced which demonstrate a classical-like behavior up to $\tau \sim 40$ for $k = 6.56$. The dependence of the initial rate of quantum diffusion on classical parameter K, shown in Fig.4, is in a good agreement with the classical dependence even for those K values where a simple theory fails. We call t_R the *diffusion* or (statistical) *relaxation time scale.*

This similarity to the classical chaos is, however, only partial. Unlike the classical one the quantum diffusion was found to be perfectly stable dynamically. This was proved in striking numerical experiments with the time reversal [50]. In a classical chaotic systems the diffusion is immediately recovered due to numerical "errors" (not random !) amplified by the local instability. On the contrary, the quantum "antidiffusion" proceeds untill the system passes, to a high accuracy, the initial state, and only than the normal diffusion is restored. An example of the time reversal in classical and quantum standard map is shown in Fig.7 [50]. The stability of quantun chaos on relaxation time scale is comprehensible as the random time scale (4.1) is

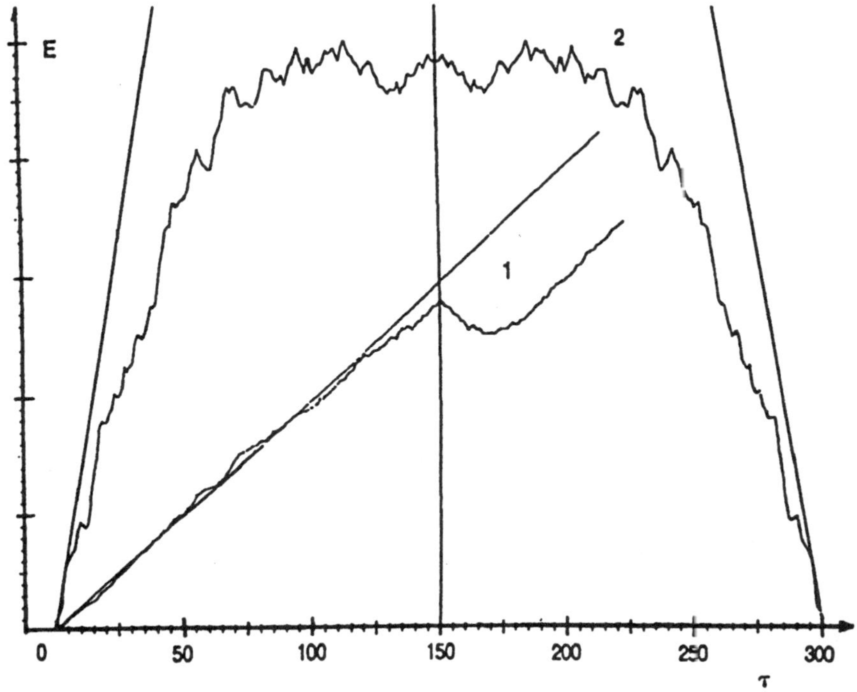

Figure 7: The effect of time reversal at $\tau = 150$ in classical (1) and quantum (2) chaos for the standard map with $k = 20; T = 0.25$. The straight lines show the same classical diffusion in different scales. The accuracy of the quantum reversal in E at $\tau = 300$ is better than 10^{-10} (!) (after Ref.[50]).

much shorter. Yet, the accuracy of the reversal is surprising. Apparently, this is explained by a relatively large size of the quantum wave packet as compared to the unavoidable rounding-off errors. In the standard map, for example, the size of the optimal, least-spreading, wave packet $\Delta\theta \sim \sqrt{T}$ [41]. On the other hand, any quantity in the computer must exceed the error $\delta < T$, hence $(\delta\theta)^2/\delta^2 \gtrsim (T/\delta)\delta^{-1} \gg 1$.

Beyond the relaxation time scale, that is for $t \gg t_R$, the quantum diffusion stops, and a certain steady state is formed which may or may not be close to the classical statistical equilibrium as will be discussed in detail below. For $k \to \infty$ $(kT = const)$ the time scale $t_R \to \infty$, again in accordance with the correspondence principle, but this quasiclassical transition is also characterized by the same double limit as for t_r above.

Thus, various properties of the classical dynamical chaos are also present in quantum dynamics but only temporarily, within finite and different time scales t_r or t_R. This is the crucial distinction of the quantum ergodic theory from the classical one which is asymptotic in t. It seems that any substantial progress in the mathematical theory requires a generalization of the existing ergodic theory to a finite time. Perhaps, it is better to say that a new nonasymptotic (finite-time) ergodic theory needs to be created.

Why the existing ergodic theory is asymptotic ? I suspect that the main reason is technical rather than physical or mathematical. Namely, the asymptotic analysis is, typically, much simpler. Remember, for example, the conception of continuous phase space in classical mechanics. One particular difficulty in a finite-time ergodic theory is the important distinction between discrete and continuous spectrum of the motion which is unambiguous only asymptotically in time.

The conception of a finite-time chaos in discrete spectrum appears unusual and even strange, indeed. Yet, in my opinion, it has no intrinsic defects or contradictions. Moreover, such a notion already exists in the rigorous algorithmic theory of dynamical systems. For a physicist, the decisive argument is that the finite-time chaos perfectly fits a broad class of quantum processes and, moreover, provides an arbitrarily close approach to the classical chaos in accordance with the fundamental correspondence principle. Also, notice that in numerical experiments on the digital computer the finite-time pseudochaos is only possible as any quantity in the computer is discrete. In computer representation any dynamical system is "superquantized" in a sense (for discussion see, e.g., Ref.[41]).

This philosophy, which has not yet many adherents, resolves also the double limit ambiguity discussed above. From the physical viewpoint there is no reason to take the limit $t \to \infty$ at all. Instead, the time should be fixed for any particular problem, the regime of quantum motion depending on the quasiclassical parameter q as outlined in Fig.8. In this picture the asymptotic classical chaos is but a *limiting pattern* to compare with the true (quantum) dynamics.

The real quantum chaos, nevertheless, is called sometimes *pseudochaos or transient chaos* to distinguish an "ugly" reality from the perfect ideal.

Of the two characteristic time scales of quantum motion discussed above the relaxation time scale t_R is most important simply because it is much longer than the other one, t_r. Peculiarity of quantum statistical relaxation is in that it proceeds in spite of the discrete energy spectrum. As is well known, the latter is always the case for the quantum motion bounded in phase space. The crucial property is a finite number of quantum states on the energy surface or, better to say, within an *energy shell.* In this case [41]

$$t_R \lesssim \rho \tag{4.7}$$

where ρ is the finite energy level density ($\hbar = 1$) (cf. Eq.(2.7)).

The physical meaning of this estimate is very simple and is related to the

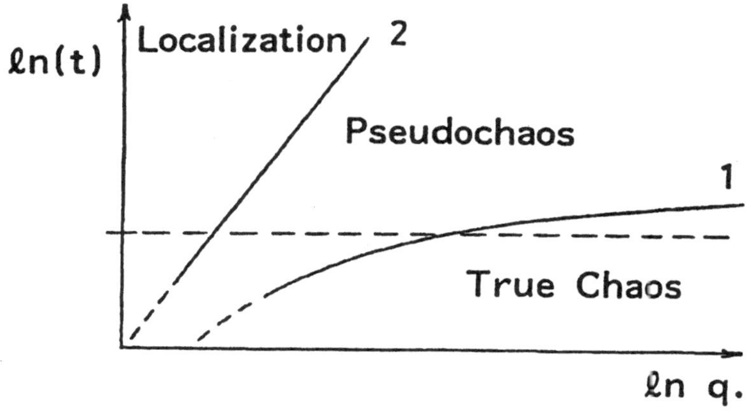

Figure 8: Classically chaotic quantum motion: 1 - random time scale $t_r \sim \ln q$; 2 - relaxation time scale $t_R \sim q^{\alpha}$; $q \gg 1$, the quasiclassical parameter.

fundamental uncertainty principle [1]. For sufficiently short time the discrete spectrum is not resolved, and a classical-like diffusion is possible, at most up to $t \sim \rho$. The same is true for the standard map on a torus which has also a finite number (C) of now quasienergy states. Since quasienergy is determined $mod\,(2\pi/T)$ the level density is

$$\rho = \frac{TC}{2\pi} = m \gtrsim t_R \tag{4.8}$$

Notice that ρ is classical parameter as is m because while $C \to \infty$ parameter $T \to 0$ in the classical limit.

The situation is much less clear for the standard map on a cylinder where the motion can be unbounded in n. In some special cases the quasienergy spectrum is, indeed, continuous, yet this does not mean chaotic motion but rather the peculiar quantum resonance. A more complicated case of continuous spectrum will be discussed below. On the other hand, all the numerical evidence indicates that typically the quasienergy spectrum is discrete in spite of infinite number of levels. Formally, the level density ρ is then also infinite. Yet, relaxation time scale t_R is finite. The point is that the quantum motion does not depend on all quasienergy eigenstates but only on those which are actually present in the initial quantum state $\psi(0)$ and, thus, control the motion. We call them *operative eigenstates* (for given initial conditions) . If their density is $\rho_0 \le \rho$ a better estimate for t_R is (cf.Eq.(4.7)):

$$t_R \sim \rho_0 \tag{4.9}$$

[1] In a different way this first principle was used in Ref.[51] to explain the Anderson localization in a random potential.

For ρ_0 to be finite all eigenfunctions have to be localized that is to decrease sufficiently fast in n. To the best of my knowledge there are as yet no rigorous results on the eigenfunction localization and/or the spectrum even for such a simple model as the standard map.

If the localization length is l, the density $\rho_0 \sim Tl/2\pi$ (for sufficiently localized intial state), and $\tau_R = t_R/T \sim l$. Actually, Eq.(4.9) is an implicit relation because ρ_0 depends, in turn, on dynamics. Consider, first, the unbounded standard map where the rate of classical diffusion has the form (2.2), and

$$D \sim k^2 \tag{4.10}$$

for $K \gg 1$ (complete classical chaos). Suppose, further, that the width (in n) of the initial state $\Delta n_0 = l_0 \ll l$. Then the final width due to a diffusion during time τ_R is $\Delta n_f \sim (\tau_R D)^{1/2} \sim l$. Since $\tau_R \sim l$, we arrive at the remarkable estimate

$$\tau_R \sim l \sim D \tag{4.11}$$

which relates essentially quantum characteristics (τ_R, l) with the classical quantity D.

Thus, the quantum diffusion in the unbounded standard map is always localized, and a certain steady state is formed which has no counterpart in classical mechanincs.

For the bounded standard map the situation is qualitatively different depending on a new parameter

$$\lambda = \frac{l}{C} \sim \frac{D}{C} \tag{4.12}$$

which we term the *ergodicity parameter*. Indeed, the quantum localization occurs for $\lambda \ll 1$ only. In the opposite limiting case $\lambda \gg 1\,(D \gg C)$ the relaxation time scale, being finite, is nevertheless long enough for the relaxation to the ergodic steady state to be accomplished. In this case the final steady state is close to that in classical mechanics. The same is true for conservative systems of two freedoms like billiards or cavities. In terms of the relaxation times $\lambda^2 \sim \tau_R/\tau_e$ (see Eqs.(2.5) and (4.11)).

5. Finite-time statistical relaxation in discrete spectrum

We turn now to a more accurate description of the quantum relaxation in the standard map. First, what are the quasienergy eigenfunctions ? We shall discuss this in detail below. So far it is sufficient to know that the quantum localization is approximately exponential with eigenfunctions

$$\varphi_m(n) \approx \frac{1}{\sqrt{l}} \exp\left(-\frac{|m-n|}{l}\right) \tag{5.1}$$

and the steady state

$$g_s(n) = \overline{|\psi_s(n)|^2} \approx \frac{1}{l_s} \exp\left(-\frac{2|n|}{l_s}\right) \tag{5.2}$$

Here the bar means averaging in time, and the initial conditions are $g(n,0) = \delta(n)$ so that g is actually the Green function. Generalization to arbitrary conditions is obvious.

Using these definitions the more accurate relations were found numerically for the standard map (see, e.g., second Ref.[41]):

$$l_s \approx 2l \approx D \tag{5.3}$$

Surprisingly, the localization lengths for eigenfunctions and for the steady state are rather different. This is due to big fluctuations around the simple exponential dependence. Generally, relations (5.3) depend also on system's symmetry [107,108].

The first attempt to describe the quantum relaxation in standard map was undertaken in Ref.[52]. The idea was very simple: the diffusion rate should be proportional to the number of quasienergy levels which are not yet resolved in time τ. This number decreases, for $\tau \geq \tau_R$, as τ^{-1}, hence

$$D(\tau) \sim D(0)\frac{\tau_R}{\tau} \tag{5.4}$$

where $D(0)$ is the classical diffusion rate. This result was corrected in Ref.[53] where, in a more sophisticated way, the so-called level repulsion was taken into account to give for the rate of energy variation

$$\frac{dE(\tau)}{d\tau} \equiv \dot{E}(\tau) \sim \dot{E}(0)\left(\frac{\tau_R}{\tau}\right)^{1+\beta} \tag{5.5}$$

where β is the repulsion parameter. Preliminary fitting of Eq.(5.5) to some numerical data looked as an agreement with $\beta \approx 0.3$.

However, recent extensive numerical similations [48] revealed a different dependence for $\tau \gg \tau_R$ (in our notations)

$$\dot{E}(\tau) \approx c\dot{E}(0)\left(\frac{\tau_R}{\tau}\right)^2 \ln\frac{\tau}{\tau_R} \tag{5.6}$$

supported by a different theory. Numerically (my fit)

$$\tau_R \approx 2l_s; \qquad c \approx 0.2 \tag{5.7}$$

in apparent contradiction with Eq.(5.5).

Still another phenomenological theory was proposed in Ref.[14] and developed in Ref.[54]. It is based on the general diffusion equation (see, e.g.,Ref.[5]):

$$\frac{\partial}{\partial\tau}g(n,\tau) = \frac{1}{2}\frac{\partial^2}{\partial n^2}Dg - \frac{\partial}{\partial n}Ag \tag{5.8}$$

The second term describes a "drift"

$$A \equiv \frac{<\Delta n>}{\tau} = \frac{dD}{dn} + B \tag{5.9}$$

Introducing this relation into Eq.(5.8), we obtain

$$\frac{\partial g}{\partial \tau} = \frac{1}{2}\frac{\partial}{\partial n}D\frac{\partial g}{\partial n} - \frac{\partial}{\partial n}Bg \tag{5.10}$$

In our problem the last term represents the so-called "backscattering", or reflection of ψ wave propagating in n. Negligible in the begining the backscattering eventually suppresses the diffusion and leads to the formation of steady state (5.2).

From Eq.(5.10) the general expression for steady state $g_s(n)$ is

$$\ln g_s = 2\int \frac{B(n)\, dn}{D(n)} \tag{5.11}$$

For homogeneous diffusion ($D = const$) g_s is given by Eq.(5.2) with $l_s = D$, hence

$$B = -\frac{n}{\mid n \mid} \tag{5.12}$$

The analysis of quantum relaxation can be performed using the two first moments of $g(n,\tau)$: $m_1 =< n > \quad (n > 0)$ and $m_2 =< n^2 >= 2E$. Notice that for initial $g(n,0) = \delta(n)$ the solution is symmetric with respect to $n = 0$, and we can consider $n > 0$ only. The equation for the moments are derived from Eq.(5.10)

$$\dot{m_1} = \frac{1}{2}Dg(0,\tau) + B; \qquad \dot{m_2} = D + 2m_1 B \tag{5.13}$$

Here $B = -1$ but we keep it for further analysis. The second equation shows that one should distinguish the rate of energy variation from the diffusion rate just because of the backscattering.

The quantity $g(0,\tau)$ in the first equation, called *staying probability* is of independent interest as a characteristic of the relaxation process.

In our case Eqs.(5.13) describe the evolution of initially spreading Gaussian distribution into the final exponential steady state (5.2). Accidentally, the ratio of moments

$$\frac{m_1^2}{m_2} \equiv \frac{\gamma^2}{2} \approx 0.5 \tag{5.14}$$

remains almost constant which allows for a simple solution

$$-t = \xi + \ln(1-\xi); \qquad \xi(0) = 0 \tag{5.15}$$

Here the new variable and time are

$$\xi = \frac{2\gamma}{D}\sqrt{E}; \qquad t = \frac{\gamma^2}{D}\tau \tag{5.16}$$

Initially, as $\tau \to 0$, Eq.(5.15) describes the classical diffusion ($\xi^2 \approx 2t$, $E \approx D\tau/2$) independent of γ. For constant D and B the relaxation $\gamma \to 1$ is exponential

$$\xi \approx 1 - e^{-t-1}; \qquad t \to \infty \tag{5.17}$$

To explain the power law relaxation observed numerically in Refs.[48,52,53] one needs to take account of the explicit time dependence for both $D(\tau) = Ds(\tau)$ and $B(\tau) = -s(\tau)$. Notice that their ratio must be independent of time, at least asymptotically, to provide the exponential steady state (see Eq.(5.11)).

The solution for the moments (see Eq.(5.13)) can be obtined by a change of time

$$\tau \to \tau' = \int s(\tau)\, d\tau \tag{5.18}$$

Then, Eq.(5.15) shows that a power-law tail is only possible for $s(\tau) \sim \tau^{-1}$. This is in accord with the first simple estimate (5.4). Assuming

$$\tau' = \tau_R \ln\left(1 + \frac{\tau}{\tau_R}\right) \tag{5.19}$$

we arrive at the following implicit dependence $E(\tau)$:

$$e^{\xi}(1-\xi)\left(1+\frac{\tau}{\tau_R}\right)^p = 1; \qquad p = \frac{\gamma^2\tau_R}{D} = 1 \tag{5.20}$$

The value of exponent p is obtained from asymptotic relation ($\tau \to \infty$): $\xi(\tau) \sim \tau^{-p} \sim s(\tau) \sim \tau^{-1}$. Hence, the relaxation time scale is

$$\tau_R = \frac{D}{\gamma^2} \approx 2D \tag{5.21}$$

The value of $\gamma^2 \approx 0.5$ was derived from the best numerical data available [48]. It is only a half of the theoretical value (5.14). Besides, Eq.(5.20) does not contain the logarithmic dependence like Eq.(5.6) [48]. The latter seems to agree better with the numerical data for large τ. The origin of this discrepancies will be discussed below.

From the first Eq.(5.13) we can derive also implicit dependence of the staying probability on time:

$$g(0,\tau) - \frac{2}{D} = \frac{e^{-\xi(\tau)}}{(\tau_R + \tau)\xi(\tau)} \rightarrow \frac{1}{e\tau} \tag{5.22}$$

which is in agreement with numerical data in Ref.[55].

Recently, an exact solution of diffusion equation (5.10) with $B = -1$ ($n > 0$) has been found [109] in the form

$$D \cdot f(z,s) = \frac{1}{\sqrt{\pi s}} \exp\left(-\frac{(z+s)^2}{s}\right) + e^{-4z} \cdot erfc\left(\frac{z-s}{\sqrt{s}}\right) \tag{5.23}$$

where the function

$$erfc(u) = \frac{2}{\sqrt{\pi}} \int_u^\infty e^{-v^2}\, dv$$

and $z = n/2D$; $s = \tau/2D$. The dependence $E(\tau)$ can be found from the equation (see Eqs.(5.13), $D, B = const$):

$$\tilde{m}_2 = 2 - D \cdot g(0,\tau) \tag{5.24}$$

Asymptotically, as $\tau \rightarrow \infty$

$$D \cdot g(0,\tau) \rightarrow 2 + \frac{D}{\sqrt{\pi}\tau[\ln(\frac{\tau}{2D})]^{3/2}}; \qquad \dot{E}(\tau) \rightarrow \frac{4D^4}{\sqrt{\pi}\tau^2[\ln(\frac{\tau}{2D})]^{3/2}} \tag{5.25}$$

where the decrease in time of D and B is taken into account as before, via the change of time variable (5.19).

Comparison of Eqs.(5.25), (5.20) and (5.6) shows that the accuracy of the "exact" solution (5.23) is logarithmic only. This is because of the original simplifying assumption of a purely exponential steady state (5.2). The exact distribution is not known but, most likely, it contains some power-law factor [41].

In many-dimensional systems or for a quasiperiodic driving perturbation the diffusion localization is typically absent besides some special cases (see, e.g., Ref.[56]). For three freedoms or two driving frequencies the localization persists but its length is exponentially large. However, the perturbative localization, mentioned above, occurs in all cases of discrete spectrum.

On the other hand, even in the lowest dimensions under consideration the so-called delocalization is possible if the motion is allowed to be unbounded. Consider, for example, the standard map on a cylinder with the perturbation $k(n)$ depending on momentum:

$$D(n) = D_0 n^{2\alpha} \tag{5.26}$$

with some constant α. To solve this problem it is essential to assume that the backscattering remains unchanged, that is $B = -1$ as before, since it does not depend on system's parameters. Then, using Eq.(5.11), we obtain the steady state distribution in the form

$$-\ln g_s(n) = \begin{cases} \frac{2n^{1-2\alpha}}{(1-2\alpha)D_0} & \alpha \neq \frac{1}{2} \\ \frac{2}{D_0} \ln n & \alpha = \frac{1}{2} \end{cases} \tag{5.27}$$

In agreement with previous results [32] the critical value of the parameter is $\alpha_c = 1/2$. For $\alpha < \alpha_c$ the localization remains exponential while for $\alpha > \alpha_c$ delocalization occurs because $g_s(n) \to const \neq 0$ as $n \to \infty$. In the critical case the steady state distribution is a power law:

$$g_s \sim n^{-2/D_0} \tag{5.28}$$

and the localization takes place for sufficiently small $D_0 < 2$ only, when $g_s(n)$ is normalizable. Notice that for the localization of energy, that is for the mean energy $< E >=< n^2 > /2$ to be finite in the steady state, a more strong condition is required, namely

$$D_0 < \frac{2}{3} \tag{5.29}$$

This result was recently confirmed numerically in Ref.[57].

In spite of all this theoretical developments no rigorous treatment of the quantum relaxation exists so far.

The analogy with disordered solids mentioned by the end of Lecture 1, being very fruitful, is nevertheless restricted since it concerns the correspondence between eigenstates only. The properties of motion in the two problems, both dynamical and even statistical, are generally different. For example, the ratio of the localization lengths for eigenfunctions and for the steady state is different: $l_s \approx 2l$ in momentum space, and $l_s \approx 4l$ in disordered solids (see, e.g., Ref.[59]).

The most striking difference is in the absence of the diffusion stage of motion in 1D solids [110]. This is because the level density of the operative eigenfunctions

$$\rho \sim \frac{l dp}{dE} \sim \frac{l}{u} \tag{5.30}$$

which is the localization (relaxation) time scale (4.9), is always of the order of the time interval for a free spreading of the initial wave packet at a characteristic velocity u. In other words, the localization length l is of the order of the free path for backscattering. On the contrary, in momentum space, for instance, in the standard map each scattering (one map's iteration) couples $\sim k$ unperturbed states, so that $\sim k^2 \gg 1$ scatterings are required to reach the localization $l \sim k^2$.

Another (qualitative) explanation of this surprising difference is in that the density of quasienergy levels for driven systems is always higher as compared to that of energy levels. The same is true for a conservative system of two freedoms as compared with the one-freedom motion in solids. Thus, the Anderson localization is the spreading, rather than diffusion, localization.

Interestingly, the asymptotic relaxation ($t \to \infty$) in solids [110] is the same as in the momentum space (5.6). Yet, the decay of the staying probability is different ([110], cf. Eq.(5.22))

$$g_s \sim t^{-3} \tag{5.31}$$

Nevertheless, the analogy in question remains very fruitful and extensively used in the studies of quantum chaos (see, e.g., Ref.[55]).

6. The quantum steady state

The quantum diffusion localization generally results in the formation of a peculiar steady state which has no classical counterpart. The statistical relaxation to this steady state is also surprising because the motion spectrum is discrete.

The ultimate origin of this steady state is in localization of all the eigenfunctions. In a homogeneous systems like the standard map on a cylinder the localization is asymptotically exponential because the equation for eigenfunctions is linear whose behavior is described by the Lyapunov exponents in n. This is the most powerful method, borrowed from the solid-state physics, to numerically calculate localization length l [58].

However, a simple exponential dependence (5.1) is only the average behavior superimposed by big fluctuations

$$\varphi(n) \approx \frac{1}{\sqrt{l}} \exp\left(-\frac{|n|}{l} + \xi_n\right) \tag{6.1}$$

By definition $< \xi_n >= 0$ while the dispersion is not only big but grows with $| \Delta n |$ as [25]

$$< (\xi_n - \xi_m)^2 >= D_\xi \mid n - m \mid ; \qquad D_\xi \approx \frac{1}{l} \approx \frac{2}{D} \tag{6.2}$$

Nevertheless, the accuracy of numerical determination of l can be fairly high:

$$\frac{\Delta l}{l} \approx \left(\frac{l}{n}\right)^{1/2} \tag{6.3}$$

for sufficiently large n. Fluctuations ξ_n have a big impact on the steady state as was already mentioned above. Namely, they double the localization length (5.3). This is essentially numerical result, no accurate theory still exists [32]. Also, it is not clear if the steady state is purely exponential asymptotically or there is a power-law factor like in solids [59].

Initial part ($| n | \sim l$) of the distribution for both eigenstates as well as the steady state must deviate from a simple exponential dependence. Again, big fluctuations impede the direct numerical measurement. Instead, two integral characteristics were studied. One is the average energy in the steady state. For exponential localization (5.2)

$$E_s = \frac{< n^2 >}{2} = \frac{l_s^2}{4} = \frac{D^2}{4} \tag{6.4}$$

and it is in agreement with numerical results within a factor of 2.

Another integral quantity – the entropy H – was introduced in Ref.[60] (see also Refs.[13, 103]) as a different measure of quantum localization. The entropy localization length, which is also called the Shannon width, is defined as

$$l_H = e^H; \qquad H = -\sum_n | \varphi(n) |^2 \ln | \varphi(n) |^2 \tag{6.5}$$

For exponentially localized eigenfunctions

$$l_{\bar{H}} = el = \frac{eD}{2} \approx 1.4D \tag{6.6}$$

where $l_{\bar{H}} = \exp(\bar{H})$, and $\bar{H}$ is the average over all eigenfunctions. Numerically, $l_{\bar{H}} \approx D$ that is less, partly due to fluctuations which decrease entropy and $l_{\bar{H}}$ by a factor of 2. Again, deviations from exponential dependence are apparently present but not very big.

Just because l_H essentially depends on the main part of the distribution its fluctuations are much bigger as compared to those for l (6.3). Namely [61]:

$$\frac{\Delta l_H}{l_{\bar{H}}} \approx 0.5 \tag{6.7}$$

Fluctuations of entropy H were numerically found [61] to be described quite well by a simple expression ($l_{\bar{H}} \gg 1$):

$$\frac{dp}{dH} = \frac{a}{\pi \cosh[a(H - \bar{H})]} \tag{6.8}$$

with $a \approx 3$. So far there is no idea as to the explanation of this distribution.

There is another class of localized eigenfunctions which we call *Mott's states.* They were conjectured by Mott [62] in the context of the Anderson localization and further studied in Refs.[55,48,63,64]. Mott's state is also called the *double-hump* state for its shape of two exponential peaks separated by distance L (in n). These states exist in pairs of the symmetric and antisymmetric superpositions of the two peaks. The mechanism of their formation can be qualitatively explained as follows. The exponential localization is the effect of resonant backscattering, that is the backscattering on a resonant harmonic of random (or sufficiently irregular) potential. Hence, the exponentially localized states are in a sense the unperturbed ones. The perturbation (nonresonant potential) mixes them. For close unperturbed states this increases still more the fluctuations. However, for distant states a new, double-hump, structure is formed. The principal parameter is the overlapping integral

$$v = \int_{-\infty}^{\infty} dn\, \varphi_1^0(n)\, \varphi_2^0(n) \approx e^{-L/l} \left(\frac{L}{l} + 1\right) \tag{6.9}$$

which determines the energy splitting in the pair: $\Delta\varepsilon \sim v$.

We studied numerically [65] the structure of Mott's states in the standard map assuming two versions of dependence $\Delta\varepsilon(L)$:

$$l_m \omega = A e^{-L/l_m} \tag{6.10a}$$

$$l_m \omega = A \left(1 + \frac{L}{l_m}\right) e^{-L/l_m} \tag{6.10b}$$

where $\omega = T\Delta\varepsilon/2\pi$, and A is a constant. The first dependence is usually accepted in literature, the second one is suggested by parameter (6.9). Our preliminary results seem to better confirm the second law with fitting parameters

$$A \approx 0.05; \qquad l_m \approx D_n \approx l_s \approx 2l \tag{6.11}$$

The fitting to the first dependence gives a close l_m but larger $A \approx 0.15$.

In disordered solids the structure of Mott's states was directly calculated in Ref.[63] via the correlation functions. The result is of the form of Eq.(6.10a) with $A \approx 5$, and $l_m = l = l_s/4$ (cf.Eq.(6.11)).

The importance of Mott's states, for which they actually were sought, is a large matrix element

$$n_{12} = \int dn\, n\varphi_1(n)\, \varphi_2(n) \approx \frac{L}{2} \tag{6.12}$$

The latter expression holds for $L \gg l_m$. The additional logarithmic dependence in the long-time relaxation (5.6) is explained just by the effect of Mott's states in the low-frequency part of the spectrum [48].

The probability for a given unperturbed (exponential) eigenstate to form the Mott pair with $L > L_1$ can be estimated as

$$p_1 = 2\alpha \int_{L_1}^{\infty} \omega(L)\, dL = 2\alpha A e^{-L_1/l_m} \tag{6.13a}$$

$$p_1 = 2\alpha A \left(2 + \frac{L_1}{l_m}\right) e^{-L_1/l_m} \tag{6.13b}$$

for two dependences $\omega(L)$ in Eq.(6.10), respectively. In both cases $\alpha \approx 1.5$ according to our numerical experiments. The total probability $p_1 \ll 1$, and this explains why multi-hump states are very rare. We have found a few states which could be interpreted as distorted three-hump eigenfunctions.

In disordered solids $p_1 > 1$ but this is not necessarily a contradiction because Eqs.(6.10,13) are asymptotic. Nevertheless, it would be interesting to analyze the structure of Mott's states in more detail.

The time-averaged density $g_s(n)$ (5.2) determines a certain invariant measure of the quantum motion which is qualitatively different from the classical measures (microcanonical, Gibbs' etc). One important distinction is in that the former depends on initial conditions as the quantum steady state results from the localization of a spreading initial state. Moreover, if the width of initial state exceeds the localization length this dependence becomes even more complicated.

Another difference is in that the relaxation of initial state into the steady state is never as full as in the classical mechanics. For example, average quantities like energy $E_s = \langle n^2 \rangle /2$ (6.4) oscillate, and can even come back, close to the initial value E_0 since the motion spectrum is discrete.

Does it make any physical sense to speak about statistical relaxation in discrete spectrum ? In my opinion, it does. First, such Poincare's recurrences are extremely rare, and their time scale has nothing to do with characteristic relaxation time scale τ_R (4.6). Second, which is even more important, those recurrences are but large fluctuations characteristic for any statistical systems.

The same occurs in classical mechanics – for trajectories, and this is the difference. In fact, the quantum density $g(n, \tau)$ plays an intermediate role between the classical density (which would never come back for chaotic motion) and the classical chaotic trajectory with its Poincare's recurrences. Namely, the quantun density which actually describes a single quantum system represents, nevertheless, a finite statistical ensemble of $M \sim l_s$ systems. Hence, finite fluctuations in the quantum steady state. For example, the energy fluctuations

$$\frac{\Delta E_s}{E_s} \approx \frac{1.5}{\sqrt{l_s}} \sim \frac{1}{\sqrt{M}} \tag{6.14}$$

in a reasonable agreement with numerical experiments (see, e.g.,Ref.[48,52] and Fig.6. Numerical factor in Eq.(6.14) is taken from our recent computer simulations [101].

One can say also that the mixing, which is responsible for relaxation, is terminated by localization, so that the quantum mixing is only partial or a *finite-time mixing*. Such a partial relaxation with persistent fluctuations is clearly seen, for example, in Fig.6 . Notice, that a big fluctuation in this run, which is a partial recurrence towards the initial state $E_0 = 0$, is approximately symmetric with respect to the minimum of E. Moreover, the growth of the fluctuation follows the "antidiffusion" law (cf. Fig.7, $\tau > 150$) while its decay is the "normal" diffusion (cf. initial part of dependence $E(\tau)$ in Fig.7). This is another manifistation of time-reversibility in the dynamical chaos.

The smooth (up to fluctuations) steady state (5.2) is formed only if localization length $l_s \gg 2\pi/T$, the period of standard map in n. In the opposite limit $l_s \ll 2\pi/T$ the quantum measure $g_s(n)$ reveals the classical resonance structure [32]. Since quantum diffusion requires both $K > 1$ (classical border) and $k > 1$ (quantum border) this regime is only possible near $K = 1$ where the diffusion in the chaotic component is very slow:

$$l_s \approx D \approx 0.3\,(\Delta K)^3\,k^2 \sim \frac{k^2}{\tau_{cr}^3} \tag{6.15}$$

and where the resonance structure is critical with characteristic time scale τ_{cr} [32,66].

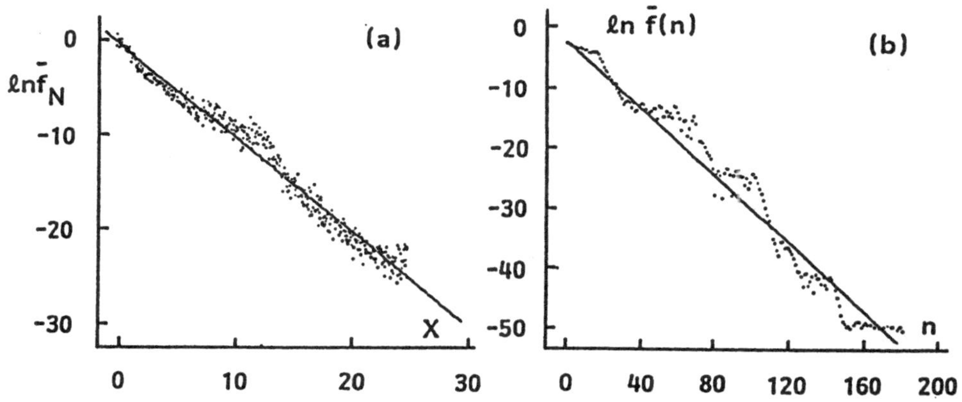

Figure 9: The quantum steady state in the standard map: a - homogeneous localization, $K = 5, k = 10, T = 0.5$, the straight line: $-\ln \bar{f}_N = x = 2n/l_s; \bar{f}_N = \bar{f}(n)2l_s/(1+x)$; b - inhomogeneous localization, $K = 1.5, k = 10, T = 0.15, l_s \approx 2, < l_s > \approx 7$, the straight line: $-\ln \bar{f}(n) = 2n/ < l_s >$ (after Ref.[32]).

The border between the two regimes is approximately at

$$l_s T \approx \frac{l_s}{k} \approx 1 \tag{6.16}$$

At the border, $\tau_{cr} \sim k^{1/3}$ as was recently confirmed in Ref.[67].

For $l_s T \ll 1$ the localization length $< l_s >$ averaged over the resonance structure is

$$< l_s > \approx \frac{k}{\sqrt{3}} \tag{6.17}$$

and the interpolation between the two regimes is approximately described by the expression [14]

$$< l_s > = \frac{l_s}{2} + \left(\frac{l_s^2}{4} + \frac{k^2}{3} \right)^{1/2} \tag{6.18}$$

Two examples of the quantum steady state are shown in Fig.9 for homogeneous (a) and inhomogeneous (b) localization, respectively.

The nature of a new time scale is a controversial question. In my understanding it characterizes the phase motion in θ rather than the excitation in n assumed in Ref.[26]. Indeed, the localization length for $l_s T \gtrsim 1$ is only $\sim k$, hence, the relaxation time scale $\tau_R \sim 1$, and does not depend on k at all. Since τ_{cr} (in my interpretation) is also of the order of local instability rise time the ratio of the two time scales $\tau_R/\tau_r \sim 1/\tau_{cr} \sim k^{-1/3} \ll 1$ in the critical structure is opposite as compared to the usual $\tau_R \gg \tau_r$.

The quantum steady state is only possible in discrete spectrum. The conditions for the latter in an unbounded quantum map remain unknown. For the standard map on cylinder the spectrum is continuous for the rational values of parameter $T/4\pi = p/q$ due to periodicity of this map in n. This results in an additional motion integral which can be termed *quasicoordinate* by analogy with the quasimomentum in spatially periodic potential [14]. By the same analogy the momentum n grows linearly in time, hence the term *quantum resonance* [15,68]. The mechanism of this resonance is especially clear in case $q = 1$ when rotation operator $\hat{R} = \hat{I}$ becomes identity (see Eq.(1.7)).

In Ref.[69] the continuous spectrum was proved to exist also for very special Liouville's (transcendental) $T/4\pi$ (see below) but if this condition is only a technical limitation remained unclear. This constitutes a very subtle mathematical problem. We shall try to discuss it using semiempirical theory of the quantum resonance [14] which leads to the expression

$$< n^2 > \approx D\tau^2 \exp\left(-\frac{q}{\pi D}\right) \tag{6.19}$$

This is the asymptotic energy growth in quantum resonance with denominator $q \gtrsim D$. A detuning $\varepsilon(q) = \mid T/4\pi - p/q \mid$ would stop the growth in time $\tau(\varepsilon)$ which we assume to satisfy the condition (see Eq.(1.7))

$$\varepsilon\tau < n^2 > = \nu \sim 1/2\pi \tag{6.20}$$

According to a few numerical results in Ref.[102] $\nu \approx 0.02$.

Consider now irrational

$$\frac{T}{4\pi} = \frac{1}{m_1 + \frac{1}{m_2 + ...}} \equiv (m_1, ..., m_i, ...) \tag{6.21}$$

$$\frac{p_i}{q_i} = (m_1, ..., m_i) \to \frac{T}{4\pi}; \qquad q_{i+1} = m_{i+1}q_i + q_{i-1}$$

where p_i/q_i are the convergents of $T/4\pi$. Comparing Eqs.(6.19-6.21) we can formulate the following conjecture: there exist infinitely many irrational values of $T/4\pi$ which provide unbounded energy growth and, hence, a continuous spectrum; moreover, $T/4\pi$ can be adjusted in such a way to achieve any desired growth rate.

Take growth law in the form

$$< n^2 > = G\tau^\gamma \tag{6.22}$$

Substituting this into Eqs.(6.19, 6.20) and excluding τ, we arrive at the relation

$$\varepsilon(q_i) = \frac{\nu}{G}\left(\frac{D}{G}\right)^{\frac{1+\gamma}{2-\gamma}} \exp\left(-\frac{q_i}{\pi D}\cdot\frac{1+\gamma}{2-\gamma}\right) \approx \frac{c}{q_i q_{i+1}} \tag{6.23}$$

where the latter expression ($c \sim 1$) follows from the continuous fraction representation (6.21) of $T/4\pi$. This relation determines a map for the construction of desired $T/4\pi$:

$$m_{i+1} \approx \frac{q_{i+1}}{q_i} \approx \frac{cG}{\nu q_i^2}\left(\frac{G}{D}\right)^{\frac{1+\gamma}{2-\gamma}} \exp\left(\frac{q_i}{\pi D}\cdot\frac{1+\gamma}{2-\gamma}\right) \tag{6.24}$$

Successive convergents determine the quantum resonances which operate in turn, each one on its own time scale

$$\tau_i = \left(\frac{\nu}{cG} q_i q_{i+1}\right)^{\frac{1}{\gamma+1}} = \left(\frac{G}{D}\right)^{\frac{1}{2-\gamma}} \exp\left(\frac{q_i}{\pi D(2-\gamma)}\right) \tag{6.25}$$

Since these time scales rapidly increase the diffusion is inhomogeneous in time, its local rate $\Gamma \equiv d < n^2 > /d\tau$ oscillating from about zero up to

$$\Gamma_{max}(\tau_i) = 2D\left(\frac{G}{D}\right)^{\frac{1}{2-\gamma}} \exp\left(\frac{q_i}{\pi D}\cdot\frac{\gamma-1}{2-\gamma}\right) \tag{6.26}$$

The ratio

$$\frac{\Gamma_{max}(\tau_i)}{<\Gamma>} = \frac{2}{\gamma} \geq 1; \qquad <\Gamma> = \gamma G \tau^{\gamma-1} \tag{6.27}$$

where $<\Gamma>$ is the mean rate from Eq.(6.22).

For maximal $\gamma = 2$ a single resonance operates according to Eq.(6.19). In the whole interval $0 < \gamma \leq 2$ the motion is unbounded, and the spectrum is (singular) continuous with a fractal structure in agreement with the rigorous results in Ref.[69]. Irrationals which are approximated by rationals to exponential accuracy, like those satisfying Eq.(6.23), are called transcendental numbers. A new conjecture is that even among those $T/4\pi$ values there are (infinitely many) such ones which provide the diffusion localization. They correspond, particularly, to $\gamma = 0$ with any finite G. A more general condition is that asymptotically

$$m_{i+1} < \exp(aq_i); \qquad a < \frac{1}{2\pi D} \tag{6.28}$$

For a particular value of $T/4\pi$ satisfying this condition the energy $E_s = G/2$ of the quantum steady state is determined by maximal G_i found from Eq.(6.23)

$$G_i^{3/2} = \frac{\nu}{c} D^{1/2} m_{i+1} q_i^2 \exp\left(-\frac{q_i}{2\pi D}\right) \tag{6.29}$$

If this $G_{max} < D^2/2$ (6.4) the resonances are irrelevant, and the usual exponential steady state is formed described by Eqs.(5.2) and(6.4). This is just the case for a typical irrational $T/4\pi$ when $G_{max} \sim \sqrt{D} \ll D$ if quasiclassical parameter $k \gg 1$ is big enough.

The change of time is a serious problem in quantum mechanics as explained above. For the steady state this problem can be solved [14] as follows. The steady state distribution is proportional to invariant measure and, hence, to (sojourn) time t. Whence, upon a change of time $t \to \tilde{t}$

$$\frac{\tilde{g}_s(n)}{g_s(n)} = \frac{d\tilde{t}}{dt} \tag{6.30}$$

the steady state distribution does change as well even though g_s does not depend on time ! Now we can change momentum n in such a way to provide $\tilde{\tilde{g}}_s(\tilde{n}) = g_s(n)$. We have

$$\frac{d\tilde{n}}{dn} = \frac{d\tilde{t}}{dt} \tag{6.31}$$

Partcularly, if $\tilde{t} = \tau$, the map time (the number of map's iterations)

$$\frac{\tilde{g}_s}{g_s} = \frac{1}{T} = \frac{d\tilde{n}}{dn} \tag{6.32}$$

where $T = 2\pi/\Omega(n)$ is map's period. If, moreover, n is action, the map momentum $\tilde{n} = E/2\pi$ is proportional to the energy (cf. Kepler map (1.10)).

7. Asymptotic statistical properties of quantum chaos

The well developed random matrix theory (RMT) (see, e.g., Refs.[70, 111, 112]) is a statistical theory which describes average properties of a "typical" quantum system. At the beginning, the object of this theory was assumed to be a very complicated, partcularly, many-dimensional quantum system as the representative of a certain statistical ensemble. With understanding the phenomenon of dynamical chaos it became clear that the number of system's freedoms is irrelevant. Instead, the number of quatum states, or the quasiclassical parameter, is of importance.

Until recently the ergodicity of eigenfunctions, that is the absence of any operators commuting with the Hamiltonian, was assumed. Of course, that is not always the case (for a very interesting and instructive review of first attempts to prove the quantum ergodicity, see Ref.[71]). One of a few rigorous results in quantum chaos is an old theorem due to Shnirelman (announced in Ref.[72] with a full proof published only now [73]). Loosely speaking the theorem states that the classical ergodicity implies the ergodicity of most quantum eigenfunctions sufficiently far in the quasiclassical region that is for sufficiently large quantum parameters. The quantum ergodicity was further discussed in Refs.[75] and well confirmed in numerical experiments with quantum billiards [21].

Shnirelman's definition of quantum ergodicity is of an integral type

$$\int dp\,dq\,W_n(p,q)\,f(p,q) \to \int dp\,dq\,g_\mu(p,q)\,f(p,q) \qquad (7.1)$$
$$n \to \infty$$

for any sufficiently smooth function f of the phase space. Here W_n are Wigner eigenfunctions, and

$$g_\mu = \delta(H(p,q) - E)\frac{dE}{dp\,dq} \qquad (7.2)$$

is microcanonical (ergodic) measure. The quantity $\rho(E) = dpdq/dE$ is the classical counterpart of the mean level density.

To understand the quantum limitations of ergodicity and the importance of the quasiclassical asymptotics ($n \to \infty$) we consider as an example the Rydberg atom in magnetic field (see Eq.(1.15)).

In Ref.[76] the eigenfunctions of this model were found, for chaotic motion in the classical limit, in the form

$$\psi_i = c\sum_m \frac{\varphi_m}{\sqrt{\Omega(m)}} \qquad (7.3)$$

Here c is normalizing constant, φ_m are some unperturbed eigenfunctions with a fixed quantum number m, and

$$\Omega(m) = 2^{3/2}\left(\omega\left(m+\frac{1}{2}\right) - E\right)^{3/2} \approx \frac{1}{n^3} \qquad (7.4)$$

is the electron longitudinal frequency depending on quantum numbers n, m. In the classical limit the ergodic measure is

$$g_\mu(m) = \int dn \frac{\delta(H(m,n) - E)}{\rho(E)} \approx \frac{1}{\rho\Omega} \qquad (7.5)$$

where $\Omega = \partial H/\partial n$.

In quantum mechanics this measure is discrete, and to satisfy ergodicity (7.1) the change in Ω must be small, hence, $\omega \to 0$, and $m \sim \mid E \mid /\omega \to \infty$. On the other hand, classical ergodicity (chaos) takes place under condition [18]

$$\varepsilon = \frac{|E|}{\omega^{2/3}} \lesssim 1 \tag{7.6}$$

Therefore, the condition for quatum ergodicity is

$$\omega \ll \varepsilon^3 \tag{7.7}$$

The RMT operates with finite matrices $N_m \times N_m$ so that expansion similar to Eq.(7.3)

$$\psi_i = \sum_j a_{ij}\varphi_j \tag{7.8}$$

is always finite, the ergodicity meaning that

$$< | a_{ij} |^2 >= \frac{1}{N_m} \tag{7.9}$$

In other words, all probabilities $| a_{ij} |^2$ are equal at average. This is not the case in a physical system whose energy shell, corresponding to the classical energy surface, is bounded. Hence, the conventional RMT is a *local theory* applicable far within a quantum energy shell. We will come back to this important question below.

Statistical properties of quasienergy eigenstates (for driven systems) were first studied in Refs.[77,78] (see also Ref.[13]) using, as a model, the standard map on a torus. Owing to condition (1.5) the parameter $T/4\pi = m/2C$ is rational. But for a finite system, with C states, the spectrum is discrete, of course, so that no delicate problems, discussed above, arise. This model represents the quantum dynamics within the energy shell of a two-freedom conservative system.

The ergodicity depends on the parameter

$$\lambda = \frac{l}{C} = \frac{D}{2C} \tag{7.10}$$

and corresponds to large values of the latter. In the quasiclassical region $\lambda \sim \frac{K}{m}k \to \infty$ ($K = kT$ and $m = CT/2\pi$ remain constant). Thus, sufficiently high quantum states are ergodic in accordance with the Shnirelman theorem.

The structure of ergodic eigenstates well agrees with the prediction of RMT, namely, the fluctuations are nearly Gaussian with the probability density

$$p(a) = \frac{\Gamma(N_m/2)}{\sqrt{\pi}\,\Gamma\left(\frac{N_m-1}{2}\right)}(1-a^2)^{\frac{N_m-3}{2}} \approx \left(\frac{N_m}{2\pi}\right)^{1/2} \exp\left(-\frac{N_m a^2}{2}\right) \tag{7.11}$$

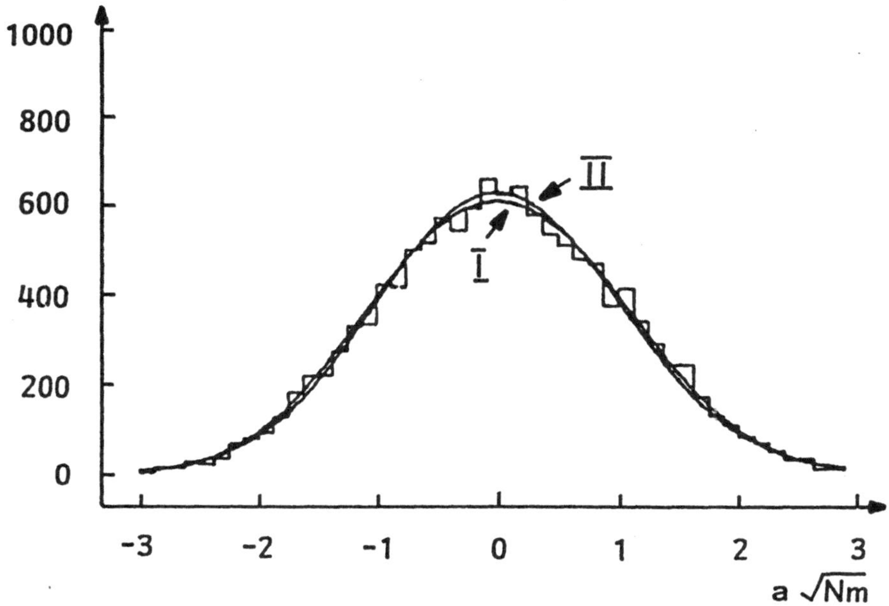

Figure 10: Fluctuations in ergodic eigenfunctions for the standard map on a torus: $K \approx 20, k = 20, T/4\pi = 4/51$; I - RMT, II - Gaussian approximation (after Ref.[78]).

Here a, assumed to be real, stand for amplitudes in expansion (7.8). Interestingly, a slight difference between the two distributions was clearly observed in Ref.[78] for $N_m = 25$ using the χ^2 criterion (Fig.10).

Big spatial fluctuations in a chaotic eigenstate are not completely random but reveal the structure of classical periodic trajectories. This interesting phenomenon had been discovered by Heller in numerical experiments with the quantum stadium billiard [79], and was subsequently confirmed by many others (see, e.g., Ref.[80]), particularly, in quantum maps. The microstructure was observed so far as some enhancements along classical periodic trajectories in both configurational and phase spaces. Such enhancements were termed *"scars"* by Heller.

A general theory of scars in conservative system with arbitrary number of freedoms N was developed by Berry [81,10] (see also Ref.[82]). He made use of the Wigner function W which is the quantum counterpart of the classical fine-grained phase space density. Notice that W is generally not positively definite.

Within a scar W forms complicated diffraction fringes, rapidly oscillating and rather extended along the energy surface. The relative width of the central fringe contracts with the quantum number n as $\sim n^{-1/2}$. In this sense the scars have essentially quantum structure which vanishes in quasiclassical region. Yet, this transition to the classical limit is not a trivial one as the fringe amplitude does not depend on n. To get rid of scars one needs a coarse-grained (averaged) density W, which is called also the Husimi distribution, and which is positively definite. Then average density of a scar vanishes $\sim n^{-(N-1)}$.

As the scars are maximally localized (essentially within one quantum cell of the phase space) they do not violate Shnirelman's integral ergodicity (7.1). However, it is not completely clear why they are not seen in the fluctuations of eigenfunctions (Fig.10).

According to Berry's theory the Wigner chaotic eigenfuntion can be approximately represented as a sum over classical periodic trajectories:

$$W(x) \approx \frac{dE}{dx}\delta(E - H(x))\times$$
$$\times \left[1 + \nu \sum_s \exp\left(-\frac{N-1}{2}\Lambda_s T_s\right) \cos(S_s + \gamma_s)\, \delta(X_s)\right] \qquad (7.12)$$

Here $x = (p, q)$ is a point in $2N$-dimensional phase space while $X = (P, Q)$ describe $2(N-1)$-dimensional Poincaré section transverse to a periodic trajectory at $X = 0$. The periodic trajectory is characterised by action S and quasiclassical phase as well as by instability rate Λ and period T. Each term in sum (7.12) represents a scar which, by the way, can be of any sign, that is it may produce both a bump or a dip in phase density W. Explicit expression for $\delta(X)$ is given in Refs.[81,10], and ν is some numerical factor.

A difficult mathematical problem in this theory is apparent strong divergence of series (7.12) since the number of periodic trajectories with $T_s < T$ grows as $\exp((N-1)\Lambda T)$ (see, e.g.,Ref.[83]). One way to approach this problem is as follows [54]. Let us try to consider Eq.(7.12) as an expansion in the basis of certain "coherent" states, the "scars"

$$W_s = \frac{1}{T_s}\delta(X_s)\delta(E - H(x)); \qquad \int W_s dx = 1 \qquad (7.13)$$

which are localized on periodic trajectories. A peculiar property of such coherent states is in that they are stationary that is they don't move in phase space, nor they are spreading. The mechanism of localization is essentially the same as for the diffusion discussed above but now it concerns the exponential spreading of a narrow wave packet prior to diffusion. The difference is in the level density which, for a scar, is $\rho_s \sim T_s$. Hence, the time scale for the localization of instability is T_s, and this is a simple explanation of the exponential factor in Eq.(7.12).

Oscillating $\delta(X)$ tails of unknown length overlap to produce somehow the average ergodic (microcanonical) distribution $\sim \delta(E - H(x))$ (see Eq.(7.12)) as well as the Gaussian fluctuations discussed above. The total number of separated scars is $\sim n^{N-1}$. Since the number of periodic trajectories grows as $\exp((N-1)\Lambda T)$ the longest period T_m of the basis scars is given by the estimate

$$\Lambda T_m \sim \ln n \tag{7.14}$$

and it coincides with the random time scale (4.1) ($q \sim n$). This is the time interval for a wave packet spreading over the whole energy surface. The scars with longer periods $T_s \geq T_m$ are not separated from each other, that is even their central fringes do essentially overlap, hence they are crucially modified. As a crude approximation one can simply drop these higher terms which makes series (7.12) trivially convergent. It is not excluded that this approach could provide some physical justification for a formal procedure of smoothing $\delta(E-H)$ [10]. It is essential that under a natural assumption of random phases in Eq.(7.12) [26] the divergence is only logarithmic in T and, hence, insensitive to the exact truncation border.

In a recent theory of the series over classical periodic orbits [105] the truncation border was found to be at $T_{max} \sim t_R$, the relaxation time scale. In this theory the series represent quantum eigenvalues. A natural conjecture is that the truncation T_{max} in both cases is of the order of the corresponding localization time scale.

Another characteristic statistical property of chaotic eigenstates is the distribution of their eigenvalues, the energies. Particularly, the spacings s between neigbouring levels are distributed, according to RMT, as

$$p(s) \approx A\, s^{\beta}\, e^{-Bs^2} \tag{7.15}$$

where A, B are obtained from normalization and condition $<s>=1$.

In the old RMT the level repulsion parameter β could take 3 values only ($\beta = 1;2;4$) depending on system's symmetry. In Refs.[77,13] this property was confirmed for ergodic quasienergy eigenstates as well.

A new problem is the impact of localization on the statistical properties of chaotic eigenstates. It was firsrt adressed in Ref.[60] for the quantized standard map on a torus to discover a new class of spacing statistics which is now called the *Izrailev distribution*:

$$p(s) \approx A s^{\beta} \exp\left(-\frac{\pi^2}{16}\beta s^2 - \left(B - \frac{\pi\beta}{4}\right) s\right) \tag{7.16}$$

where now β is a continuous parameter in the whole interval (0,4). This semiempirical relation was found using Dyson's model of charged bars on a ring. In this model the parameter β, which is the inverse bar temperature, can take any value. Yet, for the level repulsion of ergodic eigenstates only

3 values, given above, make sense. Izrailev has found that the intrermediate values describe localized eigenstates. The Izrailev distribution is also called *intermediate statistics* as contrasted to the *limiting statistics* (7.15) for ergodic states. This intermediate statistics should be distinguished from another one proposed in Ref.[84] to account for the lack of ergodicity in the classical limit. Earlier a few cases of big deviations of unknown nature from the limiting statistics (mainly in heavy nuclei) were described by a purely empirical *Brody's distribution* ($0 \leq \beta \leq 1$):

$$p(s) = A\, s^{\beta}\, e^{-Bs^{1+\beta}} \tag{7.17}$$

The next important step would be to relate parameter β in Eq.(7.16) to the localization length l or rather to the ergodicity parameter $\lambda = l/C$. Instead, Izrailev introduced a new ergodicity parameter

$$\beta_{\bar{H}} = \exp(\bar{H} - H_e) \approx \frac{2l_{\bar{H}}}{C} \tag{7.18}$$

Here $\bar{H}, l_{\bar{H}}$ are the average entropy of eigenstates and corresponding length, respectively (see Eq.(6.5)); and $H_e \approx \ln(C/2)$ is the entropy of an ergodic state which is less than maximal ($\ln C$) owing to fluctuations (7.11). Surprisingly, the new parameter $\beta_{\bar{H}} \approx \beta$ proved to be very close to the repulsion parameter β of intermediate statistics (7.16). Why this relation is so simple remains an open question.

Particularly, in case of strong localization ($\beta_{\bar{H}} \ll 1$) the spacing distribution (7.16) approaches the Poisson law

$$p(s) = e^{-s} \tag{7.19}$$

which originally was associated with the completely integrable systems and regular dynamics. Also, this limiting case shows that Eq.(7.16) is an approximation because clearly $p(0) \neq 0$ for sufficiently small $\beta_{\bar{H}}$. At most, the residual level repulsion could be exponentially small.

In any event, this limit explains the absence of repulsion for Anderson localization in infinite disordered solids. Yet, in a finite sample the repulsion must appear which is also an interesting mathematical problem.

Notice, that Poisson distribution holds only for all levels. For the operative eigestates, which determine the quantum dynamics, the repulsion reappears again. This is another difficult problem.

The level repulsion does not change the relaxation time scale (4.9) but can modify the relaxation tail (see, e.g, Eq.(5.5) and Ref.[55]). In this context an interesting question concerns the repulsion among specific Mott's states (6.10). For each pair of such states the repulsion is very strong in the sense that their spacing is bounded from below by ovelapping untegral (6.9). On the other hand, the total number of Mott's pairs increases as the spacing (ω) decreases owing to the growth of state's size L. Both effects seem to cancel, and the integral repulsion vanishes. Indeed, from Eqs.(6.13a) and (6.10a) (both versions (a) and (b) are asymptotically equivalent), we have

$$p_1 = 2\alpha l_m \omega \tag{7.20}$$

This is in apparent disagreement with numerical results in Ref.[55] where the level attraction was inferred from the asymptotic behavior of the staying probability (5.22). However, this conclusion is very sensitive to the exact relaxation law. On the other hand, our result is in agreement with another relaxation (5.6) observed in Ref.[48]. To conclude, this question certainly requires further studies.

Empirical dependence $\beta_{\bar{H}}(\lambda)$ was found in Refs.[60,85,86]. Parameter $\beta_{\bar{H}}$ was defined by Eq.(7.18) with the entropy averaged over all eigenstates. The dependence can be approximately described by two expressions

$$\beta_{\bar{H}} \approx \begin{cases} \frac{4\lambda}{1+4\lambda} & \lambda \leq 0.5 \\ 1 - \frac{1}{4\sqrt{\lambda}} & \lambda \geq 0.1 \end{cases} \tag{7.21}$$

Besides the limit $\lambda \to 0$ there is no explanation of this dependence so far, nor even the physical mechanism underlying Eq.(7.21) has been identified.

For example, we could use a simple Eq.(5.1) for localized eigenstates. On a torus it becomes

$$\varphi_m(n) \approx \left(\frac{2\lambda}{1+\lambda\sinh(1/\lambda)}\right)^{1/2} \cosh\left(\frac{|\,m-n\,|}{l}\right) \tag{7.22}$$

with the Izrailev ergodicity parameter

$$\beta_{\bar{H}} \approx \begin{cases} 2e\lambda\left(1+\frac{e^{-1/\lambda}}{\lambda}\right) & \lambda \ll 1 \\ 1 - \frac{1}{360\lambda^4} & \lambda \gg 1 \end{cases} \tag{7.23}$$

that is quite different from Eq.(7.21). Thus, the real dependence $\beta_{\bar{H}}(\lambda)$ is related to deviations from simple eigenfunction shape (5.1).

Remarkably, dependence (7.21) has the nature of a scaling in the sense that $\beta_{\bar{H}}$ and $\beta \approx \beta_{\bar{H}}$ depend on the ratio $\lambda = l/C = D/2C$ only, whatever the underlying mechanism could be.

The importance of this scaling is in that both quantities, β and λ, are invariant with respect to the rotation of the basis in Hilbert space whereas the intermediate quantities, β_H and H, are not.

The standard map on a torus can be considered also as a model for the so-called ***longitudinal localization*** in conservative systems that is one along the energy shell which destroys the ergodicity.

The ststistical counterpart of the theory of quantum localization is not only old Anderson's theory but also a new development in RMT which makes use of the so-called band random matrices (BRM, see, e.g., Refs.[87]). These have nonzero random elements within a band of width $2b$ along the main diagonal only. They are defined in a certain physically significant basis, and also are not invariant under basis rotation.

The unitary matrix in quantized standard map (1.6) is also of a band structure with $b \approx k$ but nonrandom elements. This similarity suggests that appropriate scaling parameter would be [88]

$$\lambda_r = r\frac{b^2}{N_m} \tag{7.24}$$

where N_m is matrix size, and r some numerical factor. All matrix elements are assumed to have the same statistical properties. Indeed, the scaling $\beta_H(\lambda_r)$ is similar but not identical to that for the dynamical problem (7.21). In fact, the first dependence is the same for $r \approx 1.5$ and it persists even farther, up to $\lambda_r \approx 3$. The second region ($\beta_H \approx 1$) is apparently different but it has not yet been studied in detail. Notice, that the origin of the difference can be attributed not so much to the distinction between random and regular matrix elements as to the different boundary conditions for a square matrix and a torus.

For $\lambda \ll 1$ the matrix of eigenfunctions a_{ij} (7.8) is also a band matrix with a_{ij} smoothly decreasing off the diagonal but with a much larger effective width ($\sim b^2$).

In a conservative system the BRM may represent both longitudinal as well as ***tranverse localization.*** The latter is related to a finite width of the energy shell. The relative width depends on a particular dynamics, and vanishes in the classical limit when shell becomes a surface. Transverse localization is a universal phenomenon independent of motion's ergodic properties. The type of localization depends on the structure of matrix elements. If their distribution along the main diagonal is homogeneous the longitudinal localization only is represented, generally with the intermediate statistics as described above. To account for a finite energy shell the diagonal (unperturbed) matrix elements have to grow, at average, along the diagonal. If, moreover, the eigenfunctions are ergodic the limiting statistics (7.15) persists in spite of localization as was found empirically in heavy nuclei and atoms as well as in simple dynamical models like billiards [21, 70, 111]. However, distant correlations among many levels may change on the scale of energy shell's width N_E. Generally, the old RMT describes the local quantum structure only, that is for $N_m \lesssim N_E$, even for ergodic eigenstates. The global structure is associated with band matrices. The former approximation is very good, for example, in heavy nuclei ($N_E \sim 10^6$) but not in heavy atoms ($N_E \sim 10$ only) [89].

A new type of statistical properties for the quantum chaos has been introduced recently in Ref.[90]. It is the statistics of bands (or gaps) in the fractal spectrum of a particle in the quasiperiodic critical potential. For a particular model the band "attraction" (or clustering) was found with the parameter $\beta \approx -3/2$ (cf.Eq.(7.15)) in the limit of small gaps. The attraction parameter characterizes also the fractal dimensions of the spectrum $d_f = -\beta - 1 = 1/2$ in this model. Apparently, the same statistics can be applied to the nonresonant unbounded motion in the standard map (see Eqs.(6.22)-(6.24)).

Also, I would like just to mention (and to attract attention to) a very interesting and less known theorem due to Shnirelman [91] (for the proof see Ref.[73]). It is related to the KAM integrability which is intermediate between the complete integrability with independent levels (see Eq.(7.19)) and the quantum chaos with level repulsion (7.16). The KAM structure is highly intricate as its chaotic part, being of exponentially small measure, is everywhere dense.

In quantum mechanics the beautiful Shnirelman theorem, which even does not need translation, asserts:

$$\forall N \exists C_N > 0, \quad \forall n > 1 \quad min(\lambda_{n+1} - \lambda_n, \lambda_n - \lambda_{n-1}) < C_N n^{-N} \tag{7.25}$$

where λ_n^2 are the energy eigenvalues. Thus, asymtotically as $n \to \infty$, a half of level spacings is exponentially small. A striking difference from both the complete integrability and quantum chaos !

8. Conclusion: the quantum chaos and traditional statistical mechanics

The dunamical chaos in classical mechanics seems to be a fundamentally new mechanism underlying statistical laws in physics as compared to the traditional ("old") statistical mechanics (TSM). It is indeed ! The only problem with this mechanism is in that the classical chaos does not exist, strictly speaking, as our world is quantal. Now, in quantum mechanics the chaos is waning and becoming a sort of pseudochaos which only mimics some properties of the "true" chaos and, moreover, on finite time scales only. Besides, it turns out that such a quantum chaos is rather similar in mechanism to TSM [92, 93].

Let us consider these complicated relations in some detail. The paradigm of TSM is the many-dimensional *linear* oscillator which can be described by the matrix of coefficients in its quadratic Hamiltonian. This is a completely integrable system with purely discrete spectrum. But the same is true for a broad class of quantum systems as described by Hamiltonian or unitary matrices. In both cases the main *dynamical* problem is to diagonalize the matrix, and to find its eigenvalues and eigenvectors. The principal difference is in the nature of matrix's size. In TSM it is the number of freedoms N while the quantum counterpart is that of states n.

If any of these parameters is big the *statistical* description becomes meaningful. In TSM it is achieved, in the formal theory, by taking the *thermodynamic limit* $N \to \infty$. Then the spectrum becomes continuous *if the eigenfunctions are delocalized.* This is indeed the case, under certain conditions, and not only for the simple linear oscillator but also for a broad class of completely integrable systems (see, e.g., [4] and references therein). Moreover, in the thermodynamic limit the completely integrable system (for any finite N) becomes a K system with positive (nonzero) KS entropy. This is a very strong statistical property.

In quantum mechanics we have the classical limit $n \to \infty$ with its new dynamical chaos. Yet, the main problem in quantum chaos is *finite* (no matter how large) n. This semiclassical region is characteristic for the quantum chaos. Actually, the same problem exists in TSM as well. What would be the impact of finite N on the statistical properties here? From the studies of quantum chaos we know that one still can speak about statistical relaxation in spite of the discrete spectrum. A striking example of such a process was observed in old numerical experiments [94] with the completely integrable Toda lattice of 5 freedoms only! The transition to pseudochaos and statistical relaxation in this simple model is shown in Fig.11.

Thus, a new phenomenon - quantum chaos - turns out to be the old TSM of completely integrable systems, both classical and quantal, under $N \to \infty$. Moreover, the quantum chaos provides a new insight into the old theory as to the impact of a finite N on the statistical properties. Such a reconcilation of the two apparently unrelated theories seems to be very satisfactory from the physical point of view.

The interrelation between the two mechanisms of chaos becomes especially close in a particular class of models described by the Nonlinear Schrödinger equation (NSE). The simplest NSE is known to be completely integrable

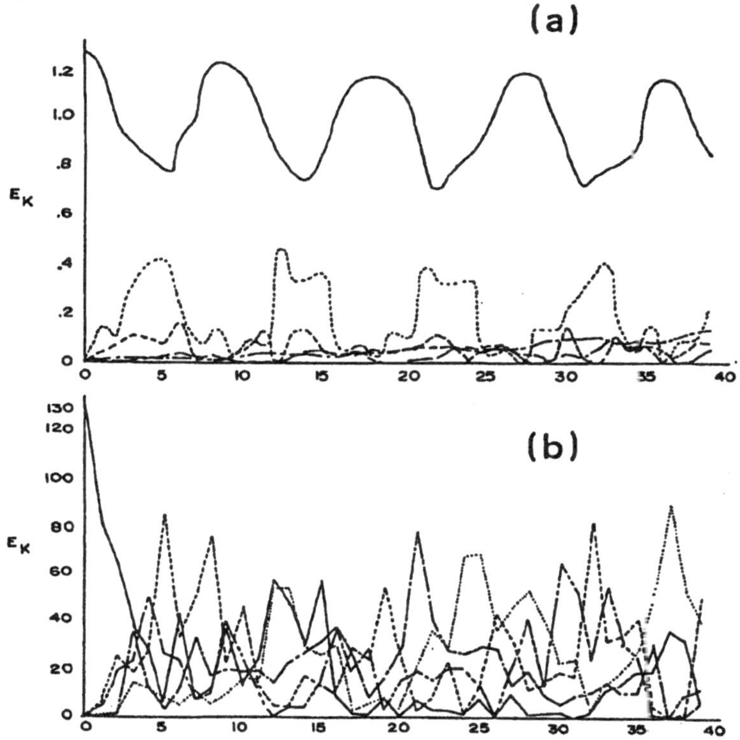

Figure 11: Classical pseudochaos in the Toda lattice: the time dependence of harmonic normal mode energies $E_k(t)$, $k = 1-5$, is shown; the total energy $E = 1.32$ (a) and 132 (b) (after Ref.[94]).

but some additional perturbation, either driving or conservative, can produce already the true (asymptotic) chaos in this quantum system [96]. Depending on the physical nature of the nonlinearity ($| \psi |^2$) it can be interpreted as the classical freedoms of motion. An example is the interaction of some quantum system with even a single mode of the electromagnetic field with infinitely many quanta. Such a system is called *semiclassical* or *partly classical* one. For any finite number of quanta the NSE is an approximation, the so-called mean-field approximation. Physically, the chaotic solutions are just the result of this approximation and hence an artifact. Yet, in the classical limit for that single freedom only the NSE becomes exact which demonstrates that even a single classical freedom is sufficient for the true chaos (an instructive analysis of such a model is presented in Ref.[97]).

On the other hand the above limit can also be interpreted as an infinite quantum system in which the old mechanism is operative. This is especially clear for the model discussed in second Ref.[96]. The interaction of many electrons is described here in the mean-field approximation by the NSE for one of them. A remarkable peculiarity of this example is in that the old mechanism for chaos is explicitly reduced to the new one, the dynamical chaos. This is also some explanation how the exponential local instability arises, in the thermodynamic limit, within a completely integrable system.

In the very conclusion I would like to make a few comments on the problem of quantum measurement. The studies in quantum chaos suggest that the latter may have close relation to this problem [95]. First, the measurement device is by purpose a macroscopic system for which the classical description is a very good approximation. Together with the measured quantum microsystem it forms a semiclassical object in which the true chaos is already possible as discussed above. Further, the chaos in the measurement device is not only possible but inavoidable because the latter has to be, by purpose again, a highly unstable system. Indeed, a microscopic interaction produces, in the process of measurement, the macroscopic effect.

The importance of chaos in the quantum measuremant is in that it destroys coherence of the initially pure quantum state converting it into the incoherent mixture. In the existing theories this is described as the effect of some external noise. In the standard map, for example, such a process was studied in Ref.[98]. Typically, a sufficiently weak noise does not affect the classical-like diffusion on relaxation time scale t_R (4.9). Yet, even arbitrarily weak noise destroys coherent localization and provides finite and permanent diffusion rate D_N where

$$\frac{D_N}{D} \sim \begin{cases} D\tilde{D} & D\tilde{D} \leq 1 \\ 1 & D\tilde{D} \geq 1 \end{cases} \tag{8.1}$$

Here $\tilde{D}$ is the diffusion rate under noise only. A sufficiently strong noise restores the permanent classical diffusion (for $1/D \leq \tilde{D} \leq D$). Notice that the critical noise level $\tilde{D}_{cr} \sim 1/D \to 0$ in the classical limit as $D \to \infty$.

A more interesting effect recently under intensive studies (see, e.g., [99] and references therein) is in that the noise of a special type substantially inhibits the quantum transitions preserving the initial state. This effect is similar to the impact of quantum measurements but unlike the latter admits the dynamical description (cf. Ref.[100] on the quantum Zeno effect).

The chaos theory allows to get rid of the unsatisfactory inclusion of external noise, and to develope a purely dynamical theory for the loss of quantum coherence. Particularly, the special type of noise in the latter example is related to the specific construction of the device for the measurement of a given quantity.

This is *almost* dynamical theory of quantum measurement except one, perhaps most difficult and important, link in the chain - the probability redistribution according to the result of the particular measurement (for discussion see Ref.[14]). The main difficulty here is in that a certain modification of the quantum mechanics appears to be inavoidable, and not simply the studies of solutions to the known fundamental equations.

This seems to be a very intriguing problem but it certainly goes far beyond the scope of my lectures and, perhaps, of the whole physics.

Acknowledgements

I would like to express my deep gratitude to the organizer of the School, Professor Dieter Heiss, for the invitation to this interesting School in the fascinating environment and conditions for both creative work and active recreation. I am very much indebted also to the participants of the School for their interest and stimulating discussions which much improved my understanding of the subject.

References

1. P.Peebles, *The Large-Scale Structure of the Universe*, Princeton Univ. Press, 1980.
2. P.Glansdorff and I.Prigogine, *Thermodynamic Theory of Structure, Stability and Fluctuations*, Wiley, 1972.
3. H.Haken, *Advanced Synergetics*, Springer, 1983.
4. I.Kornfeld, S.Fomin and Ya.Sinai, *Ergodic Theory*, Springer, 1982.
5. A.Lichtenberg, M.Lieberman, *Regular and Stochastic Motion*, Springer, 1983; G.M.Zaslavsky, *Chaos in Dynamic Systems*, Harwood, 1985.
6. B.V.Chirikov, Phys.Reports **52** (1979) 263.
7. V.I.Arnold and A.Avez, *Ergodic Problems of Classical Mechanics*, Benjamin, 1968.
8. B.V.Chirikov and V.V.Vecheslavov, in: *Analysis etc*, Eds. P.Rabinowitz and E.Zehnder, Academic Press, 1990, p. 219.
9. Proc. Les Houches Summer School on Chaos and Quantum Physics, Elsevier, 1991.
10. M.Berry, Some Quantum-to-Classical Asymptotics, Ref.[9].
11. B.Eckhardt, Phys.Reports **163** (1988) 205.
12. G.Casati and L.Molinari, Suppl.Prog.Theor.Phys. **98** (1989) 287.
13. F.M.Izrailev, Phys.Reports **196** (1990) 299.
14. B.V.Chirikov, Time-Dependent Quantum Systems, Ref.[9].
15. G.Casati et al, Lecture Notes in Physics **93** (1979) 334.
16. G.Casati et al, Phys.Reports **154** (1987) 77.
17. G.Casati et al, Phys.Rev. A **36** (1987) 3501.

18. H.Friedrich and D.Wintgen, Phys.Reports **183** (1989) 37.
19. Ya.G.Sinai, Usp.Mat.Nauk **25** #2 (1970) 141.
20. L.A.Bunimovich, Zh.Eksp.Teor.Fiz. **89** (1985) 1452.
21. O.Bohigas and M.Giannoni, Lecture Notes in Physics **209** (1984) 1.
22. P.Šeba, Phys.Rev.Lett. **64** (1990) 1855; J.Krug, ibid **59** (1987) 2133; R.Prange and S.Fishman, ibid **63** (1989) 704.
23. B.V.Chirikov, Linear Chaos, preprint INP 90-116, Novosibirsk, 1990.
24. S.Fishman et al, Phys.Rev. A **29** (1984) 1639.
25. D.L.Shepelyansky, Physica D **28** (1987) 103.
26. R.Prange, these proceedings.
27. V.M.Alekseev and M.V.Yakobson, Phys.Reports **75** (1981) 287.
28. G.Chaitin, *Information, Randomness and Incompleteness*, World Scientific, 1990.
29. B.V.Chirikov, Chaos, Solitons and Fractals **1** (1991) 79.
30. D.Umberger and D.Farmer, Phys.Rev.Lett. **55** (1985) 661.
31. M.Kac, *Statistical Independence in Probability, Analysis and Number Theory*, Math. Association of America, 1959.
32. B.V.Chirikov and D.L.Shepelyansky, Radiofizika **29** (1986) 1041.
33. A.Rechester et al, Phys.Rev. A **23** (1981) 2664.
34. V.V.Vecheslavov, private communication, 1991.
35. M.Born, Z.Phys. **153** (1958) 372.
36. G.Casati et al, Phys.Lett. A **77** (1980) 91.
37. B.V.Chirikov, Intrinsic Stochasticity, in: Proc. Int. Conf. on Plasma Physics, Lausanne, 1984, Vol. 2, p. 761.
38. M.Gell-Mann and J.Hartle, Quantum Mechanics in the Light of Quantum Cosmology, in: Proc. 3rd Int. Symposium on the Foundations of Quantum Mechanics in the Light of New Technology, Tokyo, 1989.
39. J.Ford, Quantum Chaos: Is there any?, in: Directions in Chaos, Ed. Hao Bai-Lin, World Scientific, 1987; F.Benatti et al, Lett.Math.Phys. **21** (1991) 157.
40. A.N.Gorban and V.A.Okhonin, Universality Domain for the Statistics of Energy Spectrum, preprint KIP-29, Krasnoyarsk, 1983.
41. B.V.Chirikov, F.M.Izrailev and D.L.Shepelyansky, Sov.Sci.Rev. C **2** (1981) 209; Physica D **33** (1988) 77.
42. S.Weigert, Z.Phys. B **80** (1990) 3.
43. M.Berry, True Quantum Chaos? An Instructive Example, Proc. Yukawa Symposium, 1990.
44. E.V.Shuryak, Zh.Eksp.Teor.Fiz. **71** (1976) 2039.
45. M.Berry, Proc.Roy.Soc.Lond. A **413** (1987) 183.
46. G.P.Berman and G.M.Zaslavsky, Physica A **91** (1978) 450.
47. M.Toda and K.Ikeda, Phys.Lett. A **124** (1987) 165; A.Bishop et al, Phys.Rev. B **39** (1989) 12423.
48. D.Cohen, Phys.Rev. A **44** (1991) 2292.

49. D.L.Shepelyansky, Quasiclassical Approximation for Stochastic Quantum Systems, preprint INP 80-132, Novosibirsk, 1980; V.V.Sokolov, Teor.Mat.Fiz. **61** (1984) 128.
50. D.L.Shepelyansky, Physica D **8** (1983) 208; G.Casati et al, Phys.Rev.Lett. **56** (1986) 2437.
51. P.Allen, J.Phys. C **13** (1980) L166.
52. B.V.Chirikov, Usp.Fiz.Nauk **139** (1983) 360.
53. G.P.Berman and F.M.Izrailev, On the Dynamics of Correlation Functions and Diffusion Limitation in the Region of Quantum Chaos, preprint KIP-497, Krasnoyarsk, 1988; Correlations and Diffusion Suppression in Quantum Chaos, preprint KIP-663, Krasnoyarsk, 1990.
54. B.V.Chirikov, CHAOS **1** (1991) 95.
55. T.Dittrich and U.Smilansky, Nonlinearity **4** (1991) 59; 85.
56. G.Casati et al, Phys.Rev.Lett. **59** (1987) 2927.
57. F.Benvenuto et al, Z.Phys. B **84** (1991) 159.
58. D.L.Shepelyansky, Phys.Rev.Lett. **56** (1986) 677.
59. I.M.Lifshits, S.A.Gredeskul and L.A.Pastur, *Introduction to the Theory of Disordered Systems*, Wiley, 1988.
60. F.M.Izrailev, Phys.Lett. A **134** (1988) 13; J.Phys. A **22** (1989) 865.
61. B.V.Chirikov and F.M.Izrailev, Statistics of Chaotic Quantum States: Localization and Entropy (in preparation).
62. N.Mott, Phil.Mag. **22** (1970) 7.
63. L.P.Gorkov et al, Zh.Eksp.Teor.Fiz. **84** (1983) 1440; **85** (1983) 1470.
64. R.Blümel et al, J.Chem.Phys. **84** (1986) 2604.
65. G.Casati, B.V.Chirikov and F.M.Izrailev, The Structure of Mott's States (in preparation).
66. S.Fishman et al, Phys.Rev.Lett. **53** (1984) 1212; Phys.Rev. A **36** (1987) 289.
67. J.Jensen and O.Niu, Phys.Rev. A **42** (1990) 2513.
68. F.M.Izrailev and D.L.Shepelyansky, Teor.Mat.Fiz. **43** (1980) 417.
69. G.Casati and I.Guarneri, Comm.Math.Phys. **95** (1984) 121.
70. T.Brody et al, Rev.Mod.Phys. **53** (1981) 385; F.Haake, *Quantum Signatures of Chaos*, Springer, 1990.
71. P.Pechukas, J.Phys.Chem. **88** (1984) 4823.
72. A.I.Shnirelman, Usp.Mat.Nauk **29** #6 (1974) 181.
73. A.I.Shnirelman, On the Asymptotic Properties of Eigenfunctions in the Regions of Chaotic Motion, addendum in Ref.[74].
74. V.F.Lasutkin, The KAM Theory and Asymptotics of Spectrum of Elliptic Operators, Springer, 1991.
75. M.Berry, J.Phys. A **10** (1977) 2083; A.Voros, Lecture Notes in Physics **93** (1979) 326.
76. M.Yu.Kuchiev and O.P.Sushkov, Phys.Lett. A **158** (1991) 69.
77. F.M.Izrailev, Phys.Rev.Lett. **56** (1986) 541.
78. F.M.Izrailev, Phys.Lett. A **125** (1987) 250.
79. E.Heller, Phys.Rev.Lett. **53** (1984) 1515.

80. R.Jensen, Phys.Rev.Lett. **63** (1989) 2771; R.Scharf et al, Phys.Rev. A **43** (1991) 3183.
81. M.Berry, Proe.Roy.Soc.Lond. A **423** (1989) 219.
82. E.B.Bogomolny, Physica D **31** (1988) 169.
83. Ya.G.Sinai, *Introduction to Ergodic Theory*, Princeton, 1976.
84. M.Berry and M.Robnik, J.Phys. A **17** (1984) 2413.
85. S.Fishman et al, Phys.Rev. A **39** (1989) 1628.
86. G.Casati et al, Phys.Rev.Lett. **63** (1990) 5.
87. T.Seligman et al, Phys.Rev.Lett. **53** (1985) 215; M.Feingold et al, Phys.Rev. A **39** (1989) 6507; M.Wilkinson et al, J.Phys. A **24** (1991) 175.
88. G.Casati et al, Phys.Rev.Lett. **64** (1990) 1851; J.Phys. A **24** (1991) 4755.
89. B.V.Chirikov, Phys.Lett. A **108** (1985) 68.
90. T.Geisel et al, Phys.Rev.Lett. **66** (1991) 1651.
91. A.I.Shnirelman, Usp.Mat.Nauk **30** #4 (1975) 265.
92. B.V.Chirikov, Foundations of Physics **16** (1986) 39.
93. B.V.Chirikov, Dynamical Mechanisms of Noise, preprint INP 87-123, Novosibirsk, 1987.
94. J.Ford et al, Prog.Theor.Phys. **50** (1973) 1547.
95. B.V.Chirikov, Wiss.Z. der Humboldt-Univ. **24** (1975) 215.
96. F.Benvenuto et al, Phys.Rev. A **44** (1991) R3423; G.Jona-Lasinio et al, Quantum Chaos without Classical Counterpart, 1991 (unpublished).
97. R.Fox and J.Eidson, Phys.Rev. A **36** (1987) 4321.
98. E.Ott et all, Phys.Rev.Lett. **53** (1984) 2187.
99. I.B.Khriplovich and V.V.Sokolov, Physica A **141** (1987) 73.
100. T.Petrosky et al, Physica A **170** (1991) 306.
101. G.Casati et al, The Quantum Steady State: Temporal Fluctuations in Discrete Spectrum (in preparation).
102. G.Casati et al, Phys.Rev. A **34** (1986) 1413.
103. R.Graham, these proceedings.
104. P.Koch, these proceedings.
105. U.Smilansky, these proceedings.
106. O.Agam et al, Experimental Realization of "Quantum Chaos" in Dielectric Waveguides, Technion-Phy-91, 1991 (unpublished).
107. U.Smilansky, private communication, 1992.
108. J.Pichard et al, Phys.Rev.Lett. **65** (1990) 1812.
109. G.Casati, private communication, 1992.
110. E.P.Nakhmedov et al, Zh.Eksp.Teor.Fiz. **92** (1987) 2133.
111. O.Bohigas, these proceedings.
112. H.Weidenmüller, these proceedings.

Semi-Classical Quantization of Chaotic Billiards

by

Uzy Smilansky
Department of Nuclear Physics,
The Weizmann Institute of Science
76100 Rehovot, Israel.

Abstract

The semi-classical quantization of chaotic billiards will be developed using a scattering theory approach. This will be used to introduce and explain the inherent difficulties in the semi-classical quantization of chaos, and to show some of the modern tools which were developed recently to overcome these difficulties. To this end, we shall first obtain a semi-classical secular equation which is based on a finite number of classical periodic orbits. We shall use it to derive some spectral properties, and in particular to investigate the relationship between spectral statistics of quantum chaotic systems and the predictions of random-matrix theory. We shall finally discuss an important family of chaotic billiards, whose statistics does not follow any of the canonical ensembles, (GOE,GUE,...), but rather, corresponds to a new universality class.

I. Introduction

In this series of lectures we shall deal exclusively with two-dimensional billiards. We shall motivate this choice in the following lines. While doing so, a few concepts and ideas will be introduced.

A billiard consists of a point mass confined to move freely in a connected and bounded domain, whose boundary Σ is a continuous curve. The dynamics is defined by the way the particle scatters from the boundary. We shall assume elastic, specular reflections. That is, denote by $\boldsymbol{v}_i(\boldsymbol{v}_f)$ the velocity of the particle, before (after) its collision with the wall, and by $\mathbf{n}$ and $\mathbf{t}$ the normal (pointing outside) and the tangent at the point of impact, then

$$\begin{aligned} \boldsymbol{v}_i \boldsymbol{t} &= \boldsymbol{v}_f \boldsymbol{t} \ , \\ \boldsymbol{v}_i \boldsymbol{n} &= -\boldsymbol{v}_f \boldsymbol{n} \ . \end{aligned} \tag{I.1}$$

This definition illustrates the main difference between smooth mechanical systems and the billiards. Typically, the Hamiltonian or Lagrangian function is sufficient to define the dynamics in a unique way. In billiards this is impossible since the infinitely strong forces must be replaced by reflection conditions which imply that the dynamical variables (here, normal velocity) suffer a discontinuity at the walls. In spite of this difference, the classical trajectories can be derived from a least-action principle which will be explained in the next chapter where we shall discuss the classical billiard in more detail. The resulting equations of motion will not be expressed as differential equations, but rather as a discrete mapping which specifies the conditions (angle of incidence and point on the boundary) at the "present" collision with the boundary in terms of the conditions at the previous collision. Such mappings are somewhat easier to handle, but retain all the important features and intrinsic complexities of their continuous counterparts.

The treatment of billiards has another attractive feature, namely, that the classical dynamics depends on the energy in a trivial way. The magnitude of the velocity serves only to rescale the time between collisions with the boundaries. In the quantum treatment, this translates to the fact that the energy is always scaled by $\hbar^2$ in such a way that the quantum theory depends only on the wave number k.

Because of the special way by which the billiards are defined, one has to be careful in adopting concepts and theorems which are formulated for smooth Hamiltonian systems. The famous KAM theory, for example, should be reformulated when applied to billiards. Due to the enormous progress achieved by Sinai and his co-workers, most of this work was already done with mathematical precision and rigor. As a matter of fact, at present, billiards belong to the class of mechanical systems whose dynamics is best understood, and therefore they are an excellent paradigm for the understanding and discussion of chaotic Hamiltonian systems in general. A review of the theory of chaotic classical billiards will be presented in the next chapter. It is intended to give an overview of the classical theory in a way of introduction to the quantum treatment. The interested reader is referred to the recent book edited by Sinai [1)] where the most important original articles are given in English translation.

Most of the present series of lectures will be devoted to quantum billiards. The Schrödinger equation reduces to the Helmholtz equation in two dimensions, and the reflection rules (I.1) are replaced by the boundary conditions (Dirichlet) that the wave function vanishes on the boundary. In this form the Schrödinger equation is nothing but a wave equation in the plane, and our discussion of its eigenvalues and eigenfunctions pertains to any wave problem! Thus, our results applies to quantum mechanics, (matter waves) as well as to electromagnetic, acoustic or surface waves. In other words, we are discussing here "wave-chaos" rather than "quantum chaos". To complete this generalization we should replace the classical trajectories mentioned above by rays, and remember that (I.1) is nothing but the law of specular reflection in geometrical optics.

It is interesting and important to note that among the experimental systems which were designed to study quantum chaos, a substantial fraction are of the billiard type, involving electromagnetic cavities [2,3,4)] or shallow water basins [5)]. Recently it became clear that ballistic electrons in GaAs junctions of certain shapes are excellent manifestations of quantum billiards [6,7)]. Thus, to those of us who look for the 'relevance' of their intellectual preoccupation with such toy systems, billiards supply a substantial source of justification (and support).

The main objective of this course is to discuss the billiard problem in the short wavelength (semi-classical) regime. This is the intermediate regime, where many features of classical mechanics (ray optics) are relevant to the quantum (wave) properties of the billiard, and hence it is the right regime for the search of the finger-prints of classical chaos in the quantum description. However, since we shall always assume a small but nonvanishing wave length ($\hbar$), there will be limits to the amount of information one could draw from classical mechanics. The rôle of the theory of quantum chaos is to discuss not only the relevance of classical chaos in the quantum description, but also to delineate clearly the limits where the classical information becomes irrelevant. This dual aspect of the semi-classical approach will be illustrated in the following lectures.

The general layout of this series will be as follows. The classical billiard dynamics will be described in chapter II. In the third chapter we shall introduce a somewhat unconventional quantization procedure, which will be the basis of the subsequent semi-classical theory. It will be shown that there are two distinct semi-classical steps in obtaining the final version of the semi-classical theory. The first step will provide the basic form of a secular function, whose zeroes are the eigen-energies of the billiard. It is obtained from the exact quantum condition by a truncation which is justified semi-classically. However, the dynamical information which is used to construct the secular equation is still fully quantum. For this reason the resulting secular equation will be called semi-quantal.

Chapter IV will be devoted to the derivation of a semi-classical secular equation. That is, we shall give an algorithm to construct a function, whose zeroes are the approximate eigenvalues of the billiard and which uses classical input exclusively. In Chapter V we shall derive a semi-classical expression for the spectral density,

$$d(E) = \sum_n \delta(E - E_n) \, , \tag{I.2}$$

or rather its integral, the ***number function*** $N(E)$ which counts the number of eigen-energies up to a given energy E

$$N(E) = \int_0^E \mathrm{d}E' d(E') = \sum_n \Theta_H(E_n - E) \tag{I.3}$$

We shall also have to discuss the energy averaged spectral quantities $\langle d(E)\rangle$ and $\langle N(E)\rangle$ and obtain semi-classical expressions in terms of the geometric parameters of the billiard (Weyl's formula). The final form for the semi-classical spectral density will be expressed in terms of contributions from periodic orbits of the classical billiard, and will be shown to be equivalent to the well known expression derived by Gutzwiller, and which serves as one of the foundations of quantum chaos. We shall close this chapter with a few remarks about the difficulties which are intrinsic to the Gutzwiller trace formula which will motivate ***a-posteriori*** the search for the alternative approach which was explained in the preceding chapters.

In chapter VI we shall discuss some applications, and in particular, derive semi-classical support to the empirically known relation between random matrix theory and the spectra of quantum chaotic systems. (see Professor Bohigas' notes). Finally we shall try to connect the approach developed in the present talk to the lectures by Prange and Chirikov, discussing a billiard which is composed of a linear array of cavities where nearest neighbours are connected by narrow channels. Many aspects of localization which are manifested in the quantum kicked rotor or quasi one-dimensional disordered systems can be studied in terms of such billiard models.

The semi-classical quantization of billiards was investigated in the past using a variety of techniques and approaches. One of the most profound and pioneering efforts in this field is due to Balian and Bloch[8–11]. Their approach is different from the one presented here. It is a highly recommended set of articles, fraught with original ideas and with penetrating insight. The present appproach was originally developed by the author and Eyal Doron[12], as a natural extension of their work on chaotic scattering. It was inspired by a recent theory developed by Bogomolny [13]. Bogomolny's articles are highly recommended to the reader who is interested in going deeper into this subject.

Due to space limitation no mention will be made here of "scars" in eigenfunctions of chaotic billiards. The reader is referred to a review by Heller[14] and papers cited therein. Nor will there be any attempt to discuss billiards on surfaces with negative curvature. This is a subject where the investigation of chaotic dynamics and their implications to the corresponding wave properties goes back to the classical works of Hadamard almost a hundred years ago. A review of this subject was recently written by Balazs and Voros [15]. Gutzwiller's book[16] also has a chapter on this important topic.

II. Classical Billiards

The classical trajectories in the billiard are composed of straight segments (chords) which reflect from the boundary according to the reflection rule (I.1). To describe the trajectory, one needs two parameters per segment, so that a trajectory is defined by a list of parameter pairs. A convenient parameterization is obtained by specifying the point of impact by the arc length s along the boundary. The other parameter is the projection of a unit vector along the chord on the tangent to the boundary at the point of impact p_t. The equations of motion are written now as a mapping which expresses $(s, p_t)_{n+1}$ as a function of $(s, p_t)_n$. Phase space corresponds to a truncated cylinder (s is a periodic variable whose period is the circumference of the billiard, and $|p_t| < 1$). This parameterization has the advantage that it can be used for billiards with arbitrary shape. (one should exclude points where the billiard has corners so that the tangent vector is not well defined). Moreover, the mapping is area preserving when expressed in terms of these coordinates. A pedagogical discussion of this coordinate system, with very nice illustrations, can be found in reference 17.

In these lectures we shall use a different set of variables to describe the billiard dynamics. It has the advantage of being the natural phase space in applications which will be described later. Its main disadvantage is that it can be used for convex (actually non-concave) billiards only. Choose an arbitrary point inside the billiard as the origin and an arbitrary reference direction. Then, any straight segment of a trajectory is completely specified in terms of two parameters - its distance from the origin b and the direction of the velocity vector relative to the reference direction θ. (see Fig. 1) The distance b is the impact parameter which is proportional to the angular momentum l relative to the origin. The explicit form of the mapping is given in Appendix A, where it is also shown that l and θ are canonically conjugate variables and that the mapping is area preserving.

Phase space in the (l, θ) representation is defined in the following way. For any

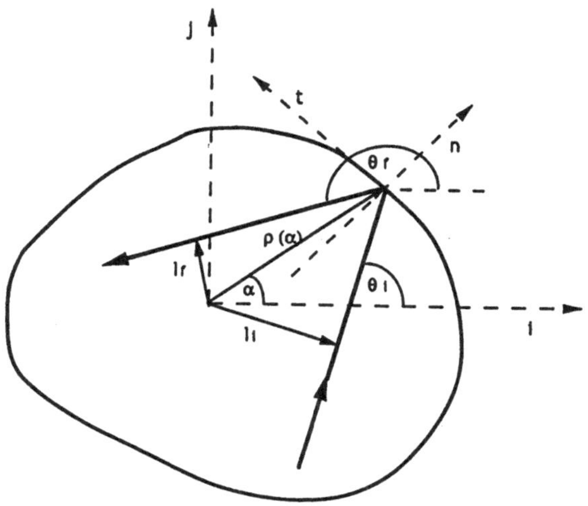

Fig. 1. A convex billiard and the parametrization of trajectories.

given direction θ construct the two tangents to the boundary Σ in the θ direction. Their impact parameters $l_-(\theta)$ and $l_+(\theta)$ define the range of l values of chords in the θ direction which the billiard can accommodate. The domain of the mapping is the section of the (l, θ) cylinder which is bounded for every θ by

$$l_-(\theta) \leq l \leq l_+(\theta). \qquad (II.1)$$

This ribbon-shaped phase space is topologically identical to the domain of the mapping in the (s, p_l) representation.

We shall now discuss the various classes of dynamics one can observe in *convex* billiards. The only smooth billiards which are completely integrable are those with elliptic boundaries. The simplest one is the circular billiard. Choosing the origin as the center, one finds that the classical mapping reads,

$$l_{n+1} = l_n \quad \theta_{n+1} = \theta_n + \omega(l_n) \qquad (II.2)$$

where $\omega(l) = \pi - 2arcsin(l)$ and we assume that the radius is of unit length. The classical orbits are periodic if $\omega(l_0)$ is a rational multiple of π. Each periodic orbit is determined by the arbitrary initial angle θ_0. The set of periodic orbits is therefore dense on the line $l = l_0$. If $\omega(l_0)$ is irrationally related to π, the classical orbit is not periodic, and it covers the line $l = l_0$ ergodically. If one draws the chords corresponding to these two classes of trajectories, one finds that periodic orbits are polygons bounded by the circular billiard. Nonperiodic trajectories cover a circular strip which occupies all the points with a distance $r \geq l_0$ from the center. The circle $r = l_0$ is a caustic- the line which separates the allowed and the forbidden regions.

For an elliptic billiard, the integral of the motion F is the product of the angular momenta with respect to the two foci of the ellipse. Taking the origin at the center of gravity of the ellipse, and the reference direction along the major axis, we find that

$$F = l^2 - d^2 sin^2\theta \ , \qquad (II.3)$$

where d is half the distance between the foci and we scale the magnitude of the velocity vector to be 1. Trajectories are confined to the curves (II.3). Of particular interest are the periodic orbits of period two, which correspond to reflections along the major axes of the ellipse. The trajectory which bounces along the major axis ($F = 0, l = 0, \theta = 0, \pi$) is unstable (sum of radii of curvatures is smaller than the diameter). The trajectory along the semi-major axis ($F = -d^2, l = 0, \theta = \pi/2, 3\pi/2$) is stable. The invariant curves (II.3) depend on the value of F. For $F > 0$ these curves never cross the line $l = 0$. If $F < 0$ the invariant curves are closed, crossing the l axis twice. As F tends to its minimal value $-d^2$, the curves shrink to a pair of points corresponding to the stable period-two orbit discussed above. The line with $F = 0$ is a separatrix distinguishing between the two kinds of trajectories. The two branches of the separatrix are the stable and unstable manifolds of the unstable period two orbit along the major axis.

An initial value (l_0, θ_0) determines the value of F and the motion can be considered to be a mapping of the closed curve $F(l, \theta) = F(l_0, \theta_0)$ on itself. One can introduce a new variable ξ which is monotonic in the arc length along the invariant curve. One can normalize ξ such that the circumference of the invariant curve is unity. In this variable, the dynamics on the invariant curve is simply $\xi \rightarrow \xi + \eta$, where η is the ***rotation number.*** If η is rational, we have periodic motion, and the set of periodic orbits cover the invariant curve. For irrational η the invariant curve is covered ergodically by a single trajectory.

We are now in a position to study what happens if the elliptic billiard is deformed. The dynamics is no longer integrable, since there exists no global invariant function which replaces $F(l, \theta)$. One expects that the loss of integrability manifests itself by the onset of chaotic motion. In the sequel we shall quote some of the results known for convex billiards in two dimensions. This theory addresses the same problems which the KAM theory does in the case of systems with smooth Hamiltonians.

Lazutkin [18] obtained the following remarkable result. Consider a convex billiard whose boundary is sufficiently smooth. (The exact requirement is that the radius of curvature of the boundary is bounded from above and below by positive constants, and as a function of the arc length, it has 553 continuous derivatives). Then, there exists a parameter $\alpha > 0$ and a set of irrationals $E(\alpha)$ with a positive Lebesgue measure, such that to every $\eta \in E(\alpha)$ corresponds an invariant curve of the billiard mapping, with a rotation number η. The set $E(\alpha)$ is defined as the set of irrationals $\eta < \alpha$ which satisfy, for any integers $n > 0, m$

$$|n\eta - m| \geq A|m|n^{-2.5} \, . \tag{II.4}$$

It can be shown that $\mu(E(\alpha)) > \alpha - C\alpha^{2.5}$ where $A > 1$ and C are constants. Lazutkin makes the important remark that "... the family of invariant curves is discontinuous: neighborhoods of rational η are excluded. This limitation is associated not with the method of proof but with the nature of the problem."

What Lazutkin's theorem says is that a smooth deformation of the elliptic billiard preserves the invariant curves with sufficiently irrational rotation numbers, and chaotic motion replaces the degenerate periodic orbits (rational rotation numbers). The important element in this theory is that it also indicates what are the crucial ingredients which should be maintained so that the deformed billiard preserves some of its invariant curves- it should remain convex, the deformation should be smooth, and should also be devoid of too flat segments. We shall now study billiards which do not conform to the above conditions, and quote some results on the conditions which are sufficient to turn the billiards chaotic.

Bunimovitch [19–20] studied a general class of (not only convex) billiards which are composed of straight segments and arcs of circles. (Polygon billiards are excluded, but billiards composed exclusively of arc segments are acceptable). Bunimovitch found the conditions under which the mapping describing the dynamics in these billiards are Bernoulli and hence K-systems:

i. Take any arc and draw the part of the trajectories which connect it to any other arc, and which can do so directly or by reflections from straight sections only. Denote by c_α the length of the chord which is the intersection of the trajectory with the circle whose arc is the starting point of the trajectory. Denote by d_α the total length of the trajectory. Then, for Bunimovitch's theorem to hold, it is necessary that the set of trajectories which satisfy the strict inequality $d_\alpha > c_\alpha$ has positive measure. In physical terms this means that the focusing elements in the billiard are so far apart from each other, that a parallel beam of trajectories which focuses at a point has sufficient distance to diverge to a size larger than its original size before it reflects again.

ii. Consider the set of trajectories A_n^α which scatter between straight segments of the billiard m times with $m > n^\alpha$. The second condition is that the measure of the sets A_n^α decrease faster than $1/n$ for some positive $\alpha > 1$. This condition can be understood as requiring that trajectories which are trapped between straight segments during n reflections are sufficiently rare.

The Bunimovitch stadium is the most famous example of chaotic billiards which satisfy the requirements above. One can think of other more exotic examples.

From the above considerations it is clear that the crucial element which leads to unstable and chaotic behavior is defocusing. In convex billiards one can achieve this by sufficient separation of the focussing elements. If one removes the restriction that the billiard is convex, and one allows concave elements, they produce the necessary dispersion of near-bye trajectories resulting in chaotic behavior. Sinai [21] was the first to consider such systems, and he proved that billiards with dispersing components are chaotic in the sense that the discrete dynamical mapping is a K system.

This short classification of billiards cannot be complete without some remarks on other classes of billiards. Rectangular billiards are integrable and so are various other polygons with rational angles, which tesselate the plane under reflections at the boundaries. The equilateral triangle is an example of an integrable billiard. Other billiards such as e.g. a 2×2 square of which a 1×1 corner square is removed, is not integrable, in spite of the fact that due to symmetry, the magnitude of the vertical and horizontal components of the momentum are constants of the motion. Richens and Berry [22)] name such systems "pseudointegrable", and show that they are not integrable in the sense that one cannot find an action -angle representation for the dynamics. The reason for this phenomenon lies in the fact that phase space does not have the topology of a torus. The motion in pseudointegrable systems has a chaotic property which is due to splitting of beams at some vertices rather than because of exponential instability. A detailed discussion of such billiards can be found in ref.22. Richens and Berry conjecture also that polygon billiards with angles which are irrational multiples of π are ergodic.

III. Quantization- The Semi-Quantal Secular Equation

We shall now enter the realm of quantum (wave) dynamics, where our main concern will be to find the spectrum of the wave analogue of the classical billiard. In other words, we look for the spectrum of the Helmholtz equation with Dirichlet conditions on the billiard boundary. We shall use a rather unconventional approach to this problem by expressing the secular function (the function whose zeroes are the eigen-energies) in terms of a related scattering problem[12)]. The advantages of this method will become clear when we introduce the semi-classical approximation. Our main purpose is to show that this approach leads to new and interesting results as well as to the well known semi-classical theory which stems from the Gutzwiller semi-classical trace formula. We shall start by treating in detail the quantization of convex billiards. Concave billiards can be quantized in the same spirit, and this will be done later.

III.a Quantization of convex billiards.

We use a polar representation of phase space in terms of the coordinates (r, ϕ) and the conjugate momenta (k_r, l) (note that classical actions are expressed in units of $\hbar$). A general scattering wave function which vanishes on the billiard boundary takes the asymptotic form

$$\Psi^{(l)}(\boldsymbol{r}) = F_l^{(-)}(\boldsymbol{r}) + \sum_m \tilde{S}_{l,m} F_m^{(+)}(\boldsymbol{r}) \qquad (III.1)$$

for $\boldsymbol{r}$ outside the smallest circle which covers the entire billiard. Here $F_l^{(\pm)}(\boldsymbol{r})$ are incoming and outgoing cylindrical waves with angular momentum l:

$$F_l^{(\pm)}(\boldsymbol{r}) = [J_l(kr) \pm iN_l(kr)] \exp(il\phi)$$

$J_l(x)$ and $N_l(x)$ are the Bessel and Neumann functions, and $\tilde{S}$ is the scattering matrix. The origin of the polar coordinates is conveniently chosen at the center of the smallest covering circle. For convex billiards it is inside the boundary Σ. $\Psi^{(l)}(\boldsymbol{r})$ as defined above describes the wave function only within the domain of convergence of the series (III.1), which is confined to the exterior of the the smallest covering circle. The divergence of (III.1) is due to the presence of the Neumann functions which diverge as $r \to 0$. If we could find a linear combination of the functions $\Psi^{(l)}(\boldsymbol{r})$ such that the coefficients of all the Neumann functions would vanish, then the resulting function will be represented by a series which converges in the *entire* plane. By construction it is a regular solution of the Schrödinger equation also inside the billiard and it satisfies the boundary conditions on Σ. In other words, it would yield an eigen-function of the billiard. Such a linear combination of the $\Psi^{(l)}(\boldsymbol{r})$ can be found only if

$$Z(E) = \det\,(I - \tilde{S}(E)) = 0\ , \qquad (III.2)$$

which is the desired secular equation for the "inside" quantized billiard.

The matrix $\tilde{S}$ is of infinite dimension. The fact that the billiard does not extend to infinity implies that we can find two angular momentum bounds, $L_+ > L_-$, such that for any l which is sufficiently far from the interval (L_-, L_+), and for any m, we have

$$\tilde{S}_{l,m} \approx \delta_{l,m} \qquad (III.3)$$

This means that the value of the determinant (III.2) is always vanishingly small, and the secular equation above is not very useful in its present form. A way to discard the contribution from the physically uninteresting range is naturally suggested in the semi-classical domain. There, the approximate form (III.3) becomes progressively more exact. Thus we can truncate $\tilde{S}$ to the domain

$$L_- \leq l, m \leq L_+ \qquad (III.4)$$

and replace (III.2) by a semi-quantal secular equation

$$Z_{sq}(E) = \det\,(I - S(E)) = 0 \qquad (III.5)$$

where S is a matrix of dimension $\Lambda = L_+ - L_-$ which is obtained by restricting $\tilde{S}$ to the range (III.4), and which is semi-classically unitary, due to (III.3).

The secular function $Z_{sq}(E)$ is the basic building block for the theory which follows here, and therefore we should reflect on it for a while. The approximation which leads to (III.5) is semi-classical in nature, since we neglected the contributions of spherical waves which are represented classically by rays which do not hit the billiard. The radial wave functions of these waves fade exponentially fast in the region occupied by the billiard. Hence we call them evanescent waves. As the energy increases, the dimension Λ increases, since $\Lambda = [kD]$ where D is a typical diameter of the billiard. However, Λ grows in a discontinuous fashion, and special care must be taken at "threshold" energies where the jumps in Λ actually occur and the secular function is discontinuous. We shall show below that in spite of the fact that threshold energies occur throughout the entire energy axis, energy intervals between successive thresholds contain increasingly larger number of eigen-values, so that the problematics due to thresholds decrease in relative importance as the energy becomes higher (or, equivalently, in the semi-classical limit). Some numerical evidence which justifies the applicability of (III.5) will be discussed at a later stage. We should emphasize that so far, the input which enters into the secular equation is *strictly quantum*. This is why we refer to (III.5) as the semi-quantal secular function. In the next chapter we shall use the semi-classical expression for the S-matrix, and obtain the desired form of the semi-classical secular equation. The truncation of

(III.2) and the assumption that the restriction of $\tilde{S}$ to the range (III.4) is unitary, can only be justified semi-classically.

As an illustration of the secular function defined above, let us consider the circular billiard. The S matrix is diagonal in the angular-momentum representation with eigen-values

$$\exp\,(i\theta_l(k)) = -(J_l(kR) - iN_l(kR)/(J_l(kR) + iN_l(kR) \qquad (III.6)$$

where R is the circle radius. To satisfy the secular equation, one of the eigen-phases $\theta_l(k)$ should be an integer multiple of 2π, which can only happen when $J_l(kR) = 0$. This is the *exact* quantization condition of the circular billiard. In the present case $L_\pm = \pm kR$, and it is known from the theory of the Bessel functions that the smallest positive zero of $J_l(x)$ is larger than l. In other words, the eigen-values with wave numbers less than k, are due to the vanishing of $J_l(kR)$ with $|l| < kR$, which shows that the semi-classical truncation which leads to (III.5) is exact in the present case.

III.b Quantization of billiards with arbitrary shapes.

We shall now show that a semi-quantal secular equation of the type (III.3) applies for the quantization of billiards of arbitrary shapes. To facilitate the discussion we shall restrict ourselves to billiards which satisfy the following requirements:

i. The billiard has to accommodate at least one chord Γ which is normal to the two tangents at its two ends (see Fig. 2). Γ divides the billiard into two parts which we refer to in the following as the "left" (L for short) and "right" (R) parts.

ii. The tangents at the end points of Γ do not intersect L in their continuation towards R, and do not intersect R in their continuation towards L.

If there exist more than one chord with this property, one can chose any chord for the purpose of the discussion and proof. There exists, however, an optimal chord, and the guide-lines for its choice will be explained.

We now turn the "inside" billiard problem into two independent scattering systems. The L system is constructed by considering the opened billiard defined by the boundary of L (to be denoted by Σ_L) and the two parallel tangents which go towards R. The two tangents form a channel which matches smoothly to Σ_L. Imposing specular reflec-

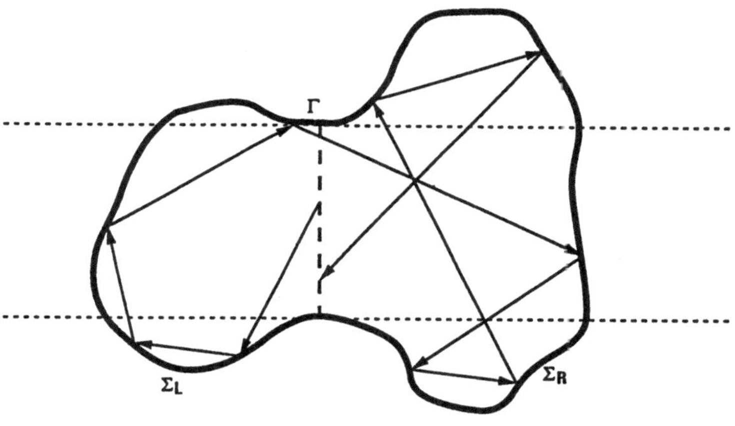

Fig. 2. Auxiliary construction for quantizing a billiard.

tion conditions classically, or Dirichlet boundary conditions for the wave problem, the L system is completely specified. The R system is defined in exactly the same manner. Note that the R (L) part of the billiard is completely discarded when we consider the L (R) scattering system. To each of the two systems we attach a coordinate system where the chord Γ coincides with the y-axis. The positive x-axis coincides with the lower channel wall for the L system. The negative x-axis for the R system coincides with the lower wall of its channel.

A scattering solution for the L (R) system must vanish on the boundary Σ_L (Σ_R), and asymptotically, for large values of x ($-x$),

$$\Psi_n^{(L,R)}(x,y) \to \sum_{l=1}^{\Lambda} \sqrt{(k_n/k_l)} \left[\delta_{n,l} e^{\mp i k_l x} + S_{n,l}^{(L,R)} e^{\pm i k_l x}\right] \phi_l(y) \qquad (III.7)$$

The scattering solutions describe a wave propagating down (up) the positive (negative) axes in the mode n. After scattering from the L (R) boundary, the Λ modes which are energetically allowed to propagate in the channel are scattered with amplitudes given by the corresponding scattering matrix $S^L(S^R)$. The number of propagating modes Λ is the integer $[Dk/\pi]$. D is the length of the chord Γ and k is the wave number. The functions $\phi_l(y)$ are a complete set of transverse mode functions with eigen-energies $(l\pi/D)^2$. The corresponding longitudinal wave numbers are k_l, for $1 \leq l \leq \Lambda$.

At this point we would like to use the scattering functions (III.7) to construct an eigen-function of the original billiard, namely, a function which vanishes on the entire boundary $\Sigma = \Sigma_L + \Sigma_R$. A possible candidate will be a function which is defined piecewise as a linear combination of the Ψ_l^L inside the L region ($x \leq 0$) and a linear combination of the Ψ_l^R inside the R region ($0 \leq x$). Such a function will automatically satisfy the boundary conditions on Σ. The requirement that the function and its normal derivative are continuous at Γ, provides the quantization condition.

The main approximation which we are now about to introduce, consists of taking the form (III.7) as a proper representation of the exact wave functions not only in the limit $|x| \to \infty$ but also in the vicinity of Γ. That is, we neglect the contribution of the evanescent modes to the scattering wave functions at the interface. Such an approximation is justified since classically, evanescent modes correspond to propagation with imaginary momenta. In the semi-classical approximation these trajectories give contributions which are exponentially small (of order $\exp(-1/\hbar)$), and are therefore discarded. This approximation is expected to be at its worst at threshold k values, namely, when the longitudinal wavenumber is very small, so that an evanescent wave takes a large distance to decay. The reason for our particular choice of the chord Γ is now becoming clear. In this way the matching of the channels to the L and R billiards is smooth, and under such conditions, the rôle played by evanescent modes is minimized. We could make the same construction but with channels which are not smoothly matched. Then, the evanescent modes play a significant role at the interface, which may limit the applicability of the semi-classical approximation.

For smooth matching at $x = 0$ we must require that there exist coefficients $a_l^{L,R}$, $l = 1, \ldots, \Lambda$, which solve the linear equations

$$\sum_{n=1}^{\Lambda} a_n^L (k_n/k_l)^{1/2} \left[\delta_{n,l} + S_{n,l}^L\right] = \sum_{n=1}^{\Lambda} a_n^R (k_n/k_l)^{1/2} \left[\delta_{n,l} + S_{n,l}^R\right] \qquad (III.8)$$

and

$$ik_l \sum_{n=1}^{\Lambda} a_n^L (k_n/k_l)^{1/2} \left[-\delta_{n,l} + S_{n,l}^L\right] = ik_l \sum_{n=1}^{\Lambda} a_n^R (k_n/k_l)^{1/2} \left[\delta_{n,l} - S_{n,l}^R\right] , \qquad (III.9)$$

resulting in the quantization condition

$$Z_{sq}(E) = \det \begin{pmatrix} I & S^R \\ S^L & I \end{pmatrix} = 0 \qquad (III.10)$$

Simple algebra brings the secular function to its final form

$$Z_{sq}(E) = \det\,(I - S^L S^R) = \det\,(I - S) \qquad (III.11)$$

where $S = S^L S^R$ is a unitary matrix of dimension Λ. Thus, we have shown that a semi-quantal secular equation can be written in the form (III.5) also for the quantization of billiards with more general shapes.

As we have done previously, we shall show that the semi-quantal secular equation applies exactly in a simple integrable case. Consider the quantization of a rectangular billiard with the four corners at $(0,0)$, $(\alpha, 0)$, $(\alpha, 1)$, $(0,1)$. We shall write the secular equation in two ways. We shall first chose the line Γ as the side of unit length along the y-axis. We construct the channels by extending the two sides which are perpendicular to it. The S matrices are of dimension $\Lambda^{(1)} = [k/\pi]$. It is easy to see that for Dirichlet boundary conditions,

$$S_{n,m}^L = -\delta_{n,m}\exp\left(2i\alpha\left(k^2 - (n\pi)^2\right)^{\frac{1}{2}}\right) \qquad (III.12a)$$

and

$$S_{n,m}^R = -\delta_{n,m} \qquad (III.12b)$$

Hence the effective S matrix is given by

$$S_{n,m} = \delta_{n,m}\exp\left(2i\alpha\left(k^2 - (n\pi)^2\right)^{\frac{1}{2}}\right) \qquad (III.12c)$$

The quantization condition can be readily shown to give the known spectrum

$$k_{n,m} = \pi\left(n^2 + (m/\alpha)^2\right)^{\frac{1}{2}} \qquad (III.13)$$

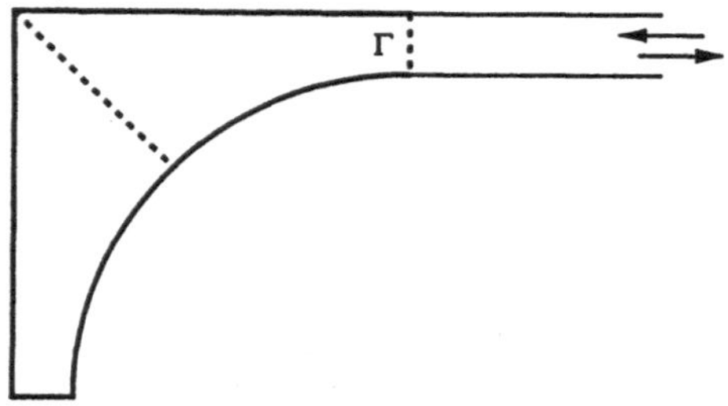

Fig. 3. Quantizing a Sinai Billiard with an auxiliary channel.

We now choose for the section Γ the side of length α along the x axis. One can repeat the calculation for this choice and find the same spectrum as was previously derived.

All the numerical studies which will be used to illustrate the semi-classical theory were performed on a model system which we shall describe now. It is the Sinai billiard, of which we treat only a symmetry irreducible eighth. (see Fig 3). To quantize this billiard we attach a wave guide and take the section Γ as the interface between the channel and the billiard. It is easy to see that in this case $S^R = -I$ and S^L is the scattering matrix of the entire cavity. The smooth connection between the billiard and the channel makes this system a particularly convenient model for our purpose.

The first problem which we would like to illustrate with this system is the applicability of the semi-quantal secular function. In Fig. 4a we compare the exact spectrum of the Sinai billiard in a certain energy interval with the zeroes of the semi-quantal secular equation. We plot the difference between the exact and approximate energies normalized to the mean level spacing. As an abscissa we use the (imaginary) momentum in the lowest evanescent mode. The figure shows clearly that the error is rather small and decreases exponentially as one gets further away from the threshold. This behaviour can be explained in the following way. The exact secular equation can be obtained from scattering solutions $\Psi^{(L,R)}$ when one does not ignore, but rather incorporates the contribution of the evanescent modes in deriving secular equation. The secular equation has the form (III.2), but now, the matrix $\tilde{S}$ is not unitary. Assume for simplicity that only the lowest evanescent mode is retained. Then, the last raw and column of $\tilde{S}$ correspond to the coupling of the propagating modes to the first evanescent mode. It can be shown by standard perturbation theory, that to lowest order in the added matrix elements, the first Λ eigen-values of $\tilde{S}$ are unimodular, and that the shift in the zeroes of the secular equation (measured in units of the mean level spacing) is of order $|\tilde{S}_{n,\Lambda+1}|^2$. (The index n stands for a typical propagating mode number). Thus, as long as the energy is sufficiently far from thresholds, the error in the semi-quantal eigen-energies is exponentially small (Fig. 4b).

The restrictions posed on the choice of the section Γ are not essential and they were necessary only to enable the most transparent presentation. One can chose Γ rather arbitrarily, and the formalism could be followed through with some more effort. For practical purposes, it helps if the channels join smoothly to the cavity (adiabatic matching) and if the sets of trapped orbits in each side is minimal.

III.c Properties of the Semi Quantal Secular Equation.

We shall end this chapter by deriving some general properties of the semi quantal secular equations. We shall prepare the ground for the semi-classical theory, by casting the secular equation in different forms which will turn out to be convenient later on.

The secular equation (III.5) is a complex valued function. It will be convenient to write it as a real amplitude times a phase factor,

$$Z_{sq}(E) = \exp\left(\frac{i}{2}\Theta(E)\right) 2^{\Lambda} \prod_{l=1}^{\Lambda} \sin\frac{\theta_l(E)}{2} \ . \qquad (III.14)$$

Where

$$\Theta(E) = \sum_{l=1}^{\Lambda} \theta_l(E) - \Lambda\pi \ . \qquad (III.15)$$

Here, the eigen-phases of the S matrix are denoted by $\theta_l(E)$ for $l = 1, \ldots, \Lambda$. $\Theta(E)$ is the phase of $\det(-S(E))$. This quantity will play a major part in our discussion. It should be pointed out that Λ is itself a function of E, but unless otherwise specified, we shall consider energy intervals between neighboring thresholds, where it is a constant.

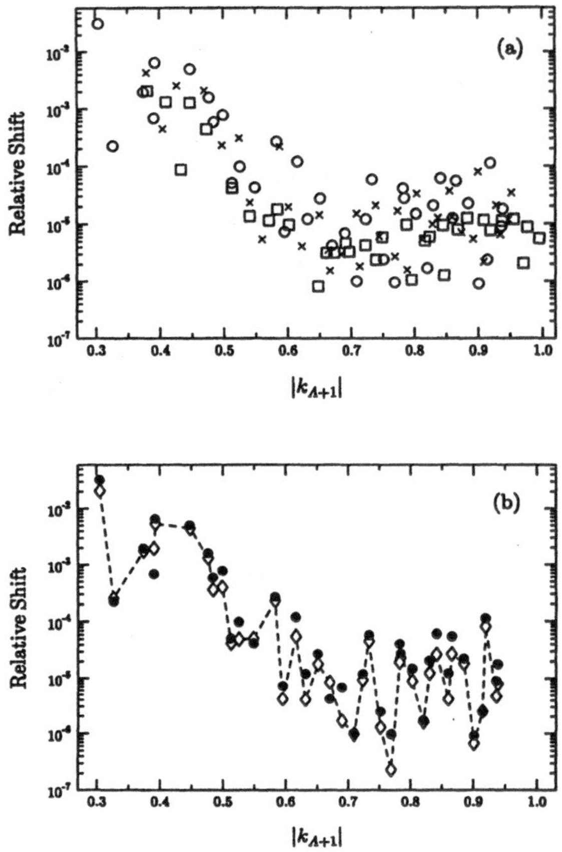

Fig. 4 (a) The deviations in the positions of the zeroes of $Z_{sq}(k)$ from the true eigenvalues. $\Lambda = 1$ (squares) $\Lambda = 2$ (crosses) and $\Lambda = 3$ (circles). The deviations are measured in units of the mean level spacing.
(b) A comparison between the deviations for $\Lambda = 3$ (full circles) and the results of perturbation theory which includes coupling to the evanescent modes.

As is clear from (III.14), an eigen-value is encountered whenever any of the eigenphases $\theta_l(E)$ is an integer multiple of 2π. Thus, the spectral density (I.2) can be written as

$$d_{sq}(E) = \sum_{l=1}^{\Lambda} \tau_l(E)\delta_p(\theta_l) \qquad (III.16)$$

where $\tau_l(E) = \frac{d\theta_l(E)}{dE}$ and $\delta_p(x)$ is the 2π-periodic delta function. Using the Fourier series for $\delta_p(x)$ we get

$$d_{sq}(E) = \frac{1}{2\pi}\sum_{l=1}^{\Lambda} \tau_l(E) + \frac{1}{\pi}\sum_{n=1}^{\infty}\sum_{l=1}^{\Lambda} \tau_l \cos(n\theta_l(E)) \qquad (III.17)$$

If the quantities θ_l and τ_l are smooth functions of the energy, one can interpret the first term on the right-hand side of (III.17) as a smooth part $\bar{d}(E)$ of the level density,

$$\bar{d}(E) = \frac{1}{2\pi}\sum_{l=1}^{\Lambda} \tau_l(E). \qquad (III.17a)$$

It is important to remember that the $\tau_l(E)$ and their sum are not guaranteed to be smooth functions of the energy. This is why we use the notation $\bar{d}$ (or $\bar{N}$) and thus distinguish them from the energy averaged quantities $< d >$ and $< N >$ which were introduced previously. However, we shall show in the next chapter that in the limit of small $\hbar$, $|\bar{d} - < d >| \to 0$ as $\hbar \to 0$. Thus, in the semi classical limit we could approximate the energy averaged quantities by their counterparts which are defined via (III.17a). An important insight is obtained by realizing that $\bar{d}(E)$ as defined above is simply related to the Wigner delay time[23] $\tau(E) = \mathrm{Tr}(T(E))$ where

$$T(E) = \frac{\hbar}{i\Lambda} S^{\dagger} S' \qquad (III.18)$$

and the prime stands for differentiation with respect to E. In other words,

$$\bar{d}(E) = \frac{1}{2\pi}\sum_{l=1}^{\Lambda} \tau_l(E) = \frac{\Lambda}{2\pi\hbar}\tau(E) \ . \qquad (III.19)$$

This result is very important since it shows the connection between three quantities- the mean level density inside the billiard, the mean delay time and the excess spectral density in the scattering problem (relative to the free space density), which is due to the presence of the scatterer.

Eq(III.19) can also be written as

$$\bar{d}(E) = \frac{1}{2\pi}\Theta'(E) \qquad (III.20)$$

as long as E does not coincide with a threshold where Λ changes by ± 1. We can obtain the corresponding counting function $\bar{N}(E)$ by integrating both sides of (III.20) over an interval $[E_1, E_2]$ which does not contain a threshold value.

$$\bar{N}(E_2) - \bar{N}(E_1) = \frac{1}{2\pi}(\Theta(E_2) - \Theta(E_1))$$

It turns out, however, that the piece-wise constant term $-\Lambda\pi$ which appears in the definition of $\Theta(E)$ allows us to rewrite the above equation in the form

$$\bar{N}(E) = \frac{1}{2\pi}\Theta(E) \qquad (III.21)$$

This relation is of crucial importance and will be used and discussed repeatedly. A semiclassical argument will be used in the next chapter to derive Weyl's expression[24)] for $\bar{N}(E)$ based on equations (III. 19-21).

It is convenient to consider the secular function at a given energy as the characteristic polynomial of the matrix S, $\det(\lambda I - S)$, evaluated at $\lambda = 1$. Thus, the secular equation is the sum of the coefficients $f_l, l = 0, 1, \ldots, \Lambda$ of the polynomial $\det(\lambda I - S) = \sum_{l=0}^{\Lambda} f_l \lambda^l$ with $f_\Lambda = 1, f_{\Lambda-1} = \mathrm{Tr}(-S), \ldots, f_0 = \det(-S)$. In general, f_l is the homogeneous symmetric polynomial of degree $\Lambda - l$ which can be constructed from the Λ eigenvalues of S. The unitarity of S implies an important symmetry among the f_l, namely

$$\exp\left[-i\frac{1}{2}\sum_{l=1}^{\Lambda}\theta_l(E)\right] f_l = (-1)^\Lambda \exp\left[i\frac{1}{2}\sum_{l=1}^{\Lambda}\theta_l(E)\right] f^*_{\Lambda-l} \qquad (III.22)$$

Using (III.15) and (III.21) we get

$$\exp\left[-i\pi\bar{N}(E)\right] f_l = \exp\left[i\pi\bar{N}(E)\right] f^*_{\Lambda-l} \qquad (III.23)$$

We shall show below that this set of equations plays the rôle of the "functional equations" which appear in the dynamical ζ function approach[25)].

Utilizing relation (III.23) one may now write for the secular function

$$\det(I-S) = \det(\lambda I - S)\Big|_{\lambda=1} = \sum_{l=0}^{\Lambda} f_l = \\ = \exp[i\pi\bar{N}(E)]\Re e\left\{\exp[-i\pi\bar{N}(E)]\sum_{l=l_0}^{\Lambda}(1+\epsilon(\Lambda,l_0))f_l\right\} . \qquad (III.24)$$

Here $l_0 = \frac{1}{2}(\Lambda+1)$ for odd Λ. If Λ is even, $l_0 = \frac{1}{2}\Lambda + 1$. $\epsilon(\Lambda, l_0) = 1$ except when Λ is even and $l = l_0$, in which case $\epsilon(\Lambda, l_0) = 0$. The advantage of writing the secular function in the form (III.24) is that it is entirely due to the unitarity of S. Thus, the removal of the contributions of terms with $l < l_0$ in (III.24) is not only a practical saving of numerical effort, but also an expression of a basic property of the system.

The f_l coefficients can be expressed in terms of $\mathrm{Tr}(S^n)$ with $n = 1, \ldots, \Lambda$. This follows from the Newton identities

$$\mathrm{Tr}\,(S^k) + f_{\Lambda-1}\mathrm{Tr}\,(S^{k-1}) + \ldots + f_{\Lambda-k+1}\mathrm{Tr}\,(S) + kf_{\Lambda-k} = 0 \ , \qquad (III.25)$$

which are valid for $1 \le k \le \Lambda$ *. This is a set of Λ linear equations for the f_l, in triangular form, with $\mathrm{Tr}(S^n)$ as coefficients. These equations can be solved using a simple iterative algorithm. The result takes the form

$$f_{\Lambda-l} = \sum_{\mathbf{p},\mathbf{r}} \phi^{(l)}_{(\mathbf{p},\mathbf{r})}(\mathrm{Tr}(S^{p_1}))^{r_1}\ldots(\mathrm{Tr}(S^{p_n}))^{r_n} \ , \qquad (III.26)$$

where $\mathbf{p}$ and $\mathbf{r}$ are vectors of non-negative integers, $p_1 < p_2 \ldots$, and the sum is on all vectors which satisfy

$$\sum_i p_i r_i = \mathbf{p}\cdot\mathbf{r} = l \ . \qquad (III.27)$$

---

* To prove (III.25), one has to compare two expressions for the logarithmic derivative of the characteristic polynomial: the first uses the explicit form $p(x) = \sum_{l=0}^{\Lambda} f_l x^l$, the other is the formal infinite series for $\log\det(Ix - S)$. Since S satisfies its characteristic equation, $\mathrm{Tr}S^{\Lambda+k} = -\sum_{l=0}^{\Lambda-1} f_l S^{k+l}$, the infinite series can be resumed. The details of the proof are left as an exercise.

The coefficient $\phi^{(l)}_{(r,p)}$ is a combinatorial factor, resulting from the triangular nature of (III.25). As an illustration, we quote the results for the first few f_l:

$$f_{\Lambda-1} = -\mathrm{Tr}\,(S)\,; \qquad (III.28a)$$

$$f_{\Lambda-2} = \frac{1}{2}\left((\mathrm{Tr}^{\,2}(S)) - \mathrm{Tr}\,(S^2)\right)\,; \qquad (III.28b)$$

$$f_{\Lambda-3} = -\frac{1}{6}\left(\mathrm{Tr}^{\,3}(S) + 2\mathrm{Tr}\,(S^3) - 3\mathrm{Tr}\,(S)\mathrm{Tr}\,(S^2)\right)\,. \qquad (III.22c)$$

In the next chapter we shall make use of the semi-classical approximation to express $Tr(S^n)$ in terms of classical quantities. This will yield the desired semi- classical quantization since it is an algorithm which gives quantum energies based on classical information.

IV. The Semi-Classical Secular Function

In the present chapter we shall introduce the second semi-classical step which will convert the semi-quantal secular function to a proper semi-classical expression. This will be achieved by deriving a semi-classical approximation for the S matrix and in particular for $Tr(S^n)$ since the latter are the building blocks of the secular equation (III.5).

We shall start by describing the classical theory of scattering for the two systems we used for the quantization of the billiard. We shall show that the intimate relationship between the motion inside the boundary and outside it, which we used for the quantization, exists also in the classical domain. Our semi-classical theory will capitalize on this duality.

Let us consider first the scattering from a convex billiard. A scattering trajectory *outside* the billiard can be parameterized in terms of the parameters (θ, l) which were used in Chapter (II) to parametrize the segments of trajectories *inside* the billiard. A single scattering event generates a correspondence between the parameters which specify the incoming ray (θ, l) and those which specify the outgoing ray (θ', l'). The domain of this correspondence can be defined in terms of the functions $l_{\pm}(\theta)$ which were introduced in Chapter (II). For any given θ there exist two limiting values of the angular momentum, $l_-(\theta)$ and $l_+(\theta)$, such that no reflection occurs for scattering trajectories which aim at the billiard at an angle θ with $l < l_-(\theta)$ or $l > l_+(\theta)$. Thus, the stripe σ appears also as the domain where the billiard affects the scattering. For (θ, l) values outside σ, there will be no deflection, and the trajectories continue as straight lines. The extreme of σ along the angular momentum axis give the semi-classical values of $L_{\pm}$ discussed in Chapter (III).

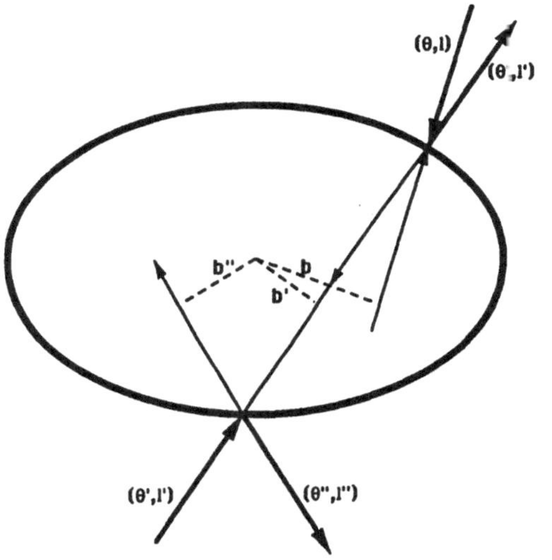

Fig. 5. The "inside-outside" duality of classical trajectories.

A further link between the "inside" and the "outside" dynamics is established by noting that any external reflection $(\theta, l) \rightarrow (\theta', l')$ corresponds to an internal trajectory which impinges on the billiard at the same point with opposite incoming and outgoing directions and the same values of the angular momenta but for a change of sign (see Fig 5.).

To make the "inside-outside" analogy complete we have to go one step further. A scattering trajectory is a one-shot event: A trajectory appears from infinity, hits the billiard and escapes back to infinity. A bounded trajectory reflects from the boundary infinitely many times, and can be described in terms of a mapping. We shall now show, that there exists a natural way to introduce a mechanism which allows scattering trajectories to be re-injected after being scattered. This will turn the *correspondence* defined above into a scattering *mapping*, mediated by scattering trajectories. As a matter of fact, such mappings occur for any scattering system and form an important tool in the study of classical scattering. One usually refers to them as "Poincaré Scattering Mappings" (PSM). A general definition of the PSM, its *raison d'être* and its relation to the semi- classical S matrix can be found in references 26, 27 and articles cited therein.

The PSM in the present case is constructed in the following way. A point $(\theta, l) \in \sigma$ corresponds to a straight line trajectory which is incident on the billiard at an angle θ and angular momentum l (see Fig. 5). The trajectory scatters into the direction θ' with an angular momentum l'. The re-injection is effected by considering (θ', l') as defining an incoming trajectory. It moves on the same line as the previous outgoing trajectory, but it impinges on the scatterer on a diametrically opposite point. The generating function which induces the mapping is given in terms of the reduced action

$$\Phi = -\left(\int r dk_r + \int \phi dl\right) , \tag{IV.1}$$

which is (but for an overall sign) exactly the same as the generating function which controls the dynamics inside the billiard (see (A.6)). Note also that this quantity can be interpreted as a WKB phase shift incurred in the incoming and outgoing branches of the trajectory. The fact that the mapping can be derived from a generating function guarantees its area preserving character.

Consider now the second method, in which we used auxiliary wave guides to quantize the billiard. Consider the 'left' scattering system first. The motion in the wave guide is integrable, since the absolute value of the transverse momentum p_y is conserved. As a matter of fact, one can express the transverse motion in the wave guide in terms of action-angle variables

$$\begin{aligned} I &= |k_y| D/\pi \\ \phi &= \pi\left[\frac{y}{D}\Theta_H(k_y) + \frac{2D-y}{D}\Theta_H(-k_y)\right] \end{aligned} \tag{IV.2}$$

(One can easily see that the quantal analogue of I is the mode number n introduced previously in the quantum version (III.7)). A trajectory in the wave-guide corresponds to a fixed value of I and a phase angle which changes periodically. As a matter of fact, trajectories with a given I fill a torus. To specify a particular trajectory one may use the value of the phase when the trajectory crosses a transverse reference line. Consider now a scattering trajectory with a given I value, and which crosses the reference line having a phase angle ϕ. It scatters in the billiard and emerges finally with action $\tilde{I}$ and phase $\tilde{\phi}$. Thus, the scattering to the left defines a correspondence $(I, \phi) \rightarrow (\tilde{I}, \tilde{\phi})$. One can turn this correspondence into a PSM by introducing the appropriate re-injection mechanism: specular reflection of the trajectory at the reference line. Thus, a mapping T_L is defined. A similar mapping can be defined for scattering in the right component of the original cavity. It is natural to choose the line Γ as the common reference line for both the left and the right sections. If we now glue again the two parts to form the original billiard, we can interpret the mapping obtained by the product $T_R T_L$ in the following way- Take a trajectory which scatters first to the left, and crossing Γ, from left to right, it defines the correspondence $(I, \phi) \rightarrow (\tilde{I}, \tilde{\phi})$. The trajectory continues into the right part and emerges back through Γ with action angle values (I', ϕ'). Thus, the product system defines a proper Poincaré section through Γ $(I, \phi) \rightarrow (I', \phi')$ which is area preserving since it is mediated by classical trajectories. The above definition introduces a 're-injection' mechanism by construction. We shall see below that this Poincaré map is the classical analogue of the unitary operator $S = S_R S_L$.

The semi-classical approximation is now introduced by the following argument: An S matrix is a unitary transformation which acts on the space of state vectors which can describe all the possible initial or final conditions of the scattering process. The PSM is the classical analogue of the S matrix, being an area preserving mapping of the section of phase space which defines the initial and final states of the scattering system. The semi-classical approximation of the S matrix makes use of the trajectories which mediate the PSM, and of the generating function (action) which is associated with the classical mapping. It reads [26)]

$$S_{nm} \cong \sum_{\alpha} A^{(\alpha)}_{n,m} \exp(i\Phi^{(\alpha)}_{n,m} + i\frac{\pi}{2}\nu_\alpha) , \qquad (IV.3)$$

where $n, (m)$ stand for the initial (final) quantum numbers which are the quantized values of the action variables which define the initial (final) states of the classical scattering system. The action Φ is the reduced action, (measure in units of $\hbar$). The absolute square of the amplitudes A_α give the classical contribution of the trajectory α to the total transition probability. ν_α is the Maslov index and it includes also the phase factor which is multiplied by (-1) for every collision with the boundary (for Dirichlet boundary conditions). The summation extends over all the classical scattering trajectories which satisfy the boundary conditions which define the transition.

We have shown, however, that for the two classes of problems which we discuss here, the classical scattering map can be identified with a mapping which describes the motion inside the billiard. Hence, from now on, we may forget the scattering origin of S in the secular equation. All we need is its semi-classical version (IV.3) which can be thought of as a semi-classical representation of a classical mapping which describes the dynamics inside the billiard. Matrix multiplication is performed by first expressing the summation over the intermediate indices as a sum of integrals by using the Poisson identity. Then, each integral is performed by the saddle-point approximation. The trace operation is also performed in the same way, and standard techniques (see e.g. ref. 16 or 28) enables us to write the semi-classical expression for $Tr(S^n)$ in terms of periodic orbits of the classical mappings which corresponds to the unitary operator S. We have shown above that these mapping are nothing but the ones describing the dynamics inside the billiards.

$$\mathrm{Tr}(S^n) \approx \sum_{\substack{\alpha \\ \eta_\alpha \rho_\alpha = n}} C_{\alpha,\rho_\alpha} \exp(i\rho_\alpha \Phi_\alpha) \qquad (IV.4)$$

The summation is over all primitive periodic orbits α of period η_α which, if repeated ρ_α times, will perform a period of length n. The action Φ_α is assumed to include the Maslov index as well as the phase factor, which is due to the reflections from the billiard boundary. The amplitudes C_{α,ρ_α} are given by

$$C_{\alpha,\rho_\alpha} = \frac{\eta_\alpha}{\sqrt{|\det(I - M_\alpha^{\rho_\alpha})|}} \,, \qquad (IV.5)$$

where M_α is the matrix of the tangent mapping (the Jacobian) calculated for the primitive periodic orbit α.

Substituting the semi-classical expression (IV.2) in (III.26) we get

$$f_{\Lambda-l} \approx \sum_{\boldsymbol{p},\boldsymbol{r}} \phi^{(l)}_{(\boldsymbol{p},\boldsymbol{r})} \prod_{j=1}^{N(\boldsymbol{p})} \left[\sum_{\substack{\alpha \\ \eta_\alpha \rho_\alpha = p_j}} C_{\alpha,\rho_\alpha} \exp(i\rho_\alpha \Phi_\alpha) \right]^{r_j} \qquad (IV.6)$$

where $N(\boldsymbol{p})$ is the number of primitive orbits (number of positive p_i) in the composite orbit $(\boldsymbol{p},\boldsymbol{r})$. This complicated sum can be viewed as a sum over objects which Berry and Keating[25] called "pseudo orbits" and which we prefer to name ***composite orbits***: For each composite period l one considers all groups of primitive periodic orbits with periods η_α and repetitions ρ_α such that

$$l = \sum_j r_j p_j \qquad (IV.7)$$

The non negative integers r_j and the corresponding $p_j = \eta_\alpha \rho_\alpha$ identify the partition labels $(\boldsymbol{p},\boldsymbol{r})$. The composite action is obtained by summing the actions of the periodic orbits together with their multiplicities, and the composite amplitude is obtained by taking the product of the amplitudes (including also the combinatorial factor $\phi^{(l)}_{(\boldsymbol{p},\boldsymbol{r})}$). Thus, expression (IV.6) for the coefficients $f_{\Lambda-l}$ can be written as a sum over composite orbits

$$f_{\Lambda-l} = \sum_s C_s^{(l)} \exp\left(i\Phi_s^{(l)}\right) \,, \qquad (IV.8)$$

where the index s labels the composite orbits.

The semi-classical expressions for f_l and $f_{\Lambda-l}$ do not manifestly obey the symmetry relation (III.23) — they are constructed from composite trajectories of different length and the corresponding semi-classical series (IV.8) are not guaranteed to satisfy the proper symmetries. These observations emphasize the basic and most severe limitations of the semi-classical approximation — it does not preserve the structural constraints which are due to a basic conservation law- namely - that the S matrix must be unitary for real energies. However, the secular function, in the form of (III.24) , already *assumes* unitarity. Inserting (IV.8) into (III.24) therefore creates a semi-classical expression for the secular function, which automatically incorporates the unitarity of $S(E)$. We get,

$$\det(I-S) = \exp\left[i\pi\bar{N}(E)\right) 2\Re\mathrm{e}\left\{\exp[-i\pi\bar{N}(E)]\sum_{l=l_0}^{\Lambda} f_{\Lambda-l}\right\} = \\ = \exp\left[i\pi\bar{N}(E)\right] 2\Re\mathrm{e}\left\{\exp[-i\pi\bar{N}(E)]\sum_{l=0}^{\Lambda-l_0}\sum_{s} C_s^{(l)}\exp\left(i\Phi_s^{(l)}\right)\right\} \qquad (IV.9)$$

This is the desired semi-classical expression for the secular function. (Note that we have written (IV.9) for the case that Λ is odd to facilitate the notation). This semi-classical secular function has the following properties:

It is constructed from composite orbits with period less than $\Lambda/2$. This limits the periods of the original periodic orbits by the same bound. Notice that so far the period is counted by the number of iterations of the PSM and not by the physical length of the period. We return to this point in the following paragraphs.

The f_l are expressed as cumulants in terms of the various powers of $\mathrm{Tr}(S^n)$. Bogomolny [13] has shown that because of the shadowing property of periodic orbits [29] there are effective cancellations between various terms in the semi-classical expression, and most of the contributions come from a subset of the possible composite orbits.

The structure of (IV.9) is similar to that of the Riemann-Siegel lookalike secular equation derived by Keating [30] by a completely different method.

For convex billiards the mean *physical* length of the periodic orbits is proportional to the number of reflections from the boundary (the period of the map). We reach the important conclusion that for convex billiards the secular equation is expressed in terms of a finite number of periodic orbits whose physical length is bounded. We can estimate the bound on the period in the following way. The dimension of the S matrix is $\Lambda = Dk$, where D is the largest diameter in the billiard. Periodic orbits which scatter $\frac{1}{2}\Lambda$ times from the wall will travel a trajectory of mean length $\lambda = \frac{1}{2}\Lambda\langle c\rangle = \frac{1}{2}D\langle c\rangle k \approx \frac{1}{2}Ak$, where A is the area of the billiard and $\langle c\rangle$ is the length of a mean chord in the billiard.

The time it takes to traverse such orbits is $\langle t \rangle = \lambda/v$. Using Weyl's formula for the mean level spacing in the billiard we get $\langle t \rangle \approx \frac{1}{2}h\,\langle d(E) \rangle$, which is the Heisenberg time (the time needed to resolve two levels which are separated by the mean level spacing). This is similar to the time bound for periodic orbits in Keating's Riemann Siegel lookalike secular equation.

The situation is much more complex when one wants to apply the same ideas in the general case. Now, there may appear periodic orbits which are localized in space and never intersect the section Γ. If, in particular the scattering into the L or the R parts of the billiard is chaotic, there is no relation between the number of times the particle is reflected from the boundary and the value of Λ. A fixed point, for example, gets contributions from infinitely many orbits, which reflect an arbitrary number of times from Σ_L and Σ_R before it hits Γ. The sums on periodic orbits (IV.9) extend over an infinity of orbits, and their convergence is not *a-priori* guaranteed. Thus, the expression (IV.9) preserves the formal structure imposed by unitarity, but might be meaningless as a numerical representation of the secular function. We shall show now that the difficulty which arises because of the presence of arbitrary long trajectories can be removed. Also here, a proper Riemann-Siegel lookalike secular function can be written in terms of periodic orbits, whose physical length is bounded by the same bound as was given for the case of convex billiards.

Following the ideas proposed by Keating[30)], we regard the semiclassical approximation (IV.9) to the f_l as formal expressions and impose on them the symmetry (III.23). We get

$$\begin{aligned} &\exp\left(-i\pi\bar{N}(k)\right)\sum_s C_s^{(l)}\exp\left(i\Phi_s^{(l)}(k)\right) \\ &= \exp\left(i\pi\bar{N}(k)\right)\sum_{s'}\left(C_{s'}^{(\Lambda-l)}\right)^*\exp\left(-i\Phi_{s'}^{(\Lambda-l)}(k)\right) , \end{aligned} \qquad (IV.10)$$

where for convenience we use the wave number k instead of the energy label. We choose a mean value $\tilde{k}$ and expand the classical actions as well as $\bar{N}(k)$ about it. At this point we make the explicit assumption that $\bar{N}(k)$ is smooth and approximately coincides with $\langle N(k) \rangle$. In other words, we assume that we are already in the semi-classical domain. The k derivative of $\bar{N}(k)$ will be denoted by λ_{av} and it is Λ times the average physical length of a trajectory between successive passages through the Poincaré section at Γ. The k derivative of the reduced actions $\Phi_s^{(l)}$ give $\lambda_s^{(l)}$ — the composite length of the composite periodic orbit. We choose an interval of the size a which is large on the scale of the mean level spacing, but smaller than the distance between successive thresholds (this is easily achieved for large $\tilde{k}$). We proceed now with formal manipulations and multiply both sides of (IV.10) by $\exp(ikx)$ (with an as yet arbitrary x) and integrate term by term both sides of (IV.10) over the interval of size a about $\tilde{k}$. Using the linearized form of the phase factors in (IV.10), and assuming that the amplitudes $C_s^{(l)}$ vary slowly with k, we get

$$\begin{aligned}\exp\left(-i\pi\bar{N}(\tilde{k})\right)\sum_s C_s^{(l)}\exp\left(i\Phi_s^{(l)}(\tilde{k})\right)\delta_a\left(\lambda_s^{(l)}+x-\lambda_{av}/2\right)=\\ =\exp(i\pi\bar{N}(\tilde{k}))\sum_{s'}\left(C_{s'}^{(\Lambda-l)}\right)^*\exp\left(-i\Phi_{s'}^{(\Lambda-l)}(\tilde{k})\right)\delta_a\left(-\lambda_{s'}^{(\Lambda-l)}+x+\lambda_{av}/2\right)\end{aligned}\qquad (IV.11)$$

where $\delta_a(x)$ is a smoothed δ function with a width a. We can use the Weyl formula (see (V.15) to write an explicit expression for λ_{av}

$$\lambda_{av}=2\pi k\frac{\hbar^2}{m}k\frac{dN_R(E)}{dE}=\Lambda v\tau(k)\qquad (IV.12)$$

where v is the velocity. The last equation is now integrated with respect to x over the interval $[-\infty,0]$. We also approximate the smoothed δ-functions by their sharp counterparts (this approximation is discussed and justified by Keating[22)]). We get

$$\begin{aligned}\exp\left(-i\pi\bar{N}(\tilde{k})\right)\sum_{\lambda_s^{(l)}\geq\frac{1}{2}\lambda_{av}} C_s^{(l)}\exp\left(i\Phi_s^{(l)}(\tilde{k})\right)=\\ =\exp\left(i\pi\bar{N}(\tilde{k})\right)\sum_{\lambda_{s'}^{(\Lambda-l)}\leq\frac{1}{2}\lambda_{av}}\left(C_{s'}^{(\Lambda-l)}\right)^*\exp\left(-i\Phi_{s'}^{(\Lambda-l)}(\tilde{k})\right)\end{aligned}\qquad (IV.13)$$

Note that the summation on the l.h.s. of the above expression is over composite orbits with length larger than $\frac{1}{2}\lambda_{av}$, whereas the summation on the r.h.s. is over composite orbits of length smaller than $\frac{1}{2}\lambda_{av}$.

Starting again from (IV.9) but now multiplying by $\exp(-ikx)$, integrating over k and x as before, we get

$$\exp\left(-i\pi\bar{N}(\tilde{k})\right) \sum_{\lambda_s^{(l)} \leq \frac{1}{2}\lambda_{av}} C_s^{(l)} \exp\left(i\Phi_s^{(l)}(\tilde{k})\right) =$$
$$= \exp\left(i\pi\bar{N}(\tilde{k})\right) \sum_{\lambda_{s'}^{(\Lambda-l)} \geq \frac{1}{2}\lambda_{av}} \left(C_{s'}^{(\Lambda-l)}\right)^* \exp\left(-i\Phi_{s'}^{(\Lambda-l)}(\tilde{k})\right) \qquad (IV.14)$$

We have thus obtained the important result that the imposition of the unitarity of the S matrix on the semiclassical approximation result in the relations (IV.13,14) which enable us to express the contribution from all the long orbits to $f_{\Lambda-l}$, in terms of the contribution from the short orbits to f_l^*, and *vice versa*. The dividing line between "long" and "short" orbits is given in terms of $\frac{1}{2}\lambda_{av}$. We can now partition expression (IV.8) for $f_{\Lambda-l}$ into its short orbits and long orbits contributions. For the long orbit sum we can use (IV.13) and write an expression for $f_{\Lambda-l}$ which is based exclusively on short (composite) orbits:

$$f_{\Lambda-l} = \sum_{\lambda_s^{(l)} \leq \frac{1}{2}\lambda_{av}} C_s^{(l)} \exp(i\Phi_s^{(l)}(\tilde{k})) +$$
$$+ \exp\left(2i\pi\bar{N}(\tilde{k})\right) \sum_{\lambda_{s'}^{(\Lambda-l)} \leq \frac{1}{2}\lambda_{av}} \left(C_{s'}^{(\Lambda-l)}\right)^* \exp\left(-i\Phi_{s'}^{(\Lambda-l)}(\tilde{k})\right) \qquad (IV.15)$$

This expression is now to be substituted in (III.24) and it provides the desired expression for the secular equation which is based on periodic orbits of a finite period, and of a finite physical length. The resulting expression is now completely equivalent to the Riemann-Siegel lookalike secular equation derived by Keating. It has a structural advantage in keeping the contributions from groups of composite trajectories with the same composite period l in the form of a cumulant. This might be exploited by using the techniques developed by Cvitanovic and Eckhardt[29,30].

As a last point, we would like to give an alternative interpretation of the bound $\frac{1}{2}\Lambda$ on the periodic orbits which contribute to the secular equation. We have shown that this limit does not restrict the physical length of the orbits when chaotic scattering prevails. However, it imposes an interesting bound on the average length of the contributing orbits. Suppose the length of the segment dividing the billiard is taken to be D, and suppose also for the sake of simplicity that the division is into two more or less equivalent pieces. The average of the physical length of a trajectory between two consecutive passes through the segment is, by simple geometrical considerations, approximately $\langle t \rangle \approx Dv/\pi A$, where v is the velocity and A is the area of the L (or R) half of the billiard. The dimension of the S matrix can be approximated by

$$\Lambda \approx \frac{kD}{\pi} \approx \frac{mvD}{\pi\hbar} \qquad (IV.16)$$

Inserting the expression for $\langle t \rangle$ into (IV.16) we get

$$\frac{1}{2}\Lambda \langle t \rangle \approx \frac{1}{2} h \langle d(E) \rangle \ . \qquad (IV.17)$$

where the average level density is $\langle d(E) \rangle = mA/2\pi\hbar^2$ by virtue of Weyl's relation. This again is reminiscent of Keating's time bound, only this time applied to the ***average*** length of the trajectories used.

The moral of the story is that to quantize semi-classically a billiard of any shape, one needs trajectories whose period are not longer than the Heisenberg time $h\langle d(E) \rangle$, which is the time needed to resolve a typical spacing between successive levels. The discussion above also helps us in making the optimal choice of the section Γ: It should be the section for which the quantities $\bar{d}$ and $\bar{N}$ give the best approximation to the properly energy averaged quantities $\langle d \rangle$ and $\langle N \rangle$. This will typically happen if the segments $\Sigma_{L,R}$ cannot trap trajectories for a long time. This has a good chance to occur, if the interface Γ is large so that escaping through it is most probable.

V. Spectral Densities

There are two complementary attitudes one can take in trying to interpret a spectrum. One can examine each level individually, and for this purpose the secular function is the appropriate tool, since by finding its zeroes, one obtains the spectrum level by level. The other approach is of more statistical nature. Here one considers average properties such as the mean density, correlations of spectral fluctuations etc. This approach will be taken in the present and the following chapters. Clearly, the secular equation is the basis for any further development since it stores the complete spectral information.

The semi-classical theory of the spectral functions which will be derived here, will start from the semi-quantal expression (III.5), or rather from the more explicit form (III.14). The latter relation is rewritten as

$$2^{\Lambda} \prod_{l=1}^{\Lambda} \sin \frac{\theta_l(E)}{2} = \exp(-\frac{i}{2}\Theta(E))\det(I - S(E)) \tag{V.1}$$

The function on the left-hand side is manifestly real on the real energy axis. Using a well known identity, we derive the semi-quantal spectral density in the form

$$\begin{aligned} d_{sq}(E) &= -\frac{1}{\pi} \lim_{\epsilon \to 0} \Im m \frac{d}{dE} \log \prod_{l=1}^{\Lambda} \sin \frac{\theta_l(E + i\epsilon)}{2} \\ &= \bar{d}(E) - \frac{1}{\pi} \lim_{\epsilon \to 0} \Im m \frac{d}{dE} \log \det(I - S(E + i\epsilon)) \end{aligned} \tag{V.2}$$

This expression is the starting point for the discussion of the two issues which we would like to clarify in the present chapter, namely, the semi-classical expression for the averaged spectral functions, and the relation of the present approach to the well known Gutzwiller theory for the spectral density.

V.a The averaged spectral density.

In the previous chapter we have shown that $\bar{d}(E)$ is proportional to the Wigner time delay. If one decomposes the S matrix into its pole (resonances) contributions, one can see that the Wigner delay time is a superposition of normalized Lorentzians, positioned at the resonance energies, and having a width which is equal to the distance of the pole from the real axis. Another term in the Wigner time may be due to a smooth overall phase of the S matrix. Both contributions, however, add up to a continuous function on the real energy axis. Hence, the main difference between $d_{sq}(E)$ and $\bar{d}(E)$ is that the former is a sum of δ spikes, whereas the latter is a series of Lorentzians superimposed on a smooth background.

We shall show now that the two densities $d_{sq}(E)$ and $\bar{d}(E)$ have a common smooth average $\langle d(E) \rangle$. This is proven by evaluating (V.2) at a distance ϵ from the real E axis. $d(E + i\epsilon)$ and $\bar{d}(E + i\epsilon)$ are the Lorentzian–weighted averages of their respective values on the real E axis. However, the S matrix decays exponentially to zero as $\epsilon \to +\infty$, and (V.2) implies that $\langle d_{sq}(E) \rangle - \langle \bar{d}(E) \rangle \to 0$.

As a matter of fact, $\bar{d}(E)$ approaches its smooth asymptote rather quickly. Consider the "worst" case when S describes chaotic scattering, and therefore is dominated by resonances. However, the resonance width is linear in $\hbar$ whereas their separation is quadratic in $\hbar$ [28]. Hence in the semi-classical limit, the resonances overlap. Moreover, the dimension of S increases as $\hbar^{-1}$. Therefore, the averaging involved in taking the trace of (III.18) is getting more effective. These two effect combine to smooth out the fluctuations in the time delay in the semi-classical domain. A comparison between the two densities and the associated counting functions, is shown in Fig. 6.

Further insight into the relationship between the two densities discussed above is obtained by coming back to the rectangular billiard which was introduced in chapter III. We have shown there that the semi-quantal secular equation, $\det(I - S) = 0$ is an exact quantization condition, and is independent of the choice of the auxiliary section Γ. Taking Γ as the section of unit length along the y axis, we get for the smooth counting function,

$$\bar{N}^{(1)}(k) = \sum_{l=1}^{\Lambda^{(1)}} \alpha \left((k/\pi)^2 - l^2\right)^{\frac{1}{2}} - \frac{1}{2}\Lambda^{(1)} \qquad (V.6a)$$

We now choose for the section Γ the side of length α along the x axis. Now,

$$\bar{N}^{(\alpha)}(k) = \sum_{l=1}^{\Lambda^{(\alpha)}} \frac{1}{\alpha} \left((\alpha k/\pi)^2 - l^2\right)^{\frac{1}{2}} - \frac{1}{2}\Lambda^{(\alpha)} \qquad (V.6b)$$

It is clear that the two functions are different. A closer inspection shows that the leading terms in the two expressions are the same,

$$\bar{N}^{(1)}(k) \approx \bar{N}^{(\alpha)}(k) \approx k^2 \frac{\alpha}{4\pi} \qquad (V.7)$$

Note that α is the area of the rectangle. This is just an example of the Weyl theorem which will be discussed shortly for billiards of arbitrary shapes.

One can write down the exact spectral counting function for the rectangular billiard [31]

$$N(k) = [\alpha \sum_{l=1}^{\Lambda^{(1)}} \left((k/\pi)^2 - l^2\right)^{\frac{1}{2}}] \qquad (V.8)$$

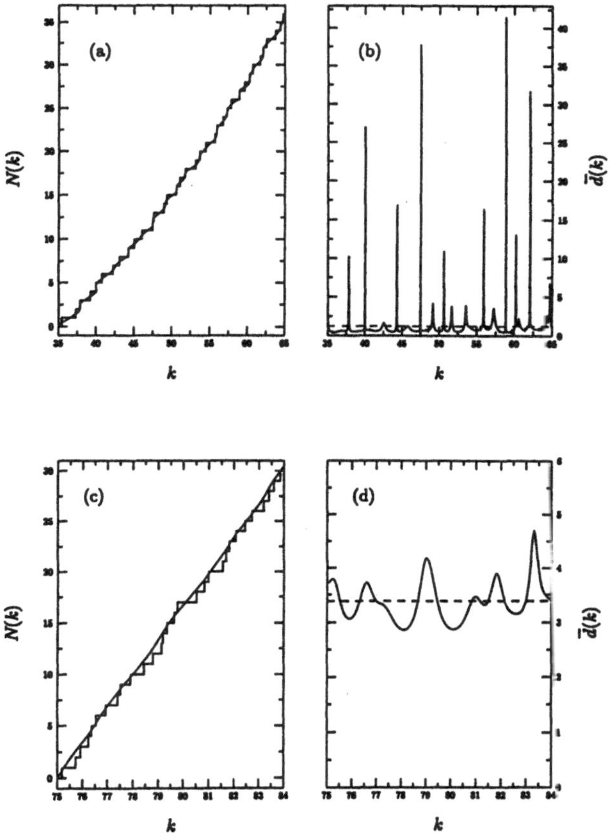

Fig. 6. (a) The counting function $\bar{N}(k)$ (smooth line) and the exact counting function $N(k)$ for $\Lambda = 1$.(*b*) The corresponding density $\bar{d}(k)$ and its average $< d(k) >$ (dashed line). (c,d) same as (a,b) but for $\Lambda = 6$.

where square brackets stand for the integer part. Thus,

$$\bar{N}^{(1)}(k) - N(k) = \sum_{l=1}^{\Lambda^{(1)}} \left(\{\xi_l\} - \frac{1}{2} \right) \tag{V.9}$$

Here $\{\xi_l\}$ is the fractional part of $\xi_l = \alpha((k/\pi)^2 - l^2)^{\frac{1}{2}}$. The fractional part is a function which oscillates periodically about its mean value $\frac{1}{2}$. Hence, the sum represents a function with a vanishing mean. A similar result can be obtained for the difference $\bar{N}^{(\alpha)}(k) - N(k)$. Thus, the smooth counting functions and the spectral counting functions share a common mean.

We can use now the relation (III.19) to derive the Weyl relation which was illustrated above for the rectangular billiard. The classical interpretation of $\tau(E)$ is that it measures the mean delay that a particle suffers due to the presence of the scattering potential. Consider first the case of a convex billiard. Here, the incoming trajectory terminates not at the centrifugal barrier as it would do in free space, but rather at the point of impact on the billiard. The same is true for the outgoing part of the trajectory. Hence the time delay (in absolute value) is the sum of the sections excluded from the trajectory (see Fig. 5) divided by the velocity. The mean value of the time delay is therefore the same as the mean chord length $\langle c \rangle$ divided by the velocity. A simple statistical argument shows that for planar billiards $\langle c \rangle = \frac{\pi A}{L}$, where A is the area of the billiard and L its perimeter. We also recall that in this case $\Lambda = [Dk]$ and D is the mean diameter so that $L = \pi D$. Substituting in (III.19) we get

$$\bar{d}(E) = \frac{Am}{2\pi\hbar^2} ,$$

where m is the mass of the particle. The above relation is the Weyl mean level density formula, which is the first term in an asymptotic series for the level density expressed in terms of the geometric properties of the billiard[32)]. It gives a partial answer to Kac's famous question[33)]: " can one hear the shape of the drum ?" in that it provides the area of the drum, in terms of it's mean spectral density.

If we consider billiards of arbitrary shapes, one has to remember that in this case S is a product of the left and the right scattering matrices. One can easily see from (III.18) that the delay time is nothing but the sum of the delay times in each of the parts of the billiard. Suppose that the particle is in the left side of the billiard. Per unit time, it has the probability $\frac{Dv}{\langle c \rangle L}$ to pass into the right side. (D is the size of the section Γ, L is the total perimeter of the left billiard (including Γ) and as before $\langle c \rangle$ is the average chord length in the billiard. The inverse of this quantity gives the mean delay time, and if one substitutes this in (III.19) together with the contribution for the delay from the right side, one reproduces the Weyl expression (V.8) for the case of billiards with arbitrary shapes.

V.b The Gutzwiller Trace Formulae for the Spectral Density

We shall now show that the semi-classical expression for the density of states $d_{sq}(E)$ (V.1) is semi-classically equivalent to Gutzwiller's original expression. This is achieved by expanding

$$\log\det(I - S(E)) = \mathrm{Tr}\log(I - S(E)) = -\sum_{n=1}^{\infty}\frac{1}{n}\mathrm{Tr}(S(E)^n)\,. \qquad (V.10)$$

In (IV.4-5) we have already given an explicit semi-classical form for $\mathrm{Tr}S^n$ in terms of periodic orbits of the Poincaré Scattering Map (PSM). We have shown that the "inside-outside" connection allows the identification of periodic orbits of the PSM with the periodic orbits in the billiard. We can also take advantage of the fact that the period of a periodic orbit is given as $T_\alpha = \frac{d\Phi_\alpha}{dE}$. Substituting in (V.10) we finally get

$$d_{sc}(E) - \bar{d}(E) = \frac{1}{\pi}\Im m\left(i\sum_{n=1}^{\infty}\sum_{\eta_\alpha p_\alpha = n}\frac{T_\alpha e^{(ip_\alpha\Phi_\alpha)}}{\sqrt{\det(I - M_\alpha^{p_\alpha})}}\right) \qquad (V.11)$$

where the inner sum is over primitive periodic orbits α as explained in chapter (IV).

Let us consider first the case of a convex billiard. The sum over periodic orbits in (V.11) is over the entire set of periodic orbits in the billiard, and therefore (V.11) is identical with Gutzwiller's expression, especially if we identify $\bar{d}(E)$ with $\langle d(E)\rangle$.

The situation is slightly more complex when we deal with billiards with arbitrary shapes. Here, the periodic orbits which appear in (V.11) are the periodic orbits of the Poincaré map which is based on the section Γ. If Γ intersects all the periodic orbits of the complete billiard (no periodic orbits are trapped in either Σ_L or Σ_R), then we are in the same situation as discussed previously for the case of convex billiards. If this is not the case, we may use a result which is due to Balian and Bloch[11]), (see also ref. 34), which shows that the trapped periodic orbits contribute to the density $\bar{d}(E)$. This is the part of $\bar{d}(E)$ which is due to the resonances in the S matrix. The set of periodic orbits in the entire billiard is the union of two distinct sets: the periodic orbits which intersect Γ, and those which do not intersect Γ. The later contribute to the first term on the l.h.s. of (V.11), and the former contribute to the second term. Hence, even for the present case, (V.11) is equivalent to the Gutzwiller sum if we allow to reorder the sum by separating the contributions of the trapped trajectories from those of the untrapped ones. One should comment at this point that the set of trapped orbits may be nontrivial (fractal) when the scattering problem is chaotic. This is the case in the example of the Sinai billiard we described above. Reordering series with uncertain convergence properties is a dubious matter. Therefore, one should consider this only when the Gutzwiller series is taken as a formal expression, and not as a series which defines a function.

The result (V.11) offers an *a-posteriori* justification of the quantization method which we pursued by showing that the semi-classical spectral density is equivalent to the Gutzwiller sum. As such, it also shares all its problematics and deficiencies.

The infinite sum in (V.11) can be reduced by first writing

$$\sqrt{|\det(I - M_\alpha)|} = \exp(\gamma_\alpha/2) - \exp(-\gamma_\alpha/2)$$

Where γ_α is the stability exponent of the orbit α. The sum over n in (V.11) is unrestricted, and therefore the ρ_α and n sums can be combined to a single sum over all the possible repetitions of primitive orbits,

$$d_{sc}(E) - \bar{d}(E) = \frac{1}{\pi}\Im m\left(i\sum_{\alpha}\sum_{r=1}^{\infty}\frac{T_\alpha e^{(ir\Phi_\alpha - r\gamma_\alpha/2)}}{1-\exp(-r\gamma_\alpha)}\right) . \qquad (V.12)$$

Expanding the resulting denominator in a geometrical series we get

$$d_{sc}(E) - \bar{d}(E) = \frac{1}{\pi}\Im m\left(i\sum_{\alpha}\sum_{r=1}^{\infty}\sum_{k=0}^{\infty}T_\alpha e^{(ir\Phi_\alpha - r\gamma_\alpha/2 - kr\gamma_\alpha)}\right) . \qquad (V.13)$$

At this point the r summation can be performed, yielding

$$\begin{aligned} d_{sc}(E) - \bar{d}(E) &= \frac{1}{\pi}\Im m\left(i\sum_{\alpha}\sum_{k=0}^{\infty}\frac{T_\alpha e^{(i\Phi_\alpha - \gamma_\alpha/2 - k\gamma_\alpha)}}{1 - e^{(i\Phi_\alpha - (k+1/2)\gamma_\alpha)}}\right) \\ &= \frac{1}{\pi}\Im m\frac{d}{dE}\log Z_{dyn}(E) . \end{aligned} \qquad (V.14)$$

Where the dynamical ζ function $Z_{dyn}(E)$ is defined as

$$Z_{dyn}(E) = \prod_{k=0}^{\infty} \prod_{\alpha} \left(1 - e^{(i\Phi_\alpha - (k+1/2)\gamma_\alpha)}\right) . \qquad (V.15)$$

Comparing (V.14) with our starting point (V.1) we find that the function

$$\tilde{Z}(E) = \exp(-i\pi \bar{N}(E)) Z_{dyn}(E) \qquad (V.16)$$

is an alternative expression for the semi-classical secular equation. We shall investigate $\tilde{Z}(E)$ more closely, and compare it to the other form which was derived in the preceding chapter.

The first issue is the domain of convergence of the infinite product (V.15). Consider first the product with $k = 0$. A product $\prod_{n=1}^{\infty}(1 + a_n)$ converges in the absolute sense if and only if the sum $\sum_{n=1}^{\infty} a_n$ converges absolutely. In the present case, we can check the convergence of the α-sum by grouping together all the contributions from trajectories with periods in a small interval $T \leq T_\alpha \leq T + \Delta T$. For chaotic billiard, the number of such trajectories proliferates as $\exp(hT)$, where h is the topological entropy. The stability exponents are also proportional to T, $\gamma_\alpha \propto \lambda T_\alpha$ and λ is the Lyapunov exponent. But, $h \approx \lambda$, and the sum with $k = 0$ will not converge absolutely on the real E axis : The factor $\exp(-T\lambda/2)$ is not sufficient to counter balance the number $\exp(T\lambda)$ of contributing trajectories! The sum will converge if we restrict the energy to the domain $\Im m(E) > h/2$, because $\Phi_\alpha \approx T_\alpha E$. By the same token, the sums with $k > 0$ converge absolutely on the real energy axis and pose no problem. The dynamical ζ-function as defined above corresponds to an analytic function only if the "entropy barrier" $\Im m(E) > h/2$ is not violated. In order to be able to reach the real E-axis where the eigen-energies are, one should find ways to analytically continue (V.15) beyond the entropy barrier.

This intrinsic difficulty with the dynamical ζ-function motivated the vigorous research into the problems of semi-classical quantization of chaotic systems. In the present lectures we presented an approach which circumvented this difficulty altogether. We shall briefly review the other approaches which are based on the study of the analytical properties of (V.15) in the concluding remarks.

It is possible to reduce the $\tilde{Z}(E)$ function further into a sum over composite orbits by using the Euler identity [35],

$$\tilde{Z}(E) = \exp(-i\pi\bar{N}(E))\sum_{c} A_c \exp(iS_c) \qquad (V.17)$$

were composite orbits with unrestricted length. This observation can be presented as an apparent paradox. Starting with the same semi-quantal secular equation, we came up with two conflicting conclusions. The semi-classical secular equation derived in chapter IV uses a finite number of periodic orbits. The dynamical ζ-function derived above uses all the periodic orbits, resulting in severe convergence problems. The reason for the contradicting results comes from the fact that the semi-classical approximation was introduced at different points. In the derivation of the secular equation (IV.9) we made use of all the constraints which were due to the unitarity of S and its finite dimension. Only then we introduced the semi-classical approximation to $\mathrm{Tr}S^n$. In the second approach we wrote down the infinite series (V.10) which is **not** an absolutely convergent series, and approximated its terms by the semi-classical expressions for $\mathrm{Tr}S^n$. No wonder that meddling with the terms of a conditionally convergent series results in a divergent series! One could cure this by making use of an important fact. For any matrix S with a finite dimension Λ one can express $\mathrm{Tr}S^n$ with $n > \Lambda$ in terms of $\mathrm{Tr}S^k$ with $k = 1, \ldots, \Lambda$. If this is implemented in (V.10) and the semi-classical approximation is introduced at this point, the above paradox is resolved.

One should also observe that if $\tilde{Z}(E)$ were extended analytically to the real axis, then (V.14) shows that it is a real function. In other words,

$$\exp(-i\pi\bar{N}(E)Z_{dyn}(E)) = \exp(+i\pi\bar{N}(E))Z^{*}_{dyn}(E) \ . \qquad (V.18)$$

This is the famous functional equation which dynamical zeta functions obey.

VI. Spectral Correlations

In the present chapter we shall emphasize an important feature of the present formalism. It enables us to introduce an interesting relationship between the distribution of the eigen-phases of the S matrix on the unit circle, and the distribution of the energies obtained as solutions of the semiclassical secular equation $\det(I - S(E)) = 0$.

As was already shown in chapter III (see (III.16)), a zero of the secular equation occurs whenever any of the eigen-phases of the S matrix, $\theta_l(E)$, equals an integer multiple of 2π. Hence, the spectral density $d(E)$ can be expressed as

$$d(E) = \sum_i \delta(E - E_i) = \sum_{l=1}^{\Lambda} \tau_l \delta_p(\theta_l(E)) \tag{VI.1}$$

where $\tau_l = \theta_l'(E)$, and the prime denotes differentiation with respect to E. Let us consider the spectral density in an interval of size Δ about a mean energy E_0. Δ is taken to be large on the scale of the mean level separation, which is given by Weyl's formula to be $2\pi\hbar^2/(mA)$, where A is the area enclosed by the billiard. Then, for $\epsilon = (E - E_0) < \Delta$,

$$d(\epsilon) \approx \sum_{l=1}^{\Lambda} \tau_l \delta_p \left(\theta_l(E_0) + \epsilon\tau_l\right) \ . \tag{VI.2}$$

If we were able to show that the fluctuations in the distribution of the τ_l about their mean value, τ, are small, we could replace all the τ_l in (VI.2) by τ. Defining $\theta = 2\pi - \epsilon\tau$ we would get

$$d(\epsilon) \approx \tau \sum_{l=1}^{\Lambda} \delta_p(\theta_l(E_0) - \theta) \ , \tag{VI.3}$$

which would have established the correspondence between the energy spectral density (the l.h.s of (VI.3)) and the eigen-phase spectral density (the r.h.s. of (VI.3)). The first part of the present chapter will be devoted to show that one can indeed substitute the τ_l by their mean value when one studies chaotic billiards. This will be done in two steps. In the first one we shall show that the eigen-phases of the S matrix at a constant energy distribute on the unit circle according to the predictions of random matrix theory for the ensemble of unitary and symmetric matrices (the COE). Further considerations will then be used to show that the distribution of the τ_l about their mean gets narrow as $\hbar \to 0$, which justify the assumption underlying (VI.3). It is known from Dyson's theory that the COE and the GOE statistics are intimately related. The validity of (VI.3) allows us to use this relation and to conclude that the COE distribution of the eigen-phases imply the GOE distribution of the energy levels in the billiard! In other words, one can deduce the billiard spectral correlations from the study of the corresponding eigen-phase spectrum of the S matrix.

VI.a S Matrix Spectral Correlations.

This subsection is a detour, whose purpose is to show that for chaotic billiards, the two-point correlation functions in the S matrix spectrum reproduce (at least asymptotically) the predictions of the theory of random matrices for the COE. We provide a semi-classical proof which is based on Berry's pioneering observation [36] and its further ramifications [37].

The S matrix spectral density is written as above,

$$d_s(\theta) = \sum_{l=1}^{\Lambda} \delta_p(\theta - \theta_l) = \frac{1}{2\pi} \sum_{n=-\infty}^{\infty} e^{-in\theta} \mathrm{Tr} S^n \tag{VI.4}$$

The two point correlations will be discussed in terms of the pair distribution function $R_2(\eta)$ and the cluster function $Y_2(r)$ which are defined in the following way.

$$R_2(\eta) = \int_0^{2\pi} d_s(\omega + \eta/2) d_s(\omega - \eta/2) d\omega - \Lambda \delta_p(\eta) \tag{VI.5}$$

It gives the number of pairs of eigen-phases which are η apart, excluding the contribution of self-correlations to the point $\eta = 0$. The pair distribution function can be rewritten in the form

$$R_2(\eta) = \frac{\Lambda}{2\pi} \left((\Lambda - 1) + 2 \sum_{n=1}^{\infty} b_2(n) \cos(n\eta) \right) \tag{VI.6}$$

where the Fourier coefficients $b_2(n)$ are given by

$$b_2(n) = \frac{1}{\Lambda} |\mathrm{Tr} S^n|^2 - 1 \,. \tag{VI.7}$$

The two-point cluster function $Y_2(r)$ is obtained from the above by measuring the eigen-phase difference η in terms of the mean spacing $\frac{2\pi}{\Lambda}$. Thus,

$$Y_2(r) = \frac{1}{\Lambda} \left(1 - 2 \sum_{n=1}^{\infty} b_2(n) \cos\left(n \frac{2\pi}{\Lambda} r \right) \right) \,. \tag{VI.8}$$

Till now, the spectral functions were defined for a single S matrix. To get smooth expressions for these functions, we must introduce some averaging mechanism. In the present context, we define the ensemble of matrices $S(E_i)$ where the energies E_i are taken from an interval where $\Lambda(E)$ is a constant, and the difference between the E_i's exceeds $\hbar\tau$, which is the energy correlation length (see ref. 7). The number of matrices in such an ensemble grows as $\hbar^{-1}$. From now on, the spectral functions defined above are considered as the ensemble averaged quantities, even if this will not be explicitly indicated. The functions defined above contain all the statistical information on two-point correlations. The more familiar Σ^2 and Δ_3 statistics are obtained from them by further integrations.

Again, the object of our discussion is $\mathrm{Tr}S^n$. This time, however, we should study the ensemble average of its absolute square. In the limit $n \gg \Lambda$, $\langle|\mathrm{Tr}S^n|^2\rangle \to \Lambda$. This can be understood by writing the explicit expression

$$|\mathrm{Tr}S^n|^2 = \Lambda + \left\langle \sum_{l \neq l'} e^{i(\theta_l - \theta_{l'})n} \right\rangle \qquad (VI.9)$$

The second term on the right averages to zero when $n \gg \Lambda$ if there are no systematic degeneracies among the θ_l. It follows that $b_2(n)$ approaches zero in this limit, which is a rather satisfactory result, since we already know from our previous discussion that this domain does not incorporate any physical information which cannot be deduced from the complementary domain $n \leq \Lambda$.

When $n \leq \Lambda$ we may use the semi-classical expression for $\mathrm{Tr}S^n$ which was derived in chapter IV. When we substitute (IV.4) in (VI.7) we have to be careful to note that because of intrinsic symmetries of the dynamics, there may exist classical orbits which are physically distinct, but whose contributions to the expression (IV.4) add coherently to the sum. Consider e.g. a mapping of a problem with time reversal symmetry. To any periodic orbit $x(j)$, $j = 1...n$ there exists a corresponding time reversed orbit $Tx(j)$, $j = 1\ldots,n$, where x denotes the phase space coordinates, and T stands for the time reversal operator. The two trajectories have the same action and the same prefactor C. We now write

$$b_2(n) \approx \frac{1}{\Lambda}\left\langle |g \sum_{\alpha} {}' C_\alpha e^{i\Phi_\alpha}|^2 \right\rangle - 1 \,, \qquad (VI.10)$$

where the sum is over primitive periodic orbits of period n, g stands for the degeneracy due to symmetry, and the prime next to the $\sum$ symbol indicates that only one orbits is included per g symmetry conjugated orbits. In the case of time reversal symmetry $g = 2$.

Eq. (VI.10) is rewritten as

$$b_2(n) \approx \frac{g^2}{\Lambda}\left(\left\langle \sum_{\alpha}{}' |C_\alpha|^2 \right\rangle + \left\langle \sum_{\alpha \neq \alpha'}{}' C_\alpha C^*_{\alpha'} e^{i(\Phi_\alpha - \Phi_{\alpha'})} \right\rangle\right) - 1 \qquad (VI.11)$$

For small enough values of n the second term (the 'nondiagonal contribution') is expected to vanish upon averaging. This is not an exact result, but it can be made plausible by the following argument: In chaotic systems action differences approach each other and get arbitrarily small for long periodic orbits due to the exponential proliferation of the orbits on the one hand, and the fact that actions are proportional to n on the mean, on the other hand. However, when n is not large, action differences are still much larger than $\hbar$ so that upon averaging the nondiagonal term vanishes.

The diagonal contribution can be written as

$$\begin{aligned} b_2(n) &= \frac{1}{\Lambda} g^2 \left\langle \sum_{\alpha}{}' \frac{\eta_\alpha^2}{|\det(I - M_\alpha^{\rho_\alpha})|} \right\rangle - 1 \\ &\approx \frac{1}{\Lambda}(gn)^2 \left\langle \sum_{\alpha}{}' \frac{1}{|\det(I - M_\alpha)|} \right\rangle - 1 \,. \end{aligned} \qquad (VI.12)$$

In writing the last line in (VI.12) we ignored the contributions from orbits which are composed of repetitions of primitive orbits. It is well known that for hyperbolic mappings of the kind we consider here, this can be justified for sufficiently large n. The sum over the periodic orbits can be given an interesting classical interpretation, which was first used in this context by Hannay and Ozorio de Almeida[38)] and was generalized in refs 37,39,40. Consider the classical mapping, which can be formally written as

$$\boldsymbol{x}_n = F(\boldsymbol{x}_{n-1}) = \cdots = F^n(\boldsymbol{x}_0) \qquad (VI.13)$$

If, instead of following the evolution of phase-space points, we want to study the evolution of phase-space distributions, we can introduce the evolution operator $U_{cl}(\boldsymbol{x}', \boldsymbol{x}; n)$ such that any initial phase space distribution $\rho(\boldsymbol{x}; 0)$ evolves into

$$\rho(\boldsymbol{x}'; n) = \int_{\Omega} U(\boldsymbol{x}', \boldsymbol{x}; n)\rho(\boldsymbol{x}; 0)\mathrm{d}\boldsymbol{x} \ , \qquad (VI.14)$$

where Ω is the phase space area. It is easy to check that the classical evolution operator is the unitary (Frobenius Perron[41)]) operator

$$U(\boldsymbol{x}', \boldsymbol{x}; n) = \delta(\boldsymbol{x}' - F^n(\boldsymbol{x})) \ . \qquad (VI.15)$$

We would like to introduce an important concept, which is the classical probability density to perform periodic motion with period n, which will be denoted by $p_{cl}(n)$. It is clear from the definition of the Frobenius Perron operator that

$$p_{cl}(n) = \frac{1}{\Omega}\int_{\Omega} U(\boldsymbol{x}, \boldsymbol{x}; n)d\boldsymbol{x} = \frac{1}{\Omega}\mathrm{Tr}U(\boldsymbol{x}, \boldsymbol{x}; n) = \frac{1}{\Omega}ng\sum_{\alpha}{}' \frac{1}{|\det(I - M_\alpha)|} \qquad (VI.16)$$

The sum is over periodic orbits of period n. It is obtained by straight forward integration, and remembering that the matrix M is the linearization of F^n cr, in other words, the Jacobian of the transformation $\boldsymbol{x}_0 \rightarrow \boldsymbol{x}_n$. Notice also that in (VI.16) we have a factor g which takes account of the contributions from the n periodic points on the orbit, and from the g symmetry conjugated orbits. Contributions from repetitions of short primitive orbits were ingnored.

We can now substitute (VI.16) into (VI.12) and get

$$b_2(n) \approx \frac{\Omega}{\Lambda} ngp_{cl}(n) - 1 \ , \qquad (VI.17)$$

which is one of the most fundamental results in the present chapter. It shows that the semi-classical expression for the two-point correlation function depends on the classical probability for period motion, and the only remnant of quantum interference is contained in the factor ng which is due to the fact that quantum mechanically, one adds amplitudes and then takes the absolute square, whereas classically one adds probabilities!

Let us come back now to the calculation of the cluster function for an S matrix whose corresponding classical mapping is chaotic. Since time reversal is conserved, and there were no other symmetries, we have $g = 2$. We notice now that if n is larger than the inverse Lyapunov exponent, the billiard dynamics is ergodic, which means that phase space is uniformly covered by trajectories, and the probability density to reach any domain in phase space is just Ω^{-1}. In particular, this is also the probability to perform periodic motion, so that (VI.17) reduces to

$$b_2(n) \approx \frac{2n}{\Lambda} - 1 \ . \qquad (VI.18)$$

This relation is not valid for all n values. In particular, if n is shorter than the mixing time n_m which is proportional to the inverse Lyapunov exponent, one still notices the particular (and sparse) distribution of periodic points, and hence one cannot use the ergodic estimate for $p_{cl}(n)$. Rather, the complete expression in terms of periodic orbits should be substituted. The special rôle of this time scale was first discussed by Berry [36] and it sets an upper limit on the scale of r for which one can expect universal behaviour of the cluster function. The other limit of validity of (VI.18) was discussed above where it was shown that $b_2(n) \to 0$ for $n > \Lambda$. The asymptotic behaviour of $b_2(n)$ in the two domains $n < \Lambda$ and $n > \Lambda$ coincides with the corresponding expression from random matrix theory for the COE. This completes the semi-classical proof that the two-point correlation functions for S matrices corresponding to chaotic billiards follow the COE statistics. It should be stressed that this relationship (or its variant for continuous time flows[36]) is ***the only*** theoretical evidence one has so far concerning the connection between chaotic dynamics and spectral fluctuations. All the other evidence, which is very impressive and convincing, comes from numerical solutions of specific models.

VI.b Energy Spectral Corelations.

We shall show now that the result discussed above implies that the energy spectrum of the billiard which is computed by solving the semi classical secular equation, must follow the GOE statistics. To this end we shall have to show that for systems which are governed by a chaotic mapping, the distribution of the τ_l is peaked about their mean, and the distribution becomes more narrow as Λ increases. We now proceed to show why such behaviour is indeed plausible.

It was shown in ref. 7) that the correlation length of the elements of the S matrix, with respect to change of energy, is approximately given by the inverse mean time delay:

$$\gamma \approx \langle \tau(E) \rangle^{-1} \approx \frac{\Lambda}{2\pi} \langle d(E) \rangle^{-1} . \tag{VI.19}$$

The correlation length is therefore, in the semiclassical limit, much larger than the mean level spacing. The correlation length of $\tau(E)$ will be similar, since it is composed of Lorentzians whose mean width is γ. We now make the assumption that the correlation length of the *individual* τ_l is also of order γ. This is reasonable, since $\tau(E)$ is the average of the τ_l, and a different result would entail strong correlations between them.

Consider the matrix $\hat{S} = \exp(-i\gamma E)S(E)$. This is in effect the original S matrix, normalized so that its average time delay vanishes. Since COE matrices exhibit eigen-phase repulsion, levels do not cross, and so $\langle \hat{\tau}_l \rangle$ must also vanish. We now make two approximations:

(i) The two matrices $S_1 = \hat{S}(E)$ and $S_2 = \hat{S}(E + \gamma)$ are statistically independent.

(ii) The $\hat{\tau}_l$ are constant over the range $(E, E + \gamma)$.

We arrange the eigen-phases $\theta_l^{(1,2)}$ of $S_{1,2}$ in ascending order over the interval $[0, 2\pi)$. Since $\langle \hat{\tau}_l \rangle = 0$, and since levels cannot cross, we can assume that the nth eigen-phase of S_1 evolved into the nth eigen-phase of S_2. According to assumption (ii) the $\hat{\tau}_l$ must therefore be

$$\hat{\tau}_l \approx \frac{\theta_l^{(2)} - \theta_l^{(1)}}{\gamma} = \left(\theta_l^{(2)} - \theta_l^{(1)}\right) \langle \tau_l \rangle . \tag{VI.20}$$

The question of the variance of τ_l has been transformed into the question of the variance of

$$\delta\theta = \theta_l^{(2)} - \theta_l^{(1)} = (\theta_l^{(2)} - l\frac{2\pi}{\Lambda}) - (\theta_l^{(1)} - l\frac{2\pi}{\Lambda} .$$

Hence $\mathrm{Var}(\delta\theta) = 2\mathrm{Var}(\theta_l - l\frac{2\pi}{\Lambda})$ which is just the fluctuation of θ_l about its mean. It was recently shown by B. Dietz[42)] that one can derive an analytic expression for the COE prediction for the variance of the $\delta\theta$ distribution. For large Λ

$$\mathrm{Var}(\delta\theta) \propto \frac{\log \Lambda}{\Lambda^2} . \tag{VI.21}$$

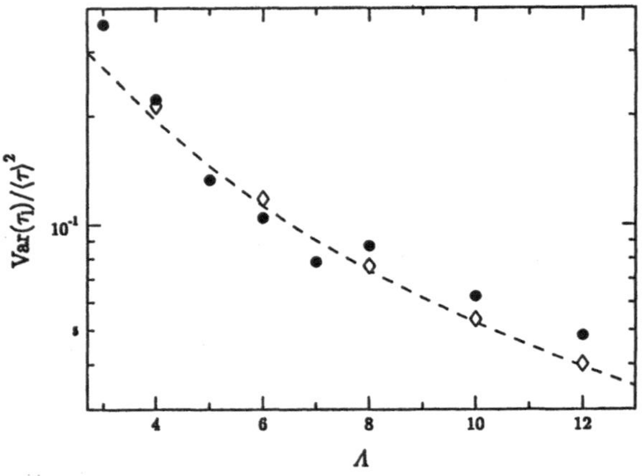

Fig. 7. $Var(\tau_l)$ as a function of Λ for the Sinai billiard. The smooth line is the prediction of random matrix theory.

This behaviour was reproduced numerically for the Sinai billiard problem which we have introduced previously. (Fig. 7).

We can thus conclude, that the COE nature of the S justifies the use of (VI.3) and therefore, the spectral distribution of the eigen-energies is given by the GOE distribution. The last statement is based on an argument of Dyson [43)] which shows that the m point correlations for the COE and GOE statistics are identical as long as m/Λ is kept small. This completes the semi-classical proof that the spectral (two-point) statistics for a quantized chaotic billiard in the limit $\hbar \to 0$ is described by the random matrix theory for the GOE.

(VI.c) Composite Billiards.

So far we have tacetly assumed that the billiard under consideration is "compact", or, in other words, cannot be decomposed in a natural way into sub systems. In the present section we shall introduce the concept of composite or extended billiards. Consider a cluster of chaotic billiards which are inter connected by narrow links. Formally, the entire system is a billiard, and the motion of a point particle in it is certainly chaotic. It differs however from the cases we discussed previously since now, ergodic coverage of phase-space by a typical trajectory follows a new pattern: The particle will perform a random walk between the various domains, and the corresponding diffusive evolution introduces a new time scale. This should be contrasted with the dynamics of a simple chaotic billiard: Once the time exceeds the inverse Lyapunov exponent, ergodic coverage is achieved and it becomes more uniform on a scale which gets finer exponentially fast.

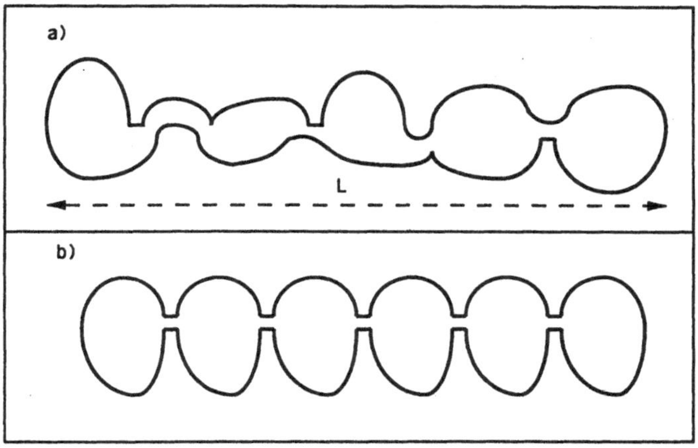

Fig. 8. Composite billiards. (a) disordered. (b) periodic.

We shall deal here with one class of composite billiards, namely, the linear chain: Take a chain of N convex chaotic billiards, which are linked by channels so that each member is connected only to its nearest left and right members. (See Fig 8). Assume also that the constituent billiards are chosen randomly and denote their mean length by a. We can define the Poincaré section Γ for this case by taking a line along the long axis of the chain. The PSM is defined in the two-dimensional phase space where y is measured along Γ and it ranges from 0 to $L = Na$. The conjugate momentum (the momentum in the direction tangent to Γ at the point of incidence) is limited by $|k_y| < k$. Due to the chaoticity of the motion, the momentum k_y will equilibrate immediately, while random walk will characterize the motion along the y axis. As long as the trajectory does not hit the ends of the chain, $\langle (y_n - y_0)^2 \rangle = Dn$. Here D is the diffusion constant, and the averaging is taken over an ensemble of trajectories. n counts the number of applications of the map.

We would like to quantize a chain billiard, and in particular to calculate its spectral two points correlation function. We can use (VI.17) to calculate $b_2(n)$. Here, $\Lambda = Lk/\pi$, $\Omega = 2kL$, $g = 2$ and we have to evaluate $p_{cl}(n)$. The classical motion is diffusive along the y axis. Hence, we can evaluate $p_{cl}(n)$ by solving the one dimensional diffusion equation in the finite interval $[0, L]$. The boundary conditions impose the condition of total reflection from the boundaries. A simple calculation gives,

$$p_{cl}(n) = \frac{1}{2kL} \sum_{l=0}^{\infty} \exp(-\frac{l^2 \pi^2 D}{2L^2} n) \ . \qquad (VI.22)$$

Note that $p_{cl}(n)$ is the probability to perform periodic motion in Ω. Hence the factor $1/2k$ in front of the sum above.

The parameter $n_D = L^2/D$ determines the behaviour of the solution with time. For $n < n_D$ the system does not have time to feel the effect of its finite size, and the evolution is diffusive. For $n > n_D$ the system reaches quickly the equilibrium value $1/L$. Thus,

$$p_{cl}(n) \approx \begin{cases} \frac{1}{(2k\pi Dn)^{\frac{1}{2}}} & \text{if } n < n_D \\ \frac{1}{2kL} & \text{if } n > n_D \ . \end{cases} \qquad (VI.23)$$

If $n_D < 1$, $p_{cl}(n) \approx 1/(2kL)$, for all times. In other words, there exists a critical length scale, $L \approx D^{\frac{1}{2}}$ for which diffusive effects are not noticeable. We shall start our discussion by considering the cases with much larger L values.

Substituting (VI.23) in (VI.17) we get

$$b_2(n) \approx (\frac{n}{\Lambda})^{\frac{1}{2}} (\frac{2L}{kD})^{\frac{1}{2}} - 1 \ . \qquad (VI.24)$$

Our billiard belongs to a class of quantum problems which usually go under the name quasi one-dimensional disordered systems. (See lecture notes by Prange and Weidenmüller.) These systems are dominated by Anderson localization, and the localization length ξ is known to have the following properties:

i. The localization length is proportional to the classical diffusion constant[44], $\xi \approx kD$.

ii. The localization length determines the critical time $n^* \approx \frac{k\xi}{2\pi}$ beyond which the semi-classical approximation which leads to (VI.24) ceases to be valid[37].

This information leads us to limit the applicability of (VI.24) to the range $n < n^*$. In localizing systems, it is natural to partition the eigen-states according to their center of localization. Roughly speaking, there are ξ states belonging to each group. They will all be strongly coupled by a local perturbation, to which the rest of the eigen-functions will hardly respond. This picture justifies the idea that the spectrum is built of independent sub-groups of $k\xi$ states, each will be statistically independent from the others. Hence, the $b_2(n)$ in such problems vanishes for $n > k\xi$ and not only after $n > \Lambda$ as is the case for noncomposite billiards. We thus propose the following expression for the two-point form factor

$$b_2(n) \approx \begin{cases} (\frac{n}{\Lambda})^{\frac{1}{2}}(\frac{2L}{kD})^{\frac{1}{2}} - 1 & \text{if } n < n^*; \\ 0 & \text{if } n > n^* \end{cases} . \tag{VI.25}$$

We substitute the above into (VI.8) and approximate the summation by an integration over the variable $\nu = 2\pi n/\Lambda$. This is the continuum limit which is commonly used in the theory of random matrices. Its existence is always assumed, and it justifies the universal applicability of random matrix theory, once energies (or phases) are scaled by their mean spacings. We get,

$$Y_2(r) = 2\gamma \int_0^1 (1 - x^{\frac{1}{2}}) \cos(2\pi\gamma r x) dx \ , \tag{VI.26}$$

where $\gamma = \frac{kD}{2L} \approx \frac{\xi}{2L}$. In the limit $L \to \infty$, $\gamma \to 0$ and $Y_2(r)$ vanishes over its entire range. This means that the spectrum of the long chain billiard tends to a Poissonian spectrum at least as far as two-points correlations are concerned! It can be shown that in the limit of short chains $L \approx \xi$, the spectrum retrieves its GOE statistics. This is done by using the explicit form of $p_{cl}(n)$. It is left as an exercise to the reader. This result is very interesting since it teaches us an important lesson: Classically chaotic billiards do not necessarily display a GOE spectrum when they are quantized. As we have shown above, the GOE–chaos connection is only true if the coverage of the billiard by classical trajectories occurs quickly, and does not involve any classical time scale which, for a fixed value of $\hbar$, may become comparable to the quantum time scale. This phenomenon was first observed and explained in the study of the spectrum of the evolution operator for the kicked rotor[37)]. The chain billiard and the kicked rotor systems are seemingly unrelated. However they have one common feature–their phase space is an elongated cylinder, along which the evolution is diffusive. This is the only ingredient which determines the two-point correlation function of the spectrum.

To illustrate the ideas presented above, we show in Fig. 9 a comparison between a numerically computed form factor and the semi-classical expression (VI.25). The model which was solved is a chain of corner billiards, each one is a quarter of a Sinai billiard, and they are connected via channels whose lengths are taken at random [44)]. The agreement between the semi-classical theory and the numerical simulation is almost too good to be true...

The last point that needs clarification is the following: suppose that our chain is composed of identical billiards, (Fig 8b) so the entire structure is periodic in the limit $L \to \infty$. The diffusion along the y-axis continues to dominate the classical dynamics. As for the quantum spectrum, we know that due to Bloch's theorem, the spectrum becomes a continuous band spectrum. For every finite L the spectrum shows the clustering of levels into subgroups which are the progenitors of the bands. This affects the spectral correlation in a nontrivial way, which can be also explained semi-classically: Due to the local translational symmetry of the billiard, a periodic orbit of period $n \ll L^2/D$ can be periodically placed along the chain. The resulting orbits are all degenerate as far as their actions and amplitudes are concerned. They are physically different, since they are positioned at different sites. The degeneracy of these orbits is of order $g = N \approx \Lambda$. Hence, the $Y_2(r)$ will be singular in the limit of long chains, reflecting the tendency of eigen-energies to cluster in bands.

Considerations of the type brought above can be extended to other composite billiards. Consider, e.g., two billiards which communicate through a narrow hole. The spectrum will not be of the GOE type if the time constant for leakage through the hole

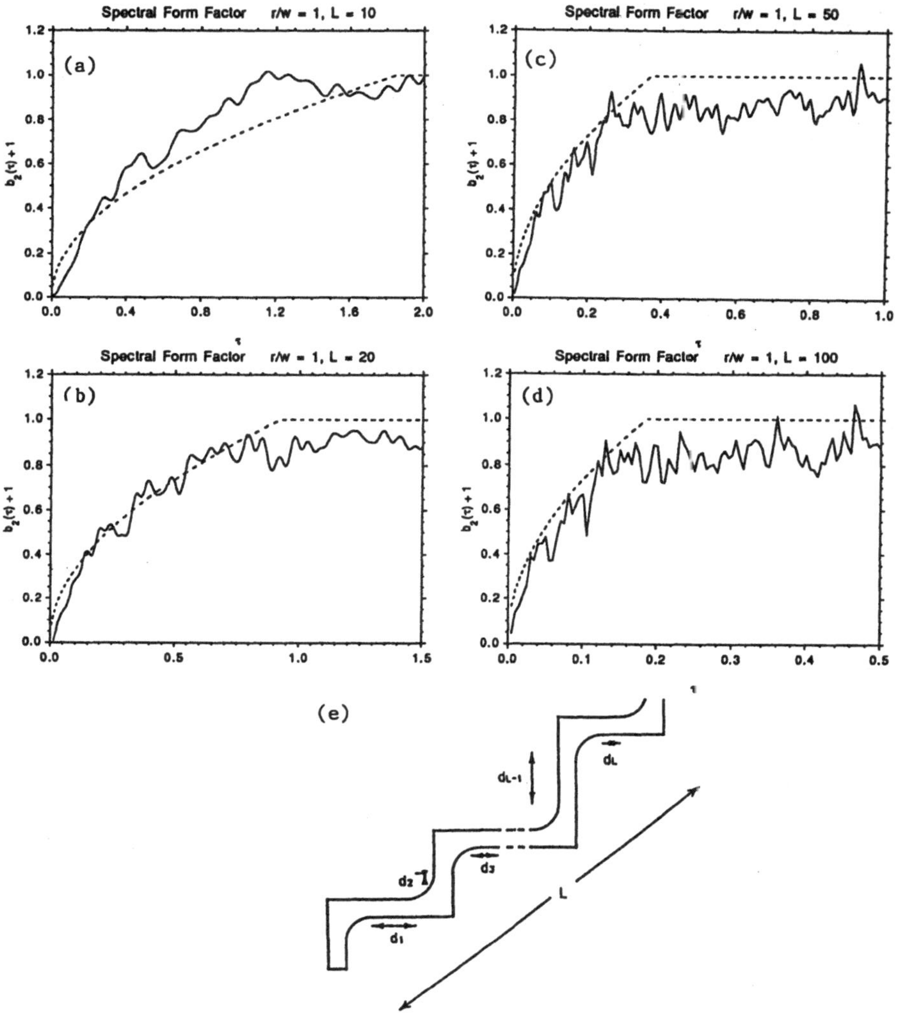

Fig. 9. The spectral form factor for composite billiards for three lengths. (a) $L = 10$; (b) $L = 20$; (c) $L = 50$; (d) $L = 100$ The dashed line is the theoretical prediction. (e) shows the composite billiard which was quantized numerically.

is large. The calculation of the corresponding two point correlation function is left as an exercise. (see also ref.39).

VII. Conclusions

In this series of lectures we have shown that chaotic billiards can be quantized semi-classically, and expressions were given for the mean level density, the secular function and some spectral two-points correlation functions. It was shown that quantum mechanics does not depend on the infinitely intricate structures which appear on the classical level. Rather, the value of $\hbar$ sets a criterion according to which one can discard irrelevant classical details. We have shown that when the problem of semi-classical quantization is properly posed, the long classical trajectories either do not appear, or can be made to disappear by imposing certain unitarity conditions on the semi-classical expressions.

Semi-classical quantization of chaotic dynamics attracted much attention recently, and the progress is very fast. In the present lectures we presented one approach to the problem. The following lines will attempt at giving a short summary of some of the other approaches. Due to space limitations it will not be possible to do justice to any of them.

As was mentioned above, the approach developed by E. Bogomolny [13)] is rather similar in spirit to the one developed here. Bogomolny was the first to realize the advantages of writing the secular equation in the form $\det(I - T(E))$ and T is a finite unitary matrix. His method of deriving T is different from the one used here, and it is strictly semi-classical, and appropriate to systems which are similar to the convex billiards which we discussed. Bogomolny made much progress in elaborating on the cummulant properties of the expansion (III.26) and used it in some numerical examples.

Most other approaches start from the semi-classical dynamical ζ-function (V.15). We have shown in chapter V that the product (V.15) or the analogous composite orbits sum (V.17) cannot converge in the absolute sense for complex energies with $\Im mE > E_h$ where $E_h = h/2$ is the "entropy" barrier. The Hamburg group [45)] studied the convergence properties of the semi-classical ζ function for $E_h > \Im mE \geq 0$ for a few cases where a large dictionary of classical trajectories was compiled. They arranged the terms of (V.17) in ascending order according to their composite length L_c which is related to the composite actions by $S_c = kL_c$. The half-plane of absolute convergence is bounded by k_h, where

$$k_h = \limsup_{c\to\infty} \frac{1}{L_c} \log \sum_{n=1}^{c} |A_n| \,. \qquad (VII.1)$$

The half plane of convergence is bounded by k_c which is defined by

$$k_c = \limsup_{c\to\infty} \frac{1}{L_c} \log |\sum_{n=1}^{c} A_n| \,. \qquad (VII.2)$$

Their numerical results confirmed the relation $k_h = h/2$. The surprise came from their evaluation of k_c which, in spite of its erratic behaviour, stays well bellow 0 (Fig.10). This means that the series converges conditionally for $\Im m(E) = 0$, and the series can be used to compute the secular function, even without resummation á lá Riemann-Siegel.

The Riemann-Siegel approximation for the Riemann ζ-function inspired Berry and Keating in their attempts to quantize chaotic systems [25]. Their research culminated in discovering a new approximation to the Riemann ζ function which is superior in some ways to the Riemann-Siegel theory [46]. They showed that the same approach can be used to find an approximant to the semi-classical dynamical ζ. One of the stages of the research was the introduction of the technique which was used in chapter (IV) in the present notes.

An important attempt to reduce the dynamical ζ-function was initiated by Cvitanovic and Eckhardt[29]. Their cumulant expansion works very well for uniform hyperbolic systems with rather special properties. They used this approach to calculate poles of the scattering matrix for scattering from three reflecting discs. Other applications of Riemann Siegel "lookalike" secular equations were recently applied to some practical problems, and proved superior to other approximate methods in practical problems[47].

Finally, we would like to list some open problems.

The applicability of the semi-classical approximations for chaotic systems is not entirely understood–we do not really know how to estimate the error and how to relate it to parameters which characterize the classical chaos. One suspects that a possible

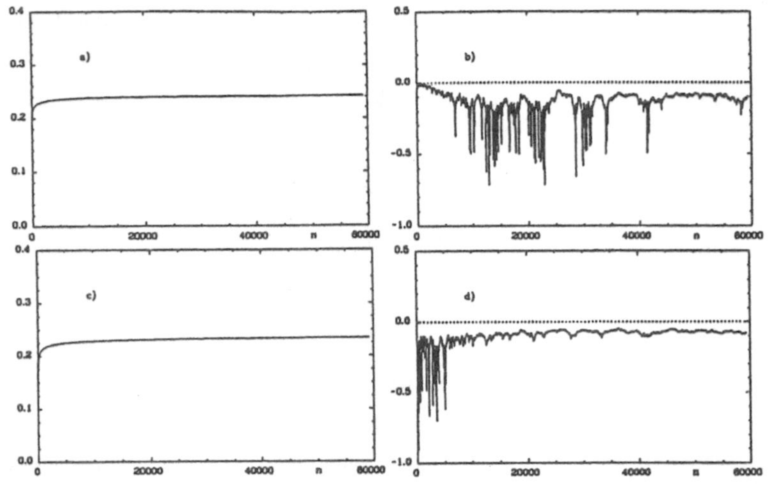

Fig. 10. An evaluation of the series (VII.1,2) as a function of $n = L_c$ for the hyperbola billiard. a) Abscissa of absolute convergence, b) Abscissa of conditional convergence for the secular function for states which are symmetric with respect to reflection about the hyperbola axis, c) and d) the same for anti-symmetric states (ref (45)).

culprit is the proliferation of unstable orbits as time increases. This conjecture has not been confirmed in any way. In particular, the suspected deterioration of the approximation for long times is not fully understood. A related problem is the justification of ignoring the "non diagonal" terms in the semi-classical expressions of the type (VI.11). It is believed that the non-diagonal terms take over from the diagonal part and describe the transition between the semi-classical domain and the strictly quantum domain. How this is done is not understood. In the last chapter we mentioned Anderson localization in chaotic elongated billiards. A semi-classical theory for this phenomenon is yet completely unknown.

Acknowledgements

Much of the theory discussed here was developed together with Mr. Eyal Doron. Many thanks are extended to him for stimulating discussions and ideas and for his masterful performance of numerical simulations. The treatment of composite billiards was developed in collaboration with Dr. Thomas Dittrich. The numerical work presented here was carried out by him, and I am obliged for his help. Finally, discussions and conversations with Drs. J. Keating, E. Bogomolny and Prof. M. Berry are acknowledged. The research reported here was supported by grants from the Israeli Academy of Sciences, the Minerva foundation and the US Israel bi-national science foundation.

Appendix A

The boundary of the convex billiard is conveniently described in terms of the distance $\rho(\alpha)$ of the point on the boundary in the direction α. Denote by $\mathbf{1}_r$ and $\mathbf{1}_\alpha$ the unit vectors along the radius vector and the perpendicular, respectively,

$$\begin{aligned} \mathbf{1}_r &= \boldsymbol{i}\cos\alpha + \boldsymbol{j}\sin\alpha \\ \mathbf{1}_\alpha &= -\boldsymbol{i}\sin\alpha + \boldsymbol{j}\cos\alpha \end{aligned} , \qquad (A.1)$$

and define the angle ψ by the relation:

$$tg\psi = \rho'/\rho , \qquad (A.2)$$

where the prime denotes differentiation with respect to α.

The radius of curvature $R(\alpha)$ is given by

$$R(\alpha) = \left(\frac{\rho(1+(\rho'/\rho)^2)^{3/2}}{1+2(\rho'/\rho)-\rho''/\rho} \right)$$

The tangent and normal to the boundary at α are

$$\begin{aligned} \boldsymbol{t} &= \sin(\psi-\alpha)\cdot\boldsymbol{i} + \cos(\psi-\alpha)\boldsymbol{j} \\ \boldsymbol{n} &= \cos(\psi-\alpha)\cdot\boldsymbol{i} - \sin(\psi-\alpha)\boldsymbol{j} \end{aligned} . \qquad (A.3)$$

Using the reflection conditions (I.1) one can easily get the relation

$$\theta_i + \theta_f = \pi + 2(\alpha - \psi(\alpha)) . \qquad (A.4)$$

The incoming and outgoing angular momenta are:

$$\begin{aligned} \ell_i &= \boldsymbol{r} \times \boldsymbol{k}_i\big|_z = \rho \sin(\theta_i - \alpha) \\ \ell_f &= \boldsymbol{r} \times \boldsymbol{k}_f\big|_z = \rho \sin(\theta_f - \alpha) \end{aligned} \tag{A.5}$$

The generating function for the mapping $(\ell_i\theta_i) \rightarrow (\ell_f\theta_f)$ is

$$S(\theta_i\theta_f) = \rho\big[|\cos(\theta_i - \alpha)| + |\cos(\theta_f - \alpha)|\big]$$

if $(\theta_i - \alpha) < \pi/2$ then $(\theta_f - \alpha) > \pi/2$ so that

$$S(\theta_i, \theta_f) = \rho[\cos(\theta_i - \alpha) - \cos(\theta_f - \alpha)] \tag{A.6}$$

One can easily check that

$$\ell_i = -\frac{\partial S}{\partial \theta_i}, \quad \ell_f = \frac{\partial S}{\partial \theta_f}, \tag{A.7}$$

which proves that the (ℓ, θ) variables are canonically conjugate and that the mapping is area preserving. The last fact can be checked directly by calculating the Jacobian of the transformation

$$J = \begin{pmatrix} \frac{\partial \theta_f}{\partial \theta_i} & \frac{\partial \ell_f}{\partial \theta_i} \\ \frac{\partial \theta_f}{\partial \ell_i} & \frac{\partial \ell_i}{\partial \ell_i} \end{pmatrix} \tag{A.8}$$

To evaluate the entries in J, one can take the derivatives of (A.4) and (A.5), extract $d\alpha$ and get

$$\begin{aligned} d\ell_i &= A d\theta_i + B d\theta_f \ , \\ d\theta_f &= -\frac{A}{B} d\theta_i + \frac{1}{B} d\ell_i \ , \\ d\ell_f &= -d\ell_i + C d\theta_i + D d\theta_f \ , \end{aligned}$$

with

$$\begin{aligned} A &= -\frac{R(\alpha)}{2} \cos(\psi + \theta_i - \alpha) + \rho \cos(\theta_i - \alpha) \ , \\ B &= -\frac{R(\alpha)}{2} \cos(\psi + \theta_i - \alpha) \ , \\ C &= \rho \cos(\theta_i - \alpha) \ , \\ D &= \rho \cos(\theta_f - \alpha) \ . \end{aligned} \tag{A.9}$$

After some algebra one finds

$$J = \frac{1}{B} \begin{pmatrix} -A & CB - DA \\ 1 & D - B \end{pmatrix} \tag{A.10}$$

so that

$$\det J = \frac{A-C}{B} = 1\,. \tag{A.11}$$

The stability is determined by $Tr J$

$$Tr J = -2\left(1 - \frac{2\cos\theta_i(1-\psi')}{\cos(\psi+\theta_i)\cos\psi}\right).$$

Consider now the simplest periodic orbit, namely, one which reflects from two points on the boundary whose normals are on the same line. Choose the origin on this line and the $\boldsymbol{i}$ vector along it. Hence, for the first reflection, $\theta_i^{(1)}$=0, $\alpha^{(1)}$=0, and $\theta_f^{(1)}$=π. For the other reflection, $\theta_i^{(2)} = \pi$, $\alpha^{(2)} = \pi$, $\theta_f^{(2)} = 0$. One can show that

$$J = J_1 J_2 =$$
$$\frac{4}{R_1 R_2}\begin{pmatrix} -\rho_1 + R_1/2 & 1 \\ \rho_1^2 - \rho_1 R_1 & -\rho_1 + R/2 \end{pmatrix}\begin{pmatrix} -\rho_2 + R_2/2 & 1 \\ \rho_2^2 - \rho_2 R_2 & -\rho_2 + R_2/2 \end{pmatrix}$$

and

$$Tr J = 2 + \frac{4}{R_1 R_2}\Big[(\rho_1+\rho_2)^2 - (R_i + R_2)(\rho_1+\rho_2)\Big]\,.$$

This periodic orbit is unstable if $Tr J > 2$ or $(\rho_1 + \rho_2) > (R_1 + R_2)$. This means that the chord length between the two reflection points exceeds the sum of the radii of curvature – a condition which is reminiscent of Bunimovich's condition $c_\alpha > d_\alpha$, which we discussed in Chapter II.

References

1) Ya. G. Sinai (editor) Dynamical Systems (collection of papers) World Scientific, Singapore (1991).
2) H.J. Stöckmann and J. Stein, "quantums chaos in billiards studied by microwave absoprtion", Phys. Rev. Lett. **64** (1990) 2215.
3) E. Doron, U. Smilansky, and A. Frenkel, "Experimental demonstration of chaotic scattering of microwaves", Phys. Rev. Lett. **65** (1990) 3072.
4) S. Sridhar "Experimental Observation of Scarred Eigen-functions of Chaotic Microwave cavities", Phys. Rev. Lett. **67** (1991) 785-788.

5) R. Blümel, I.H. Davidson, W.P. Reinhardt, H. Limand, and M. Sharnoff, Phys. Rev. A. in press (1992).

6) R. A. Jalabert H. U. Baranger and A. D. Stone, "Conductance fluctuations in the ballistic regime: A probe of quantum chaos?" Phys. Rev. Lett. **65** (1990) 2442.

7) E. Doron, U. Smilansky, and A. Frenkel, "Chaotic scattering and transmission fluctuations", Physica **D50** (1991) 367.

8) R. Balian and C. Bloch Ann. Phys. NY **63** (1971) 592-606.

9) R. Balian and C. Bloch Ann. Phys. NY **64** (1971) 271-307.

10) R. Balian and C. Bloch Ann. Phys. NY **69** (1972) 514-45.

11) R. Balian and C. Bloch, "Solution of the Schrödinger equation in terms of classical paths", Ann. Phys. **85** (1974) 514.

12) E. Doron and U. Smilansky, "Chaotic spectroscopy", Weizmann Institute preprint WIS-91/11/Mar-PH, 199. PPhys. Rev. Letters in press (1992);
E. Doron and U. Smilansky, "Semiclassical quantization of chaotic billiards – a scattering theory approach", Weizmann Institute preprint WIS-91/63/Sep-PH, 1991, Nonlinearity, in press (1992).

13) E. B. Bogomolny, "Semiclassical quantization of multidimensional systems", *Comments on Atomic and Molecular Physics* **25** (1990) 67 and E.B. Bogomolny, Nonlinearity in press (1992).

14) E. Heller in Proceedings of the 1989 Les Houches Summer School on "Chaos and Quantum Physics", Ed. M.J. Giannoni, A. Voros and J. Zinn-Justin, (North Holland) 1992.

15) N. Balazs and A. Voros, 'Chaos on the Pseudosphere' Physics Reports, **143**, (1986) 109.

16) M. C. Gutzwiller "Chaos in Classical and Quantum Mechanics" Springer Verlag New York (1990).

17) M. V. Berry "Regularity and Chaos in Classical Mechanics, Illustrated by Three Deformations of a Circular "Billiard". Eur. J. Phys. 2 (1981) 91-102.

18) V.F. Lazutkin, Math of the USSR Izvestiya (1) 7 (1973) 185-214.

19) L.A. Bunimovich, Comm. in Math. Phys. 65 (1979) 295-312.

20) L.A. Bunimovich, Chaos 1 (1991), 187-193.

21) Ya.G. Sinai, Russian Math. Surveys (2) 25 (1970) 137-189.

22) P.J. Richens and M.V. Berry, Physica D2 (1981) 495-512.

23) M.L. Goldberger and K.M. Watson, Collision Theory John Wiley & Son Inc. New York, (1964).

24) See ref. 16) section 16.2 p 257.

25) M. V. Berry and J. P. Keating, "A rule for quantizing chaos?", J. Phys. A **23** (1990) 4839.

26) U. Smilansky, "The classical and quantum theory of chaotic scattering". In Proceedings of the 1989 Les Houches Summer School on "Chaos and Quantum Physics" Eds. M.J. Giannoni, A. Voros and J. Zinn-Justin, (North Holland) 1992.

27) C. Jung, "Poincaré map for scattering states", J. Phys. A **19** (1986) 1345.

28) M. Tabor, "A semiclassical quantization of area-preserving maps", Physica **D6** (1983) 195.

29) P. Cvitanovič and B. Eckhardt, "Periodic-orbit quantization of chaotic systems" Phys. Rev. Lett. **63** (1989) 823.

30) B. Eckhardt, "Periodic Orbit Theory" in Proc. of the International School of Physics "Enrico Fermi" on Quantum Chaos (Varenna 1991). Eds. G. Casati, I. Guarneri and U. Smilansky, in press.

31) J. P. Keating and M. V. Berry, "False singularities in particular sums over closed orbits", J. Phys. **A20** (1987) L1139.

32) K. Stewartson and R. T. Waechter, Proc. Camb. Phil. Soc. **69** (1971) 363.

33) M. Kac "Can one hear the shape of a drum?" Amer. Math. Monthly 73,no 4 part II (1966), 1-23.

34) P. Gaspard in Proceedings of the International School of Physics "Enrico Fermi" on Quantum Chaos (Varenna 1991). Eds. G. Casati, I. Guarneri and U. Smilansky. in press.

35) L. Euler, Introduction in Analysis in Infinitorum (1748).

36) M. V. Berry, "Semiclassical theory of spectral rigidity", Proc. Roy. Soc. London A **400** (1985) 229.

37) T. Dittrich and U. Smilansky "Spectral Properties of Systems with Dynamical Localization: I. The Local Spectrum". Nonlinearity 4 (1991) 59-84. "II. Finite Sample Approach", ibid. 85-101.

38) H. J. Hannay and A.M. Ozorio de Almeida, J. Phys. **A17** (1984) 3429.

39) U. Smilansky, S. Tomsovic and O. Bohigas, "Spectral Fluctuations and Transport in Phase Space". J. Phys A in press (1992).

40) N. Argaman, Y. Imry and U. Smilansky, in preparation.

41) see, e.g., "Deterministic Chaos" by H.G. Schuster VCH (second edition), pages 29-30.

42) B. Dietz, private communication.

43) F. J. Dyson, J. Math. Phys. **3** (1962) 140.

44) T. Dittrich, E. Doron and U. Smilansky, Physica B, in press.

45) M. Sieber and F. Steiner Phys. Rev. Lett. 67 (1991), 1941. M. Sieber, Ph.D Thesis University of Hamburg (1991). C. Matthies and F. Steiner, "Selberg's zeta function and the quantization of chaos", DESY preprint 91–024, April 1991. R. Aurich and F. Steiner, "From classical periodic orbits to the quantization of chaos", DESY preprint 91–044, May 1991.

46) M.V. Berry and J. Keating "A new Asymptotic Representation for $\zeta(1/2 + it)$ and Quantum Spectral Determinants", Proc. Roy. Soc. Lond., in press. (1992).

47 G. Tanner, P. Scherer, E. B. Bogomolny, B. Eckhardt, and D. Wintgen, Phys. Rev. Lett. 67 (1991) 2410.

Stochastic Scattering Theory
or
Random-Matrix Models for Fluctuations in Microscopic and Mesoscopic Systems

Hans A. Weidenmüller,
Max-Planck-Institut für Kernphysik, Heidelberg, Germany

Abstract.

In these lectures, I consider stochastic fluctuations in microscopic and mesoscopic systems. In microscopic systems (atoms, molecules, atomic nuclei) such fluctuations arise as a consequence of chaotic dynamics. In mesoscopic systems, where the inelastic mean free path for electrons is larger than the system size, such fluctuations originate in disorder and/or chaotic scattering. In both cases, the complex Hamiltonian dynamics can be simulated in terms of random-matrix models having the correct symmetry properties. From such models, mean values and variances of observables can be calculated with the help of a technique that uses a generating function written as an integral over commuting and anticommuting variables. The following examples are discussed. (i) Statistical nuclear cross-sections; (ii) Chaotic quantum scattering; (iii) Conductance fluctuations in mesoscopic systems.

1. Motivation: The Phenomena.

Two concepts lie at the root of all that follows: *Stochasticity* and *Resonance*. In a physical system, stochasticity has either of two causes: chaotic dynamics or disorder. In both cases, it may be useful or even necessary to construct a stochastic model for the Hamiltonian which obviates the need to construct solutions of the actual Hamiltonian of the system. Such stochastic modelling forms the topic of these lectures. In this context, resonances play a particularly significant role. Indeed, most experiments on chaotic or disordered systems can be viewed as scattering experiments. Resonances are long-lived, quasistationary states populated in a scattering process. In a long-lived intermediate state, the system has the chance to fully explore the available phase space, and to fully develop stochasticity. Hence, chaotic or disordered systems with a large number of resonances are particularly suitable for generic stochastic modelling. In contradistinction, scattering events with a short staying time in the domain of chaotic motion or disorder show system-specific features and do not allow for such generic modelling. Since time representation and energy representation are connected by Fourier transformation, it follows that generic stochastic modelling is expected to be possible and useful on the scale of the mean level spacing d between resonances, while system-specific features (which are amenable to semiclassical analysis) appear on a larger energy scale. It is useful to keep these remarks in mind as we proceed.

A prime example for a real physical system with chaotic dynamics is provided by the Hydrogen atom in a constant, homogeneous magnetic field B. This system is almost fully chaotic for energies close to the ionization threshold if the perturbation caused by B is comparable in strength with the mean spacing between neighbouring Rydberg states. This example has been dealt with in the lectures by Oriol Bohigas and Peter Koch. Chaotic dynanics in general and in this specific case was defined and the connection to properties of the quantum case was made. None of this is repeated here, save for the following point.

In the chaotic regime, the spectral fluctuation properties of eigenvalues belonging to eigenfunctions with identical symmetry properties coincide with those of the Gaussian Orthogonal Ensemble (GOE). The latter is an ensemble of matrices N (considered in the limit $N \to \infty$). The matrix elements $H_{\mu\nu}$ (with $\mu, \nu = 1, \ldots, N$) are real, obey the relation $H_{\mu\nu} = H_{\nu\mu}$, and are uncorrelated random variables with a Gaussian probability distribution, with first and second moments given by

$$\begin{aligned} \overline{H_{\mu\nu}} &= 0 \quad , \\ \overline{H_{\mu\nu} H_{\mu'\nu'}} &= \frac{\lambda^2}{N} (\delta_{\mu\mu'} \delta_{\nu\nu'} + \delta_{\mu\nu'} \delta_{\nu\mu'}) \quad . \end{aligned} \tag{1.1}$$

The bar indicates the average over the ensemble. The parameter λ has the dimension energy and fixes the mean level spacing of the GOE. The fluctuations about the mean are therefore specified in a parameter-independent fashion.

The coincidence of the spectral fluctuation properties (nearest-neighbour spacing distribution, Σ_2 and Δ_3 statistics and other fluctuation measures discussed by O. Bohigas) of the hydrogen atom in the chaotic domain and of the GOE is not accidental, but generic. Indeed, it is now understood that such coincidence holds for the spectral fluctuation properties of fully chaotic systems in the semiclassical domain, subject to some provisos. Of these provisos, I mention but two: (i) The chaotic system must be invariant under time reversal and under rotation. Otherwise, the GOE must be replaced by another one of Wigner's ensembles as listed in O. Bohigas' lectures. (ii) All of classical phase space must be equally accessible. Otherwise, the spectral fluctuation properties may coincide with those of several coupled GOE's. An example for such a situation was given by O. Bohigas.

This generic coincidence furnishes the justification for an essential aspect of these lectures: The modelling of quantum systems with classically chaotic dynamics in terms of an ensemble of random matrices.

A second class of systems considered in these lectures is the class of disordered systems. The passage of light waves through a medium with an index of refraction which varies spatially in a random fashion, and the passage of electrons through a metal or semiconductor with a random distribution of impurities, furnish examples for such systems. For this class of systems, it is the disorder which justifies stochastic modelling. Symmetry arguments again lead to the three matrix ensembles (GOE, GUE and GSE) listed by O. Bohigas. Examples for the modelling of disordered systems in terms of these ensembles are given in section 2.

In the study of chaotic and disordered systems, wave coherence is a central issue. Suppose that a Schrödinger wave packet representing an electron passes through a disordered medium where it undergoes a large number of scattering events on randomly distributed elastic scattering centers (i. e., impurities). Will the wave packet retain its coherence properties (and if so, in which sense)? Answering this question leads to an understanding of the occurrence of random fluctuations in some of the scattering processes which form the topics of these lectures. In particular, one understands why even systems of mesoscopic size with typical length scales in the μm region do not self-average. Another phenomenon related to wave-coherence in disordered media is Anderson localisation, a topic mentioned in the lectures by R. Graham and R. Prange.

Similar questions arise in the scattering of waves in situations where the classical scattering is chaotic ("Quantum chaotic scattering"). In his lectures, U. Smilansky has given examples of chaotic scattering processes. Further examples are given below. A general definition is not attempted here. The close similarity between quantum properties of disordered and chaotic systems is demonstrated, for instance, by the occurrence of dynamical localization in chaotic systems where not all parts of classical phase space are equally accessible.

After these introductory remarks, I present a number of examples of wave scattering by classically chaotic or disordered systems. Taken all by themselves, chaos and disorder may not seem terribly interesting objects of physical study:

What can we hope to learn beyond the statement that wave coherence does persist, and that it leads to random cross-section fluctuations? To answer this question, I will in each example define a physical question which goes beyond the mere statement of randomness. The reader is invited to judge for himself the interest and relevance of these questions.

1.1 Microwave Scattering in Cavities.

Several groups [1, 2, 3] have reported on such experiments. The connection with chaotic scattering is this. Cavities can be built in the shape of one of the well-studied two-dimensional chaotic billiards (the Sinai billiard [2] or the Bunimovich stadium [1, 3]), with size parameters in these two dimensions large in comparison to the third. Then, with a proper choice of microwave frequency, only a single mode is excited in the third dimension, and the microwave scattering in such a cavity (with several antennas for incoming and outgoing waves) becomes mathematically the same problem as quantum-wave propagation in the corresponding two-dimensional billiard, in spite of the vector nature of electromagnetic radiation and the scalar nature of Schrödinger waves. Microwave scattering therefore is a tool to study indirectly quantum scattering in classically chaotic systems. Similar problems arise in the passage of electrons through mesoscopic systems of sufficiently low impurity concentration. (Fabrication of such devices has become possible in the last few years). The electrons pass through such systems without being scattered by impurities (hence the expression "ballistic electrons"), but do experience scattering on the boundaries. This may again lead to chaotic scattering.

Figure 1.1 shows the first experiment on chaotic microwave scattering reported in the literature. The upper part gives the reflected power versus frequency. The insert shows the form of the cavity. The minima of the reflected intensity indicate resonances; they are related to the eigenvalues of the closed cavity (without antennas). The nearest-neighbour spacing distribution of eigenvalues (histogram) is compared with the Wigner distribution based on the GOE (full line). The insert shows the corresponding distribution for a rectangular billiard (integrable case).

Some of the questions which arise in this context are the following. Do the elements of the scattering matrix for chaotic microwave scattering (or for chaotic scattering of ballistic electrons) display generic features independent of the actual system? What are they? Can we experimentally study the localization transition by constructing a chain of such cavities? In the framework of semiclassical theory, some of these questions have been tackled in U. Smilansky's lectures. Here, we shall address them from the point of view of random-matrix theory.

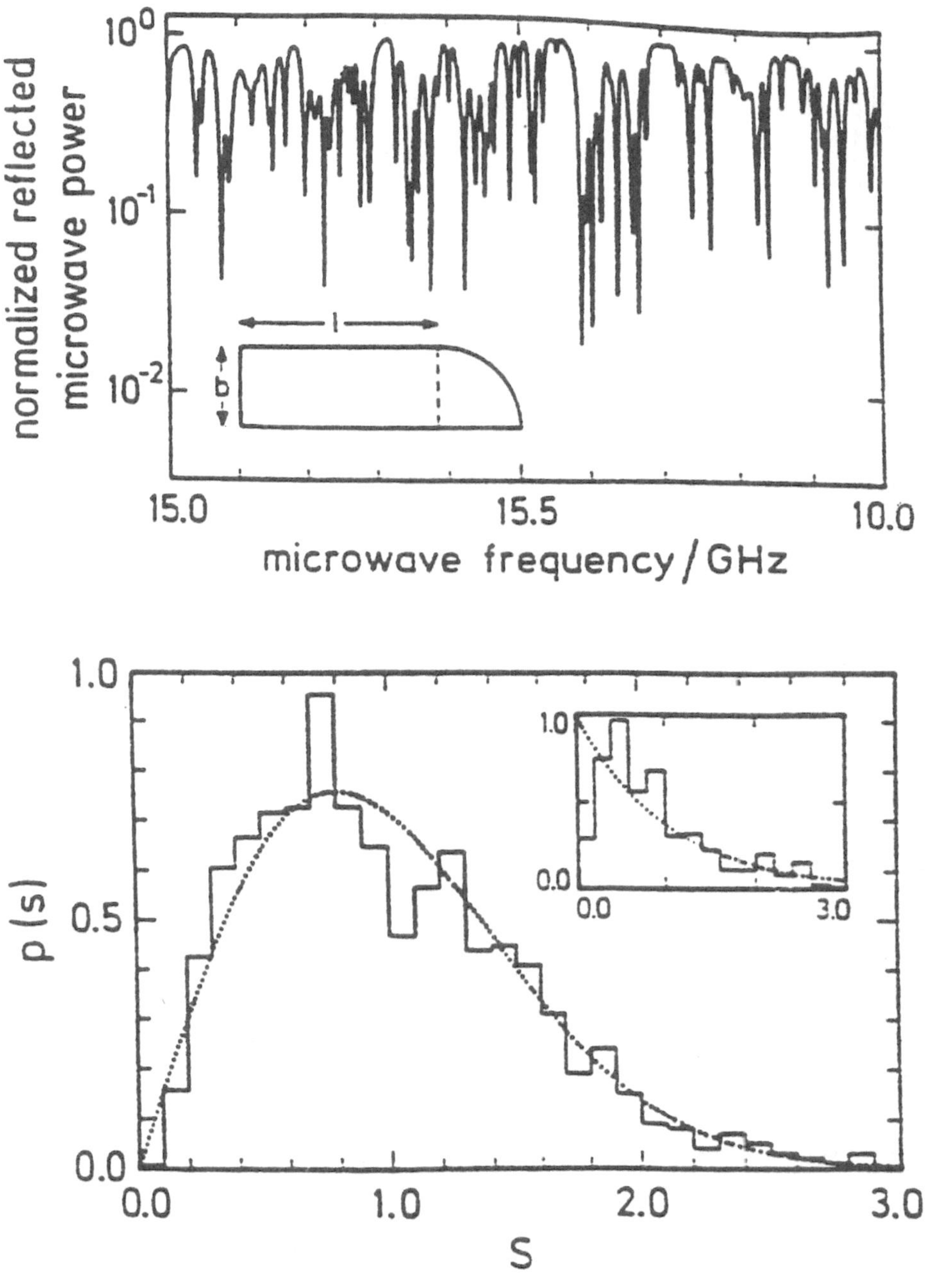

Fig. 1.1. Upper part: Microwave intensity reflected from a stadium versus frequency. Lower part: Nearest-neighbour spacing distribution $\rho(s)$ of eigenvalues versus s, the actual spacing in units of the mean spacing. Taken from ref. [1].

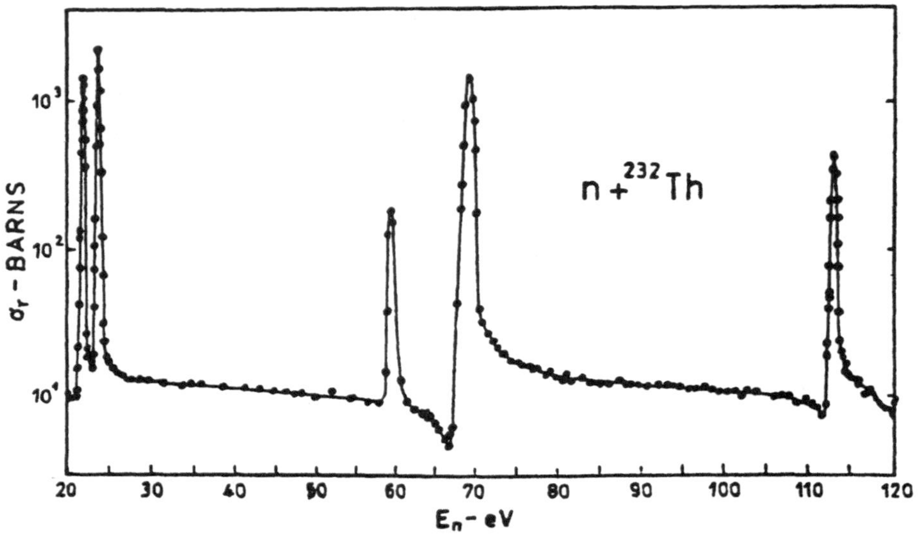

Fig. 1.2. The total neutron cross-section on ^{232}Th versus neutron energy. Taken from ref. [4].

1.2 Compound-Nucleus Scattering in the Domains of Isolated and of Overlapping Resonances.

Data on s-wave neutron scattering on heavy nuclei displaying isolated resonances (cf. fig. 1.2) yielded the first statistically significant example of a physical system for which the spectral fluctuations coincide with GOE predictions. We now attribute this coincidence to chaotic nuclear dynamics. In the present context, we pose the following questions. The very fact that the resonances show GOE statistics implies that the nuclear S-matrix in its energy dependence is a random process. Suppose that we replace the nuclear Hamiltonian by a GOE. Is it possible to correctly predict salient features of this random process like variance, autocorrelation function and higher moments? Is there a connection between this stochastic modelling of compound-nucleus reactions and quantum chaotic scattering as described under (i)? And can we work out an universal approach covering both the case of isolated resonances (fig. 1.2) and the case of overlapping resonances? An example for overlapping resonances is shown in figure 1.3. At each value of the energy, the S-matrix element for the reaction shown is the coherent superposition of *many* resonances; the latter are governed by GOE statistics.

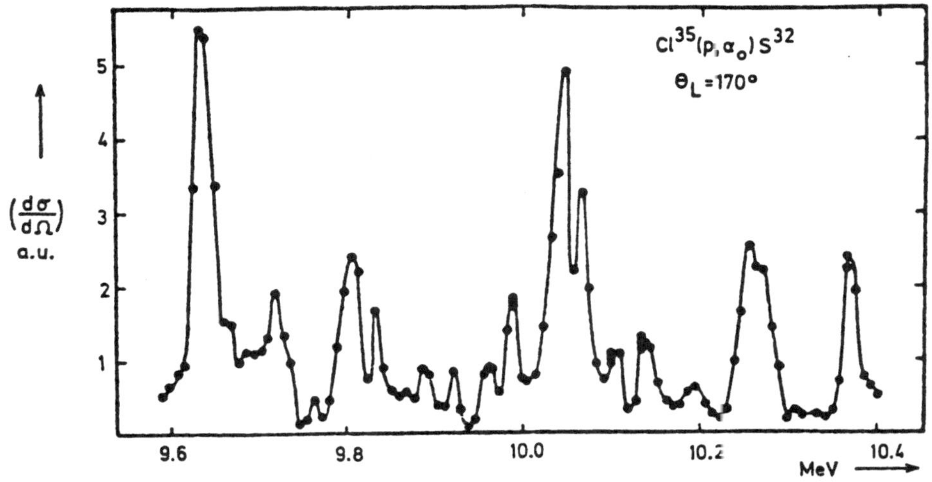

Fig. 1.3. The cross-section for the reaction ^{35}Cl$(p, \alpha_0)^{32}$S at a lab scattering angle of 170° versus proton energy. Taken from ref. [5].

In spite of the introductory remarks, it may be useful to re-iterate at this point the significance of stochastic modelling, and its relation to data on chaotic or compound-nucleus scattering. First and foremost, data of the type shown in figs. 1.1 to 1.3 correspond to a fixed Hamiltonian and therefore are deterministic and reproducible. So what is stochastic about them, and which of their features can be modelled as being universal? Clearly, if we want to reproduce *exactly* the data shown in the figures, there is only one way: To solve the equations of motion. It is well to remember, however, that in classical mechanics, chaos means instability. And this instability carries over, to a considerable extent, to the quantum situation. Therefore, the calculation of the eigenvalues even of a stadium (the minima of fig. 1.1) requires considerable effort, while the calculation of the curves in figures 1.2, 1.3 is beyond any hope of practical realization. In such a situation, some theoretical information, even if incomplete, is desirable and useful, if that information can be obtained easily, and particularly if it turns out to be generic.

Stochastic modelling can be used to calculate *averages* over observables, and these averages can be compared with the corresponding averages taken from the data themselves. Examples are the nearest-neighbour spacing distribution of eigenvalues (this quantity does not imply knowledge of the individual

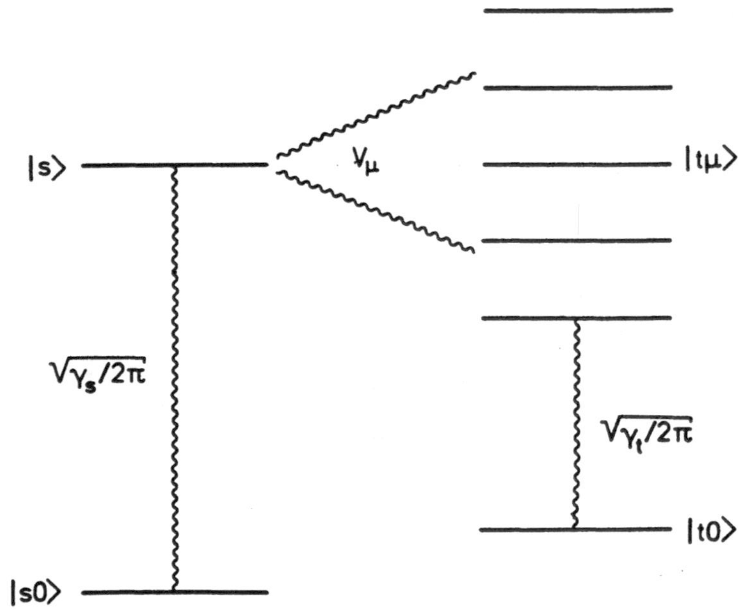

Fig. 1.4. The laser experiment on Methylglyoxal (schematic). Taken from the first of refs. [7].

eigenvalues!), the Δ_3 statistic or, in the case of the curves in figures 1.1 to 1.3, the average intensity (cross-section), the variance, and the energy autocorrelation function of the cross-section. In comparing averages obtained from stochastic modelling with averages computed from the data, we rely on what is referred to as ergodicity. Indeed, stochastic modelling introduces a statistical *ensemble* of Hamiltonians (example: the GOE), and all averages are ensemble averages. To be applicable to a data set (which, of course, always yields only a running average over energy or another parameter), the stochastic model must be ergodic, i. e. guarantee equality of ensemble average, and of running average over a single member of the ensemble. Ergodicity has been proved for several (but not all of) the cases described in these lectures.

1.3 Chaotic Motion in Molecules.

To understand chemical reactions and energy transport in molecules, it is important to address the question whether in a given range of excitation energies, the level spacing fluctuations do or do not signal classical chaos. A pertinent experiment [6] is schematically showm in figure 1.4. A singlet state at an excita-

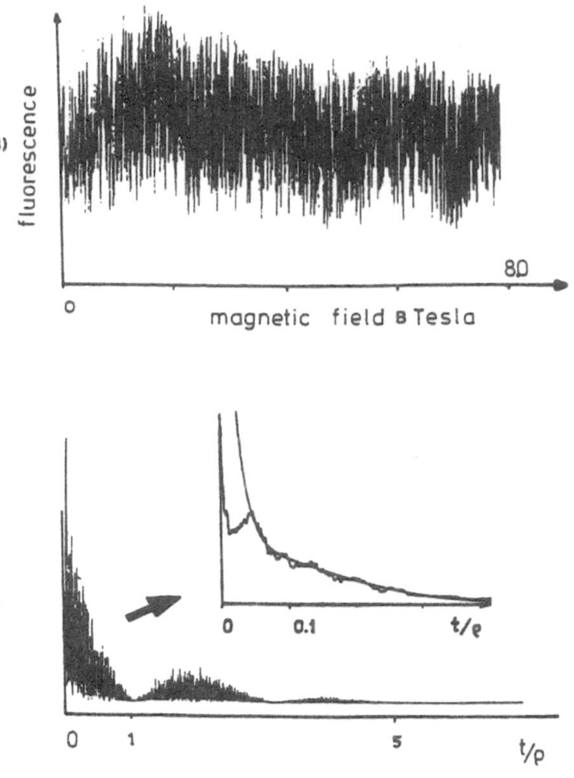

Fig. 1.5. Upper part: Resonance fluorescence yield $I(B)$ versus applied magnetic field B. Lower part: The Fourier transform $\int_{-\infty}^{\infty} \mathrm{d}B \cdot \int_{-\infty}^{\infty} \mathrm{d}b \exp(ibt) \cdot I(B+b/2)I(B-b/2)$ of the autocorrelation function versus t. The insert shows a correlation hole for small values of t. Taken from ref. [6].

tion energy of about 26 000 cm^{-1} in Methylglyoxal is laser-excited; the resonance fluorescence yield from this level is observed. The s-state is embedded in a dense background of triplet-states. Changing an externally applied magnetic field, one shifts the $m = \pm 1$ components of the t-states past the s-state, therebye inducing many Landau-Zener crossings. At each crossing, the s-state wave function mixes with one of the t-state wave functions; the mixing enables the s-state to gamma decay into other modes and reduces the resonance fluorescence yield. The upper part of figure 1.5 displays many hundred such crossings. The resonances overlap at least partially. This makes a direct determination of the level spacing distribution from the data impracticable. The insert in the lower part of the figure shows that the Fourier transform of the intensity-autocorrelation function displays a "correlation hole". It was suggested [6], that this hole indicates GOE statistics and thereby chaos in the classical limit. This statement needs to be verified in terms of a random-matrix calculation modelling the situation shown in figure 1.4. The problem is of interest not only in its own right, but also because the signal (the correlation hole) is obtained without tediously analysing the data with regard to individual resonances and their spacing distribution. To sum up: The physical questions are whether the signal in fig. 1.5 indicates molecular chaos, and whether such a signal is universal.

1.4 Passage of Light Through a Medium with a Spatially Randomly Varying Index of Refraction.

The passage of laser light through such a medium produces a random pattern on a screen (top part of right-hand side of figure 1.6). As the angle of incidence changes (by rotation by 10 mdeg in each step shown in figure 1.6) the pattern does, too (lower parts on right-hand side of figure 1.6). But the change is gradual, as seen by inspection and demonstrated by the correlation functions on the left-hand side, all taken with the first pattern. This experiment [8] demonstrates beautifully wave coherence in random impurity scattering. Physically relevant questions are: What determines the shape of the autocorrelation function? Can one observe localization in these experiments?

1.5 Universal Conductance Fluctuations.

In the last few years, it has become possible to manufacture metallic wires of μm length, and semiconductor devices in which electrons move in a virtually two-dimensional layer. At sufficiently low temperature, the conductance in both kinds of samples is due to elastic impurity scattering of the electrons. (It is commonly assumed that the electron density is sufficiently small to render the electron-electron interaction unimportant). The conductance G is measured as a function of an external magnetic field H or, in the case of MOSFET's, also as function of the Fermi energy E_F (the gate voltage). The dependence on H

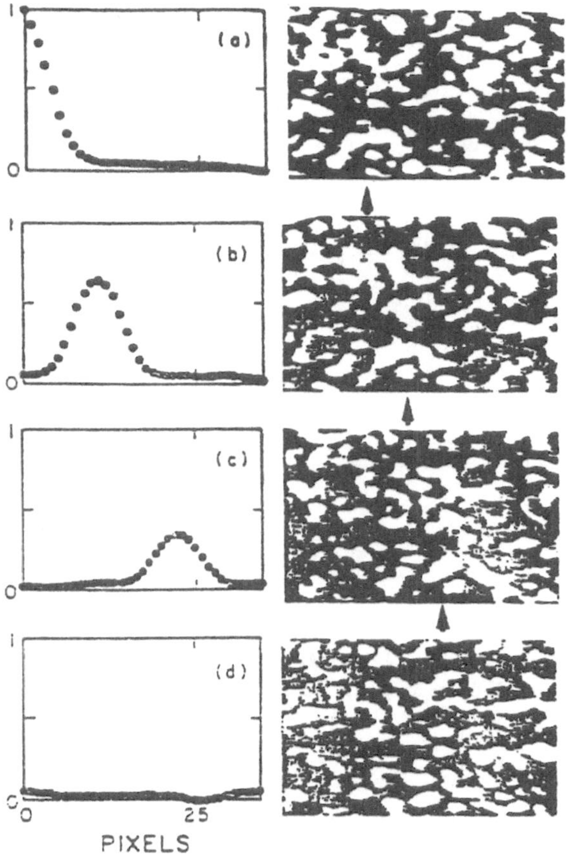

Fig. 1.6. Speckle patterns and correlation functions in the passage of light through a disordered medium. Taken from ref. [8].

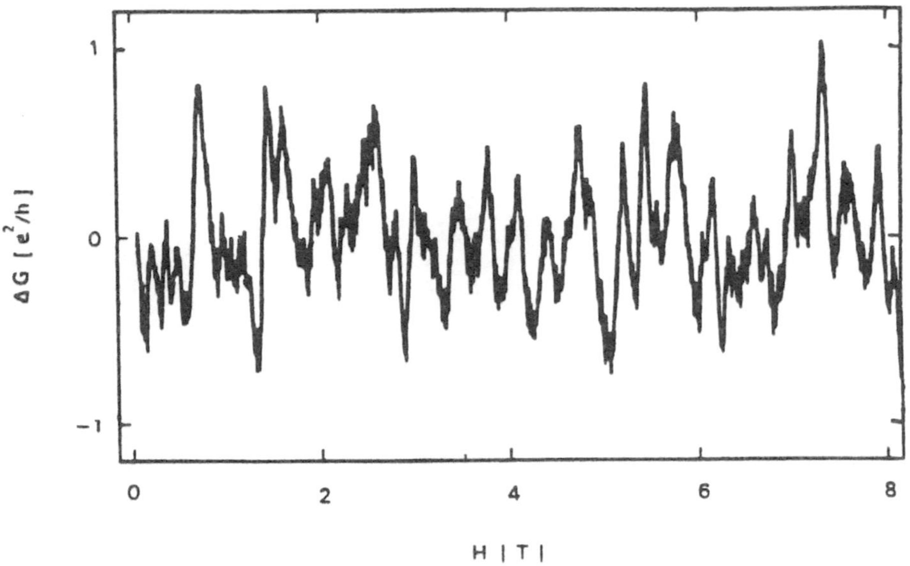

Fig. 1.7. The fluctuations of $\Delta G = G - \overline{G}$ in units e^2/h versus an external magnetic field H in tesla. The wire was 310 nm long and 25 nm wide, the temperature 0.01 K. Taken from ref. [9].

or on E_F is supposed to simulate the fluctuations of G over an ensemble of macroscopically identical probes. The intuitive, classical argument is that in changing H (or E_F), one changes the curvature (the velocity) of the electron which therefore experiences a different set of random scatterings on its way through the sample. With $g = Gh/e^2$ a dimensionless variable, the remarkable finding of such experiments is this: The average $\overline{g}$ of g over the fluctuations depends on the material and the geometrical dimensions of the wire, and changes much with either. However, the size of the fluctuations, quantified by $\overline{g^2} - \overline{g}^2 = \mathrm{var}(g)$, *is always of order unity*, see figure 1.7. The phenomenon has been named "Universal Conductance Fluctuations". It is of obvious interest for applications in microelectronics. More importantly, the fluctuations relate to the effect which disorder has on the coherence properties of wave functions. The importance of coherence is manifest in figure 1.8 which shows the fluctuations of G for a ring geometry rather than for a simple wire. The rapid oscillations are Aharanov-Bohm interference patterns superposed on the longer-wavelength random fluctuations.

To sum up: Physical questions relating to conductance fluctuations concern their universality, the dependence of the correlation function on external para-

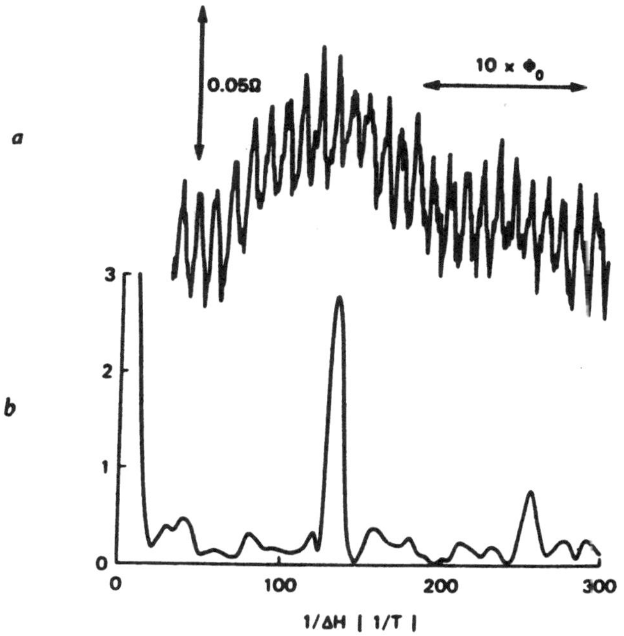

Fig. 1.8. Fluctuations of G versus H for a ring geometry. Part a shows the data, part b the Fourier transform.Taken from ref. [9].

meters, and the transition to localization. As an additional point of interest, we mention a recent experiment on isolated rings. This experiment forms the content of section 6.

2. Stochastic Modelling.

Because of lack of space and time, I confine myself to two cases: The stochastic modelling of chaotic scattering and compound-nucleus scattering (examples (i) and (ii) of section 1 for which the modelling is identical), and conductance fluctuations (example (v) of section 1). Other cases, partly described in the published literature, can be modelled along very similar lines. For brevity, I always use the GOE. An external magnetic field breaks the time-reversal symmetry of the intrinsic Hamiltonian in the case of conductance fluctuations; I do not pay attention to this fact which, of course, must be taken into account in actual calculations. The two examples which follow are intended to demonstrate the general approach rather than detail.

2.1 Chaotic and Compound-Nucleus Scattering.

Let $a, b, c, \ldots$ denote the open channels, defined by a fixed transverse mode in the antenna in the case of microwave scattering, and by a fixed fragmentation, channel spin, relative orbital angular momentum, and total spin and parity for compound-nucleus scattering. These channels feed resonances (compound states of fixed spin and parity) labelled $\mu, \nu, \ldots$. In the absense of a direct dynamical coupling between channels and under omission of elastic phase shifts (both of which can be easily included), the elements of the scattering matrix have the form

$$S_{ab}(E) = \delta_{ab} - 2i\pi \sum_{\mu\nu}^{N} W_{a\mu} \left[D^{-1}\right]_{\mu\nu} W_{\nu b} \quad . \tag{2.1}$$

Here, E is the Energy and $W_{a\mu}$ are matrix elements coupling channels and levels. The $W_{a\mu}$ typically change very little when E ranges over a mean level spacing and are taken as constants. The inverse propagator is

$$D_{\mu\nu}(E) = E\delta_{\mu\nu} - H_{\mu\nu} + i\pi \sum_{c} W_{\mu c} W_{c\nu} \quad . \tag{2.2}$$

To simulate stochastic fluctuations of S_{ab} and, thereby, of the cross-section (which is proportional to $|S_{ab}|^2$), we consider an ensemble of S-matrices by keeping $W_{\mu a}$ fixed and by letting $H_{\mu\nu}$ run over the GOE. The parameters in this model are λ (the GOE parameter) and the $W_{\mu a} = W_{a\mu}$ (time-reversal symmetry). Because of orthogonal invariance of the GOE, the distribution function of the S-matrix elements depends only on the dimensionless orthogonal invariants $\sum_{\mu} W_{a\mu} W_{b\mu}/\lambda$. The number of these latter input parameters is equal to the number of average S-matrix elements $\overline{S_{ab}}$. Given the average S-matrix, the model therefore describes the fluctuations of S in a parameter-free fashion.

The form (2.1), (2.2) of the S-matrix is generic to resonance theory; it is obtained from any of the standard nuclear reaction theories; it can be read as an N-level Breit-Wigner formula which preserves the unitarity of S. Aside from the resonance energies determined by $H_{\mu\nu}$, the denominator (2.2) also contains an imaginary width term which accounts for the coupling to the open channels.

To repeat: A description of the actual data of figure 1.1 should make use of the actual Hamiltonian of the stadium. Replacing this Hamiltonian by the GOE can give no more than characteristic average properties. The same remark applies to compound-nucleus scattering.

In both cases, the statistical input consists of the *average* S-matrix elements. It is in terms of these that the model (2.1) and (2.2) aims at predicting the fluctuations. Average S-matrix elements relate to short time scales. Therefore, they are system-specific and cannot be obtained from stochastic modelling. On the other hand, it is usually easy to model $\overline{S}$ or the running (i. e. energy) average $\langle S \rangle$ dynamically: Short time scales imply that only few degrees of freedom are involved. In the case of compound-nucleus scattering, it is the complex optical model which serves this purpose, and a few parameters in this potential are fitted to experimental values of $\langle S \rangle$. For systems with few degrees of freedom, it is possible to calculate $\langle S \rangle$ directly dynamically from the relation $\langle S \rangle = S(E+iI)$, where I is the width of the averaging interval. And $S(E+iI)$ is determined from the dynamical equations, solved at complex energy $E+iI$. This calculation avoids the complexity of the many resonances which occur at energies with *negative* imaginary part.

2.2 Conductance Fluctuations.

Although experiments on mesoscopic wires usually employ a four-lead geometry, I consider for simplicity here a two-lead setup: A disordered wire of length L_x connected at either end to an ideal lead. In each lead, channels $a, b, c, \ldots$ are defined by the transverse modes of the electron. The number of such channels at the Fermi energy is given by $\Lambda = L_x L_z K_F^2/\pi$ where $L_x L_z$ is the cross-section area of the wire and K_F the Fermi wave number. For typical experiments, Λ is at least several thousand. With $G = (e^2/h)g$, we use the many channel approximation to Landauer's formula which reads $g = 2\sum_{ab} \left|S_{ab}^{LR}\right|^2$. Here, S_{ab}^{LR} is the element of the scattering matrix connecting channel a on the left with channel b on the right (figure 2.1); $\sum_{ab} \left|S_{ab}^{LR}\right|^2$ is obviously the total transmission through the disordered sample; the factor 2 is due to spin. To determine S_{ab}^{LR}, we consider the longitudinal motion of the electron in the leads as free. Then, the S-matrix has a structure similar to eqs. (2.1), (2.2): The Hamiltonian $H_{\mu\nu}$ can be thought of as an Anderson tight-binding Hamiltonian for the electron in the disordered sample in the absence of any coupling to external leads. To implement a random-matrix model, we use a different construction. The disordered sample is thought to be divided (see figure 2.1) by transverse cuts into K equal slices of length l, the elastic mean free path. (We obviously assume $L_x = Kl$ with K integer.) In each slice the Hamiltonian for that slice is replaced by a GOE; GOE's in different slices are uncorrelated but have the same variances. States in neighbouring slices are coupled by real matrix elements that are Gaussian distributed uncorrelated random variables with zero mean and fixed variance (the same for all matrix elements, and all pairs of neighbouring slices). States in two slices separated by at least one further slice have zero coupling matrix elements. The resulting Hamiltonian has block structure, each diagonal block corresponding to one of the slices. Labelling the slices by running indices $j, l = 1, \ldots, K$ and the states in each slice by by running indices $\mu, \nu = 1, \ldots, N$ we can write the Hamiltonian as $H_{\mu\nu}^{jl}$. For $j = l$, this is a GOE; for $j = l \pm 1$, the matrix elements are Gaussian; for $|j - l| > 1$, the matrix elements vanish.

This model for $H_{\mu\nu}$ encompasses the idea of a mean free path for random impurity scattering. It applies to probes with transverse dimensions of the order of l. It contains two parameters: The GOE parameter λ which is adjusted to the mean level density, and the variance of the block non-diagonal elements $H_{\mu\nu}^{jj+1}$. Dividing the latter variance by λ^2 we obtain a dimensionless quantity which, for $K \gg 1$, is related to the diffusion constant D.

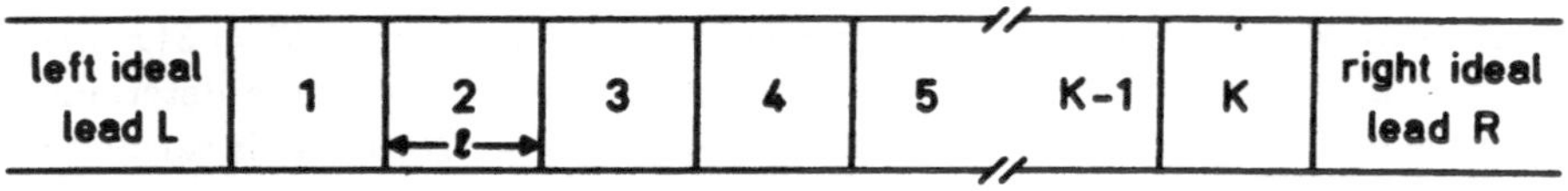

Fig. 2.1. A random matrix-model for conductance fluctuations in a two-lead experiment (schematic). L and R stand for left and right, respectively. The disordered sample is divided into K slices of length l each.

To construct the S-matrix, it is consistent with this model to assume that an electron entering from the left (right) lead does not travel beyond the first (the last) slice before it undergoes its first random scattering. We accordingly assume that the matrix elements $W^L_{a\mu}(W^R_{b\nu})$ are zero unless $\mu(\nu)$ is a state in the first (the last) slice. Then the S-matrix has the form

$$S^{LR}_{ab} = -2i\pi \sum_{\mu\nu} W^L_{a\mu} \left[D^{-1}\right]^{1K}_{\mu\nu} W^R_{\nu b} \quad . \tag{2.3}$$

The propagator is given by

$$D^{jl}_{\mu\nu} = E\delta_{jl}\delta_{\mu\nu} - H^{jl}_{\mu\nu} + i\pi\,\delta_{jl}\left[\delta_{j1}\sum_a W^L_{a\mu}W^L_{a\nu} + \delta_{jK}\sum_b W^R_{b\nu}W^R_{b\nu}\right] \quad . \tag{2.4}$$

We work in the metallic regime $K_F l \gg 1$ so that $N \gg \Lambda \gg 1$. Orthogonal invariance in the end slices reduces the $2N \cdot \Lambda$ input parameters $W^{L,R}_{a\mu}$ to the quantities $\sum_\mu W^L_{a\mu}W^L_{b\mu}$ and $\sum_\nu W^R_{a\nu}W^R_{b\nu}$; without loss of generality it may be assumed that the latter quantities are diagonal in the channel indices; it is physically reasonable to postulate that the remaining quantities $\sum_\mu (W^L_{a\mu})^2$ and $\sum_\nu (W^R_{b\nu})^2$ are all equal. This reduces the model to a three-parameter model: It depends on the mean level spacing, on the diffusion constant, and on the strength of the coupling between the two end-slices and the adjacent lead.

So far, we have described a stochastic model for the S-matrix describing scattering of a single electron on a disordered piece of wire as in figure 2.1. How does the S-matrix relate to the observable, i. e. the conductance? The answer, briefly mentioned above, is given by the many-channel generalization of Landauer's formula. With G the conductance, it reads

$$G = \frac{e^2}{h} g \quad \text{and} \quad g = 2\sum_{ab} \left|S^{LR}_{ab}(E_F)\right|^2 \quad . \tag{2.5}$$

Here, E_F is the Fermi energy. At first sight, this formula is obvious: The dimensionless conductance is, including a factor two for spin, given by the total transmission from left to right through the disordered wire. And g is bounded from above by 2Λ, the total number of channels. On second thought, eq. (2.5) is not obvious at all: How can a quantal transmission coefficient determine a *dissipative* coefficient (the conductance)? Where does dissipation occur? This is not a simple question and has been much discussed in the literature. (A recent review is ref. [10].) The answer briefly is this: Dissipation happens in the reservoirs, located at infinite distance to the right and to the left in figure 2.1. These reservoirs prevent the electron from bouncing back from infinity and make viable a description of the passage of the electron as a scattering process. Even after accepting this answer, one is left marvelling how a formula like (2.5), encompassing the entire scattering process, can yield a dissipative coefficient. This formula clearly embodies the long-range coherence of the wave function which is so important in mesoscopic physics.

The model of eqs. (2.3) to (2.5) has limitations: It does not allow a transition to the case of ballistic electrons because $K \geq 1$, and the size of the slices is somewhat arbitrary. Replacing this model by assuming a white-noise random impurity potential in the disordered region removes these limitations but leads to identical results for $K \geq 3$ or so [10, 11].

In closing this section, I mention that stochastic modelling as exemplified in cases (i) and (ii) is not confined to the examples (i) to (v) presented in section 1. Further applications are to precompound reactions [12] and to symmetry breaking in chaotic scattering: Parity violation is much enhanced in compound-nucleus scattering and is presently a topic of intense study, both experimentally and theoretically [13]. Finally, the Hamiltonian in eq. (2.4) is kin to a banded random matrix. Such matrices have found some interest recently in the study of quantum chaos.

3. Methods of Averaging.

How can we calculate averages of observables from stochastic models as presented in the examples of section 2? Typically, the stochastic Hamiltonian (which contains a large number $\sim N^2$ of random variables) appears in the propagator, i. e. in the denominator of a scattering amplitude. We are interested in average values of the scattering matrix, of $|S_{ab}|^2$ (this yields the average cross-section, or the scattered average intensity, or the average conductance), of $|S_{ab}|^4$ (this yields the variances of the entities just mentioned), and even of $|S_{ab}(E_1)|^2 \, |S_{ab}(E_2)|^2$ (this yields the correlation functions). The stochastic approach is viable only if such averages can be calculated.

Three methods are basically available for this purpose; they are briefly sketched in this section.

3.1 Monte-Carlo-Simulation [14].

This method is entirely numerical. Its essence is this. With the help of a random-number generator, the Gaussian distributed random variables in the Hamiltonian are drawn at random and a member S_{ab} of the stochastic ensemble of S-matrices is calculated. This procedure is repeated $M \gg 1$ times to make it possible to calculate averages with small statistical error.

This method is very important to establish trends, gain physical insight into modes of behaviour of physical systems, and for order-of-magnitude results. Limitations result from the need to keep both M (the number of repetitions) and N (the dimension of H) finite and, in fact, rather small to save CPU time. In my experience, the method is good for mean intensities and variances, reaches its limit for correlation function, and cannot be used to study parameter dependeces, except very qualitatively.

3.2 Disorder Perturbation Theory [15].

Here, the scattering matrix is expanded in a Born series in the random part of H. To calculate the ensemble averag $\overline{S}$ of S, each term in the Born series is averaged, and the result resummed. To average a given term, the following theorem is utilized. Let $z_1, \ldots, z_K$ be independent random Gaussian variables with zero mean values. With $i_1, i_2, \ldots, i_L$ integers taken from the set $1, \ldots, K$, we have

$$\begin{aligned} \overline{z_{i_1} z_{i_2} \cdots z_{i_L}} &= 0 \quad \text{for } L \text{ odd}, \\ \overline{z_{i_1} z_{i_2} \cdots z_{i_L}} &= \overline{z_{i_1} z_{i_2}} \cdot \overline{z_{i_3} z_{i_4}} \cdots \overline{z_{i_{L-1}} z_{i_L}} \\ &\quad + \text{all other pairwise contractions of } z_{i_1} \cdots z_{i_L} \quad . \end{aligned} \tag{3.1}$$

Applying (3.1) to a given term in the Born series generates a lot of terms, and it is impossible to resum the Born series when all these terms are kept. Therefore, this method requires a small parameter which can be used to identify terms of leading order. In the context of solid state physics, this parameter is $(K_F l)^{-1}$, with l the elastic mean free path and K_F the Fermi wave number. In the metallic regime, $K_F l \gg 1$. In the present context, the small parameter is N^{-1}. A second small parameter is Λ^{-1}, with Λ the number of open channels and $\Lambda \gg 1$. (If this condition is not met, impurity perturbation theory fails, and the methods under (iii) must be used.) Keeping only the terms of lowest (zeroth) order in N^{-1}, and of the few lowest orders in Λ^{-1}, we can resum the Born series and obtain an asymptotic expansion in inverse powers of Λ. The same method can also be applied to calculate averages of $|S_{ab}|^2$, variances etc., although the complexity of the calculation increases rapidly with increasing powers of S. To keep track of terms, it is useful to use a diagrammatic representation [15].

Because of the condition $\Lambda \gg 1$, this method does not work for chaotic scattering when the number of open channels is of the order of 10 or less, see section 4. Isolated mesoscopic rings form the subject of section 6. Here, $\Lambda = 0$, and at zero temperature impurity perturbation theory fails. For disordered quasi-onedimensional conductors (the example (v) of section 2), the asymptotic series works less well as the length L_x of the disordered piece increases; the expansion breaks down as L_x approaches the localization length, and other methods must be used to study the localization transition. This last point is not discussed any further.

3.3 The Generating Functional.

The amplitude (S-matrix element) is written as the logarithmic derivative of a suitably chosen generating function,

$$S_{ab} = \frac{\partial}{\partial J_{ab}} \ln Z(J)|_{J=0} \quad . \tag{3.2}$$

This procedure has the advantage that the random variables in H appear in an exponent in Z. Averaging an exponential of a Gaussian random variable is trivial, and likewise the calculation of $\overline{Z}$ (or of powers of Z, like $\overline{Z^n}$ with n integer) is straightforward. Unfortunately, the calculation of $\overline{S_{ab}}$ requires the calculation of $\overline{\ln Z}$ and this is not straightforward at all. Two paths out of this impasse exist.

(a) The replica trick. We write

$$\ln Z = \lim_{n\to 0} \left[\frac{1}{n}(Z^n - 1)\right] \tag{3.3}$$

and calculate $\overline{\ln Z}$ by calculating $\overline{Z^n}$ for integer n, taking the limit $n \to 0$ at the end. This procedure is not exact because n is restricted to integer values. It is now known to apply in the same regime as disorder perturbation theory. In comparison with that expansion, the replica trick is simpler to handle because the calculation is more compact. This trick can also be used to calculate $\overline{|S_{ab}|^2}$ and terms of higher order in S.

(b) Grassman integration [16]. If in eq. (3.2) we had $Z(0) = 1$ identically, then the logarithmic derivative were to become a normal derivative, and $\overline{S_{ab}} = \frac{\partial}{\partial J_{ab}} \overline{Z(J)}|_{J=0}$ could be calculated straightforwardly. The normalization $Z(0) = 1$ can be achieved by introducing integration over anticommuting variables. I wish to illustrate this point without going too much into detail. For the case of the model of eqs. (2.1) and (2.2), the generating function $Z(J)$ in eq. (3.2) can be chosen as

$$Z(E, J) = \text{const.} \int_{-\infty}^{\infty} (\prod_{\mu}^{N} \mathrm{d}S_\mu) \exp\{\mathcal{L}(\phi)\} \quad . \tag{3.4}$$

The S_μ with $\mu = 1, \dots, N$ are real integration variables. The "Lagrangian" $\mathcal{L}$ is given by

$$\mathcal{L} = \frac{1}{2} i \sum_{\mu,\nu} S_\mu D^J_{\mu\nu} S_\nu \tag{3.5}$$

with

$$D^J_{\mu\nu} = D_{\mu\nu} - 4\pi \sum_{ab} W_{\mu a} J_{ab} W_{\nu b} \quad . \tag{3.6}$$

The last term in eq. (3.6) is the "source term"; logarithmic differentiation with respect to the source of $Z(E, J)$ yields the observable. Different observables for the same ensemble are obtained by keeping the formal structure of eqs. (3.4) to (3.6) the same and by changing the source term appropriately. Aside from the constant factor in eq. (3.4) which is immaterial because of the logarithmic derivative in eq. (3.2), the normalization of $Z(E, 0)$ contains the factor $[\det(D_{\mu\nu})]^{-1/2}$. This is the standard factor which appears in the evaluation of a Gaussian integral of the form (3.4). This factor depends on the random variables in a very non-trivial way.

Now let us see how anticommuting (or Grassmann) variables can indeed be used to normalize $Z(E, J)$ to unity at $J = 0$. We define an algebra of anticommuting variables $x, y, z, \ldots$ with $xy = -yx$ so that $x^2 = y^2 = \ldots = 0$. Differentials of these variables are also anticommuting, $\mathrm{d}x\mathrm{d}y = -\mathrm{d}y\mathrm{d}x$, $x\mathrm{d}x = -\mathrm{d}x \cdot x$ and $y\mathrm{d}x = -\mathrm{d}x \cdot y$. Integration over anticommuting variables is defined by the two conventions

$$\int \mathrm{d}x = 0 \quad \text{and} \int x\mathrm{d}x = (\sqrt{2\pi})^{-1/2} \quad . \tag{3.7}$$

These conventions imply that

$$\int \exp\left\{\sum_{i,k}^{N} \chi_i^* A_{ik} \chi_k\right\} \mathrm{d}\chi_1^* \mathrm{d}\chi_1 \cdots \mathrm{d}\chi_N^* \mathrm{d}\chi_N = (-2\pi)^{-N} \cdot \det A \tag{3.8}$$

where $\chi_1, \ldots, \chi_N$ and $\chi_1^*, \ldots, \chi_N^*$ are independent anticommuting integration variables. The reader is invited to verify eq. (3.8) in the simplest nontrivial case $N = 2$; the calculation is simple and straightforward.

If now we multiply the square of the function (3.4) (which at $J = 0$ according to the above is proportional to $(\det D)^{-1}$) with another function formally defined in exactly the same way but with anticommuting instead of commuting integration variables, the result is a function $Z^G(E, J)$ (where G stands for Grassmann) in which the factor $(\det D)^{-1}$ is cancelled by the factor $(\det D)$ coming from integration over anticommuting variables, see eq. (3.8), and which is therefore trivially normalized to unity for $J = 0$. A suitable derivation of this function yields $S_{ab}(E)$ by a relation analogous to eq. (3.2). To each level $\mu, 1 \leq \mu \leq N$, there are now associated four integration variables which we write as $S_\mu, T_\mu, \chi_\mu, \chi_\mu^*$. Here, S_μ and T_μ are commuting and χ_μ, χ_μ^* are anticommuting variables.

In this way, an exact (non-perturbative) calculation of averages becomes possible. The method has a price: It is necessary to learn analysis for anticommuting variables. This task is simplified when one uses the symmetries of the effective Lagrangian in Z. Such symmetries lead to the study of graded Lie groups, where "graded" indicates the presence of both commuting and anticommuting variables.

To proceed further, a cumbersome calculation is necessary. I refer to [17] for details and sketch the main points. After averaging over the ensemble, one finds that the exponent in Z depends only on certain bilinear forms in the integration variables S_μ, T_μ, χ_μ and χ_μ^* which appear both linearly and quadratically. Introducing these bilinear forms as new integration variables and integrating over the remaining independent variables reduces the number of integrations to a value independent of N. The resulting "effective action" in the exponent is proportional to N. This suggests the use of the saddle-point approximation.

Minimzing the effective action one finds not a single saddle point but rather a saddle-pint manifold. The integration variables can be arranged in such a way that one part defines coordinates within and the other, coordinates normal to the manifold. Integration over the latter is easy in Gaussian approximation, justified by the factor N. This is the origin of the asymptotic expansion in powers of N^{-1}. Integration over the former (the "Goldstone modes") may also be done approximately in terms of an asymptotic expansion, or a loop expansion, if a further small parameter in the problem exists ($(K_F l)^{-1}$ or Λ^{-1}). In the absence of a small parameter, the integration has to be done exactly (chaotic and compound-nucleus scattering in the case of few open channels; mesoscopic isolated rings; localization transition).

In summary, we see that powerful methods are available to yield reliable approximations to or, if necessary, exact expressions for the quantities of interest. These methods have been used to obtain the analytical or numerical results used in the next three sections.

4. Chaotic Scattering and Compound-Nucleus Reactions.

Irregular or chaotic scattering has actively been studied in the framework of classical mechanics since the middle of the 1980s. Prior to this period, chaos in classical dynamics was almost exclusively understood as a property of bounded systems. Since then, many papers have shed light on the connection between irregular scattering and chaotic dynamics.

During the last few years, interest has grown in the *quantum* properties of systems exhibiting classically chaotic scattering. This interest is not purely theoretical, but has also been stimulated by applications: As explained in section 1, quantum chaotic scattering is encountered in atoms, molecules, atomic nuclei, and more recently, also in the ballistic regime of the conductance of mesoscopic semiconductor devices, and in microwave reflection by specially shaped cavities.

Historically, the first encounter of physicists with what now is termed quantum chaotic scattering happened in nuclear physics. Niels Bohr introduced the concept of the compound nucleus in the '30s and drew an analogy between a nuleon-induced nuclear reaction and the collision of an ensemble of hard spheres. The theme was developed by Bethe, Weisskopf, Wigner, Feshbach, Moldauer and others, and many ideas which are pertinent to chaotic scattering in general were developed. (A review of this field, mirroring the situation at the end of the 1970s, is given in ref. [18]). In particular, the stochastic model of eqs. (2.1,

2.2) originated in nuclear physics, and the relevant averages were worked out in this context [17]. Later, it was recognized that the model of eqs. (2.1, 2.2) and the results obtained from it apply to chaotic scattering in general [19], in the energy intervals referred to in section 1.

A second approach to quantum chaotic scattering [19] uses the semiclassical approximation, well-established for bounded phase-space problems. In the context of billiards, this approximation has been the topic of the lectures by U. Smilansky at this school. The central concept in this approach is that of the unstable classical periodic trajectories. For chaotic systems the density of such trajectories grows exponentially with period. Using the argument of the Fourier transform in the same way as in section 1, we are led to expect that the semiclassical approach, in which the (few) short periodic trajectories are the easiest to calculate, gives straightforwardly the long-range correlations in energy of the cross-section and is therefore complementary to the stochastic approach. Moreover, the exponential proliferation of periodic unstable trajectories restricts the semiclassical approach to the calculation of *mean* values of observables. For technically different reasons, this approach is thus restricted in a similar way as the stochastic approach.

A further restriction of the semiclassical approach lies in the condition $\hbar \to 0$ which for chaotic scattering entails $\Lambda \gg 1$ where Λ is the number of open channels. This consequence is obvious for compound-nucleus scattering where $\hbar \to 0$, the semiclassical limit, is attained with increasing excitation energy, i. e. a growing number of open channels. For microwave scattering, the condition $\Lambda \to \infty$ follows from $\hbar \to 0$ via the observation that with decreasing $\hbar$, the number of transverse (open channels) in the antenna(s) increases. There are physically interesting cases where the condition $\Lambda \gg 1$ does not hold. Here, it is of interest to compare the stochastic approach (which is not in this way restricted) with data, and with the semiclassical one.

I shall proceed as follows. I will present results of the stochastic approach and elucidate their physical content. Then I will compare these results to what is found semiclassically, and to relevant data.

As explained in section 1, in the stochastic approach all averages are given in terms of the average S-matrix elements $\overline{S_{ab}}$ which define the input of this approach. In keeping with the model of eqs. (2.1, 2.2) (where a dynamical coupling between channels was neglected), I assume that $\overline{S_{ab}}$ is diagonal in the channels, $\overline{S_{ab}} = \delta_{ab}\overline{S_{aa}}$. (The theory has also been worked out for cases where this restriction is not met). The unitarity deficit of the average S-matrix is then given by the "sticking probabilities" $T_a = 1 - |\overline{S_{aa}}|^2$. These coefficients measure that part of the flux in channel a which is not re-emitted speedily but rather spends a substantial time in the interaction region. (Here, I used again ergodicity, $\overline{S_{aa}} = \langle S_{aa}\rangle$, and the argument involving the Fourier transform as in section 1). This is the part which explores details of the complex dynamics, populates the long-lived resonances, and is responsible for fluctuations. Obviously, $0 \le T_a \le 1$. For $T_a = 1$ (all a), the fluctuations attain maximal size, while for $T_a = 0$ (all a) we have no fluctuations because $S_{aa} = \overline{S_{aa}}$.

The central object of interest is the correlation function of two S-matrix elements taken at different energies. It is defined by

$$C_{abcd}(\epsilon) = \overline{S_{ab}(E)S^*_{cd}(E+\epsilon)} - \overline{S_{ab}}\,\overline{S_{cd}} \quad . \tag{4.1}$$

For the stochastic model of eqs. (2.1, 2.2), this function has been calculated. I give here the complete result because I want to draw attention to some of its features, but also because I want to make it clear that the conclusions drawn below are firmly established. The result is [17]

$$\begin{aligned} C_{abcd}(\epsilon) = \\ &\frac{1}{8}\int_0^\infty \mathrm{d}\lambda_1 \int_0^\infty \mathrm{d}\lambda_2 \int_0^1 \mathrm{d}\lambda \frac{(1-\lambda)\,\lambda\,|\lambda_1-\lambda_2|}{[(1+\lambda_1)\,\lambda_1\,(1+\lambda_2)\,\lambda_2]^{1/2}(\lambda-\lambda_1)^2\,(\lambda+\lambda_2)^2} \\ &\times \exp\Big\{-i\frac{\pi\epsilon}{d}(\lambda_1+\lambda_2+2\lambda)\Big\} \prod_e \frac{(1-T_e\lambda)}{(1+T_e\lambda_1)^{1/2}\,(1+T_e\lambda_2)^{1/2}} \\ &\times \Big\{\delta_{ab}\delta_{cd}\,\overline{S_{aa}}\,\overline{S^*_{cc}}\,T_aT_c\Big(\frac{\lambda_1}{1+T_a\lambda_1}+\frac{\lambda_2}{1+T_a\lambda_2}+\frac{2\lambda}{1-T_a\lambda}\Big) \\ &\quad\times\Big(\frac{\lambda_1}{1+T_c\lambda_1}+\frac{\lambda_2}{1+T_c\lambda_2}+\frac{2\lambda}{1-T_c\lambda}\Big) \\ &\quad+\Big(\delta_{ac}\delta_{bd}+\delta_{ad}\delta_{bc}\Big)\,T_aT_b \\ &\quad\times\Big[\frac{\lambda_1\,(1+\lambda_1)}{(1+T_a\lambda_1)(1+T_b\lambda_1)}+\frac{\lambda_2\,(1+\lambda_2)}{(1+T_a\lambda_2)(1+T_b\lambda_2)} \\ &\quad\quad+\frac{\lambda\,(1-\lambda)}{(1-T_a\lambda)(1-T_b\lambda)}\Big]\Big\}\,, \end{aligned} \tag{4.2}$$

where d is the mean level spacing of the GOE. In comparing eq. (4.2) with data, d has to be taken as the mean resonance spacing. This is in keeping with our general attitude – mean values like $\overline{S_{aa}}$ or d serve as input for the stochastic model.

The right-hand side of eq. (4.2) consists of four building blocks. They are typical for the results obtained from Grassmann integration. This is why they are briefly discussed here. First, there is a threefold integral over real variables $\lambda, \lambda_1, \lambda_2$ and a rational expression involving the λ's. The integral stems from the integration over the saddle-point manifold, the weight factor essentially is an invariant measure of a graded Lie group with both commuting and anti-commuting elements. The exponential factor gives the dependence on ϵ. We note that ϵ is measured in units of the mean level spacing d, in keeping with the remarks in section 1. The product over all channels c contains the coupling to the open channels in the form of the sticking probabilities T_c. The curly bracket contains the terms which relate to the channels a, b, c, d which appear on the left-hand side of eq. (4.2). We note that the expression is zero unless the channels are pairwise equal. This is a consequence of our assumption $\overline{S_{ab}} = \delta_{ab}\overline{S_{aa}}$. When this assumption is dropped, only the curly bracket in eq. (4.2) is changed, while the T_a must be taken to be the eigenvalues of the Hermitean matrix $P_{ab} = \delta_{ab} - (\overline{S}\,\overline{S}^\dagger)_{ab}$.

Further insight into the structure of eq. (4.2) is gained by an asymptotic expansion, valid if $\sum_c T_c \gg 1$. The leading term in this expansion is

$$C_{abcd}(\epsilon) = (\delta_{ac}\delta_{bd} + \delta_{ad}\delta_{bc}) \frac{T_a T_c}{\sum_f T_f - i\pi\epsilon/d} \quad . \tag{4.3}$$

For $\epsilon = 0$, the correlation function $\overline{|S_{ac}|^2} - |\overline{S_{ac}}|^2$ factorizes into a term T_a which gives the probability of formation of long-lived intermediate resonances, and another factor $T_c/\sum_f T_f$ which gives the relative probability for decay of the intermediate system into final channel c. This factorization corresponds to N. Bohr's idea of independence of formation and decay of the compound nucleus, and eq. (4.3) with $\epsilon = 0$ is referred to as the Hauser-Feshbach formula in nuclear physics.

For $\epsilon \neq 0$, eq. (4.3) yields for the autocorrelation function a Lorentzian with width

$$\Gamma = \frac{d}{2\pi} \sum_f T_f \quad . \tag{4.4}$$

This is the energy interval over which the S-matrix is correlated with itself. The expression (4.4) for the correlation width confirms a simple semiclassical argument due to Weisskopf. Let us consider the time-evolution of a wave packet $\Psi(t)$ which at time $t = 0$ starts in the interaction regime (i. e. in the domain of Hilbert space spanned by the levels $|\mu\rangle$). We ask for the probability per unit time of this wave packet to escape to infinity. An estimate for the time it takes the wave packet to return to the same point in Hilbert space is given by the Heisenberg time d/h. Therefore, the partial width Γ_c for escape into channel c is given by $\Gamma_c = (d/2\pi)T_c$, and the total width by eq. (4.4). This semiclassical argument is not expected to hold unless $\Lambda \gg 1$, and we shall indeed see below that the autocorrelation width differs from Γ for the case of few open channels. – The argument just given refers actually to the average *life-time* of the compound system, and to the associated *decay width*, while eq. (4.4) is the *correlation width*. The equality of both quantities again follows from Fourier-transforming the autocorrelation function.

For the case of equivalent channels (all T_a equal), eq. (4.3) yields an elastic cross-section ($a = b = c = d$) which is twice as large as the inelastic one ($a = c$, $b = d \neq a$). This amplification is referred to as "elastic enhancement factor" in nuclear physics. In mesoscopic systems, a related effect carries the name "weak localization correction".

We expect semiclassical theory to work when many channels are open. This is why we compare the results of this approach with eqs. (4.3, 4.4). We note that semiclassically, all channels are coupled equally strongly to the chaotic domain so that in this limit, $T_a = 1$ for all a. Factorization of the cross section then reduces to the statement that the cross section is the same for all channels except for the elastic one, where it is twice as large. This is exactly what the semiclassical theory yields. If $T_a = 1$ (all a), eq. (4.3) yields an average inelastic cross section, $\overline{|S_{ab}|^2} = \Lambda^{-1}$. We are not aware of a semiclassical derivation of this simple formula, although it follows trivially from unitarity

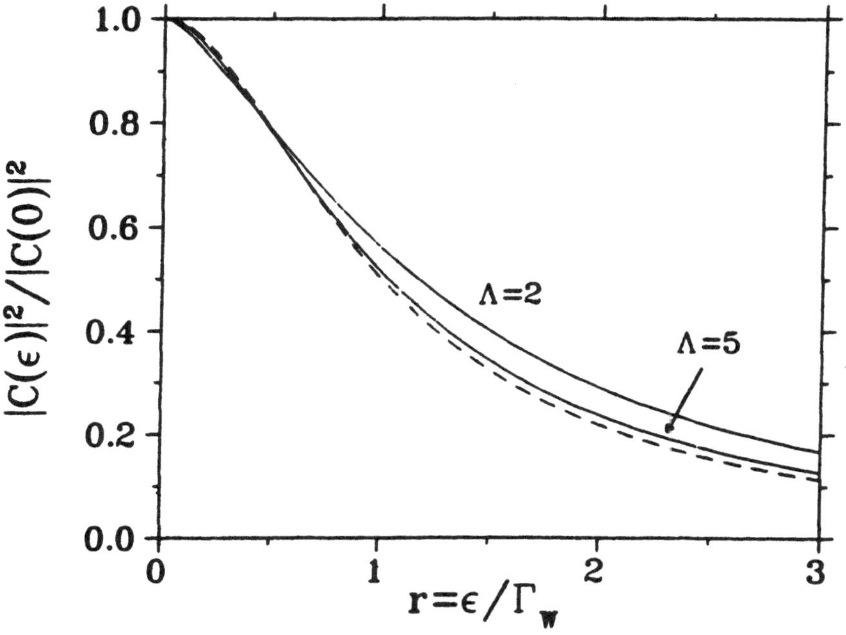

Fig. 4.1. The modulus squared of the normalized S-matrix autocorrelation function as given by eq. (4.2) for different numbers of open channels (Λ) compared with the Lorentzian prediction. The r-axis is rescaled by the Weisskopf estimate (Γ_W). All channels are equivalent with $T = 1$.Taken from ref. [20].

and from the independence (of the cross section) of the channel indices. The form of the autocorrelation function found semiclassically coincides with that of eq. (4.3), including the Weisskopf expression (4.4) for the correlation width, if we put all $T_a = 1$. Finally, both eq. (4.3) and the semiclassical approach predict zero correlation between S-matrix elements pertaining to different channels. In summary, we note that eq. (4.3) coincides with the semiclassical result if we put all $T_a = 1$. As it stands, eq. (4.3) is valid without this specification. Its range of validity therefore extends beyond that of semiclassics. It is interesting to compute not only the second, but also higher moments of the S-matrix in both approaches. We return to this point below.

We now leave the domain of validity of the asymptotic expansion (4.3) and study cases with a moderate number of open channels. To relate our findings to the case $\Lambda \gg 1$, we compare the actual correlation width Γ to the asymptotic estimate (4.4) which we denote by Γ_W. And we compare the time-dependence of the Fourier transform of the autocorrelation function $C(\epsilon)$ to the exponential decay of $P_{ab}(E,t)$, the semiclassical average probability of connecting channels a and b with trajectories that have a delay time in the interval $[t, t+dt]$. Our results are obtained by numerical integration of eq. (4.2). Figure 4.1 shows the normalized autocorrelation function plotted versus ϵ (in units of Γ_W) for different numbers of open channels, but with $T_a = 1$ in all open channels. (This quantity does not depend on the channel indices a and b.) We observe that as the number of channels increases, the correlation width approaches the Weisskopf estimate from above. Moreover, near $\epsilon = 0$ the curves for the smaller Λ values fall off more steeply than a Lorentzian would.

The origin of this behaviour emerges as we consider the Fourier transform $F_{abcd}(t)$ of the autocorrelation function $C_{abcd}(\epsilon)$. It is given by

$$\begin{aligned} F_{abcd}(t) &= \frac{1}{8}\int_0^\infty \mathrm{d}\lambda_1 \int_0^\infty \mathrm{d}\lambda_2 \int_0^1 \mathrm{d}\lambda\, \delta\Big[\frac{t}{\hbar} - \frac{\pi}{d}(\lambda_1 + \lambda_2 + 2\lambda)\Big] \\ &\quad \times \prod_e \frac{1 - T_e\lambda}{[(1+T_e\lambda_1)(1+T_e\lambda_2)]^{1/2}}\, J_{abcd}(\lambda, \lambda_1, \lambda_2) \end{aligned} \tag{4.5}$$

(where J_{abcd} is the content of the curly bracket in eq. (4.2)), and corresponds semiclassically to the function $\langle P_{ab}(t, E)\rangle_E$ which is independent of both a and b (save for the elastic enhancement factor).

For small values of time (such that $\Gamma_W t/\hbar < 1$), the product over the channels in eq. (4.2) can be approximated by an exponential; combining this with the delta function we find immediately $F_{abcd}(t) \sim \exp(-\Gamma_W t/\hbar)$, in keeping with the semiclassical result where $\Gamma_W/\hbar$ is replaced by γ, the classical escape rate. However, the terms remaining under the integral also depend on time and might jeopardize the approximation just made. On the other hand, the relevant time scale decreases with increasing Γ_W, so that we expect the exponential form to become an ever better approximation as Γ_W increases.

For very long times ($t \gg \hbar/\Gamma_W$), $F_{abcd}(t)$ decays with a power law:

$$F_{abcd}(t) \sim t^{-2-\Lambda/2} \quad , \tag{4.6}$$

where $t^{-\Lambda/2}$ originates from the product over the channels and the factor t^{-2} comes from the function J_{abcd}. In the one-channel case, this gives $t^{-5/2}$.

Numerical results for $F_{abcd}(t)$ are shown in fig. 4.2. Again, we have $T_a = 1$ in all channels; this makes the (normalized) function $F_{abcd}(t)$ independent of channel indices and causes us to denote it by P (because of its similarity to the semiclassical case). We observe that as Λ increases, the exponential approximation becomes better over an increasing number of decades. And the point of departure from the exponential law moves to ever shorter times in keeping with the comments made before.

Large time scales correspond to small energy scales. The arguments given in section 1 suggest that, in this domain, the stochastic approach should reproduce the generic features of chaotic scattering. Therefore, we attribute general significance to the differences between the stochastic and the semiclassical approaches displayed above. In our opinion, these differences reflect genuine quantum effects which are beyond the scope of the standard semiclassical approach.

Serious deviations from the semiclassical regime occur when we deal with nonequivalent channels, i. e., cases where the transmission coefficients T_a differ

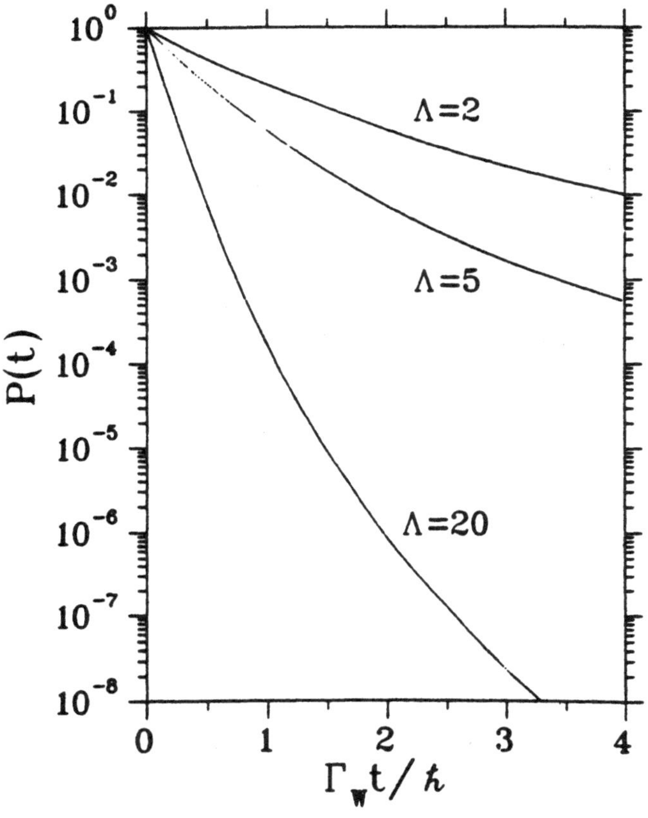

Fig. 4.2. Fourier transform of the S-matrix autocorrelation function (P) for different numbers of open channels. Taken from ref. [20].

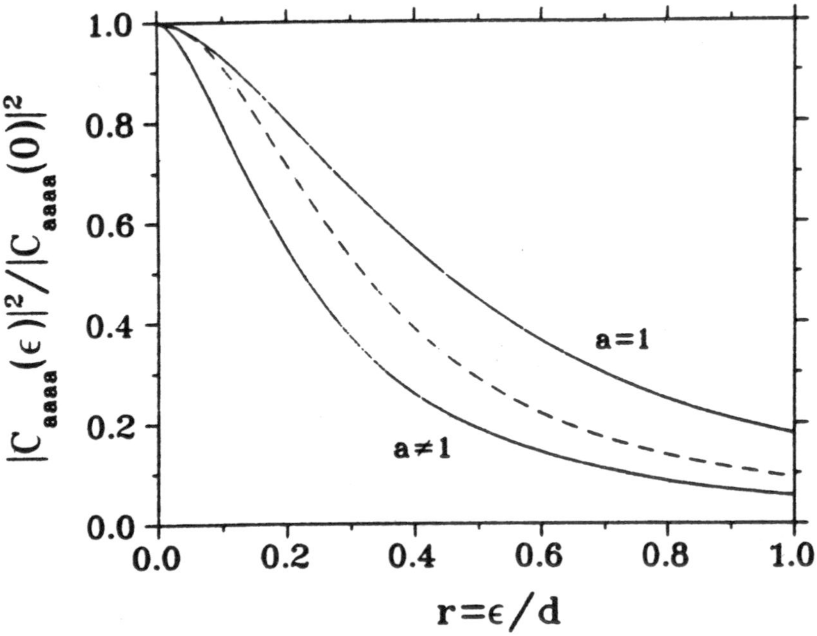

Fig. 4.3. Modulus squared of the normalized S-matrix autocorrelation function for non-equivalent channels: 11 open channels, $T_a = 1$ for $a = 1$ and $T_a = 0.1$ for $a \neq 1$. The solid lines represent the numerical integration of eq. (4.2) and the dashed one represents the Lorentzian with Γ_W for $\sum_a T_a = 2$.Taken from ref. [20].

from each other. In fig. 4.3 we show two normalized autocorrelation functions versus $r = \epsilon/d$. Both are calculated for 11 open channels, with $T_i = 0.1$ for $i \neq 1$ and $T_1 = 1.0$. The upper curve gives $|C_{abcd}(\epsilon)/C_{abcd}(0)|^2$ for $a = b = c = d = 1$ and the lower one corresponds to the choice $a = b = c = d = i$ with $i \neq 1$. We show this example to emphasize that not only average cross sections, but also autocorrelation functions behave in a non-universal fashion.

Much attention has been devoted in nuclear physics to the elastic enhancement factor W_a. It is defined by

$$C_{abab}(0) = \frac{T_a T_b}{\sum_c T_c} \left[1 + (W_a - 1)\, \delta_{ab}\right] \quad . \tag{4.7}$$

For equivalent channels ($T_a = T_b$), this coincides with eq. (4.3). The elastic enhancement factor can be computed from eq. (4.2). Before this equation was available, extensive Monte Carlo simulations of the GOE model described in section 2 were undertaken which led to fit formulas for W_a that were later found to provide a very good approximation to the exact result obtained from eq. (4.2). There is no need for us to go into details here. Suffice it to say that

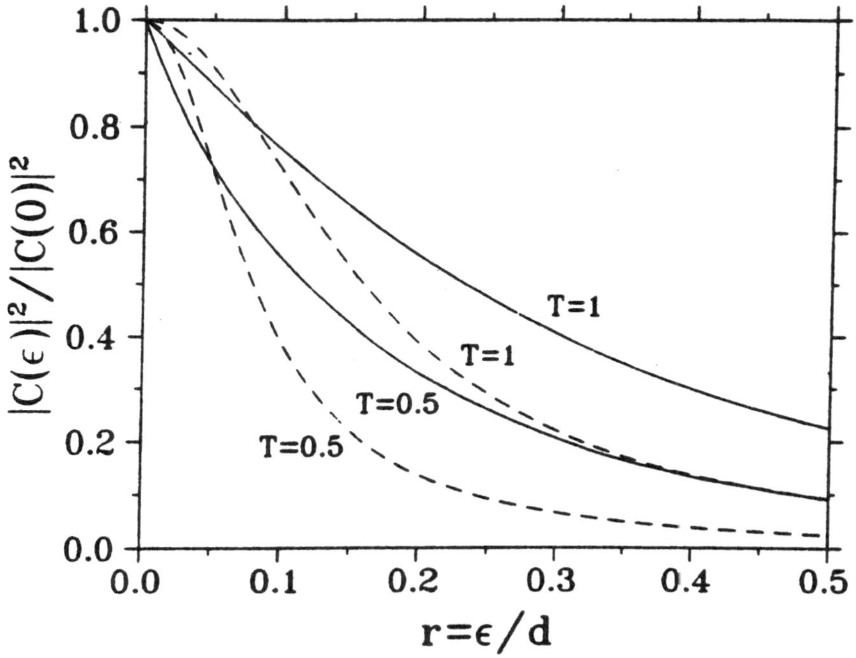

Fig. 4.4. The modulus squared of the normalized autocorrelation function $C(\epsilon)$ of eq. (4.2) versus $r = \epsilon/d$ for $T = 1/2$ and $T = 1$ (full curves), and two Lorentzians (dashed curves) for comparison. Taken from ref. [20].

in the few-channel case, W_a can significantly deviate from two. It is generically larger than this value and may become as large as three in the limit $\sum_c T_c \ll 1$ of extremely isolated resonances. The situation for the elastic enhancement factor is therefore very similar to that of the correlation function: The semiclassical limit $W_a = 2$ is attained for many equivalent channels and if $T_a = 1$, all a, for any number of channels; otherwise there are significant deviations which again reflect generic quantum features.

The differences between the semiclassical and the stochastic approaches are most pronounced in the case of a single open channel. It is here that the semiclassical treatment fails, at least in the channel region, although it may still apply in the interaction region. Moreover, this case is not academic. It applies in many instances in atomic and nuclear physics. Recent experiments on ballistic electrons in semiconductor devices realized a few-open-channel (albeit not a single-open-channel) situation. A recent microwave experiment claims to correspond to a one-open-channel situation (although absorption by the cavity may be thought of as being due to one or several other channels).Figure 4.4 shows the normalized one-channel correlation function for $T = 1/2$ and $T = 1$. The dashed curves are Lorentzians based on the Weisskopf estimate. A full account of the similarities and differences between the stochastic and the semiclassical approach, which includes other aspects not discussed here, can be found in ref. [20].

How do these theories compare with experiment? Or with explicit calculations for chaotic scattering problems with few degrees of freedom? As for data on microwave scattering, it is possible, of course, to realize experimentally a situation with one or two open channels. The difficulty is, however, that microwave absorption at the walls of the cavity amounts to several more "fictious" open channels. This is why really stringent tests, especially concerning the difference between semiclassical and stochastic theory in the few-channel case, are not available yet. Hope must be placed in the next generation of experiments which use superconducting materials.

The situation is better for a comparison with *calculated* results, especially for the billiards of U. Smilansky's lectures. Here, semiclassical theory has been fully established. Again, I am not aware of decisive tests in a situation of one or two open channels to distinguish between semiclassical and stochastic quantal theory.

With this proviso, we conclude that the stochastic approach to chaotic scattering yields the universal behaviour on long time scales, or short energy scales, taking full account of quantum effects. It does so in terms of few input parameters which in turn relate to short time scales, or energy averages over large intervals. The approach is sufficiently flexible to allow for any number of open channels and arbitrary transmission coefficients. This is of interest for applications, for instance in nuclear physics.

The stochastic approach will eventually fail when the energy interval considered becomes too large, since then the GOE assumption fails and system-specific nonuniversal features become relevant. The semiclassical approach, while failing in the quantum regime, has the advantage that via the shortest peeriodic orbits it contains information on this limit of universal behaviour. Moreover, it is capable of incorporating such system-specific effects into the theory.

The length of the energy interval beyond which the stochastic approach fails is given by the inverse of the period of the shortest unstable classical periodic trajectory. For dynamical systems with few degrees of freedom, this interval is very important, and its influence on the scattering process cannot be neglected. For systems with many degrees of freedom like atoms, molecules, or nuclei, this interval becomes so large (in units of the mean level spacing) that we do not expect important modifications of the stochastic approach. In these systems, however, another effect (not considered here) may be important: Not all parts of phase space may be equally accessible, and this fact may necessitate a modification of the stochastic approach. In the domain of nuclear physics, such modifications go under the name of "precompound" or "preequilibrium" processes.

In a different context, we have encountered an example of such a situation in the model of eqs. (2.3) to (2.5). Here, it is no longer possible to simulate the dynamics in the interaction region in terms of a single GOE. We turn to this case in the next section.

We have answered some of the questions on chaotic scattering raised in section 1. It may be argued, of course, that the similarity between chaotic and compound-nucleus scattering comes about by construction, and that I have not deduced it from the dynamics. This is true. On the other hand, the analysis of compound-nucleus resonances does show that their spacings obey GOE statistics and this is the input used in section 2. (The same statement applies to chaotic scattering). Hence whatever the cause of GOE statistical behaviour – the model of section 2 is applicable to both cases. And in the case of compound-nucleus scattering, this statistical model is indeed in complete agreement with the data.

5. Universal Conductance Fluctuations.

Results obtained from the stochastic model of eqs. (2.3, 2.4) depend on three parameters. It is useful to discuss these parameters first before we look at the results of the model. We recall that L_x is the length of the disordered sample.

The first of these parameters (the variance of $H^{jj}_{\mu\nu}$) is uniquely related to the mean level spacing d in the disordered sample. We obviously have $d \sim L_x^{-1}$. The second parameter (the variance of $H^{jj+1}_{\mu\nu}$), when expressed in units of the first, relates to the diffusion constant D for electron propagation through the disordered sample. This constant D in turn appears in the diffusion equation for the probability density $P(x,t)$ to find an electron at time t at the longitudinal position x in the sample,

$$\frac{\partial}{\partial t} P(x,t) = D \frac{\partial^2 P}{\partial x^2} \quad . \tag{5.1}$$

Dimensional arguments show that the diffusion time T_{diff} through the sample is proportional to $[D/L_x^2]^{-1}$. The energy E_c associated by the uncertainty principle with T_{diff} (the Thouless energy) is given by

$$E_c = \frac{\pi^2 \hbar D}{L_x^2} \quad . \tag{5.2}$$

Working out E_c from the model one sees that E_c is (except for a factor of order unity) what nuclear physicists call the spreading width. It is the FWHM of the probability to find an eigenfunction of the unperturbed Hamiltonian in the disordered sample (defined as $H^{jj+1}_{\mu\nu} = 0$ for all j) mixed into an eigenfunction of the full problem. The third parameter (the strength $\sum_a W^2_{a\mu}$ of the coupling to external channels) determines another energy scale, the decay width Γ of the sample. It is given by

$$\Gamma = \frac{d}{2\pi} 2\Lambda\alpha \tag{5.3}$$

where the coefficient α obeys the inequalities $0 \leq \alpha \leq 1$. Eq. (5.3) is intuitively obvious: By the Weisskopf argument, $d/2\pi\hbar$ is the frequency with which a time-dependent quasiperiodic wave packet in the disordered sample returns to its original position; α is the average probability with which it escapes into one of the open channels; $(d/2\pi\hbar)2\Lambda\alpha$ is the total escape rate into all channels. For good conductors (good coupling to the leads) we expect $\alpha \simeq 1$. In this case, $\Gamma \gg d$: We deal with the case of strongly overlapping resonances. We can interpret $T_{dec} = \hbar/\Gamma$ as the decay time: An electron localized at time $t = 0$ within the sample will, for $t \to \infty$, attain equal probability density everywhere (in leads and disordered sample); since the leads are infinitely extended by definition, this implies that the probability for finding the electron within the disordered region goes to zero; the rate of the associated exponential decay is given by T_{dec}. We note the close similarity to the discussion in the previous section. We also note that $\Gamma \sim L_x^{-1}$ because Λ and α do not depend on L_x.

The standard approach to the problem of conductance fluctuations does not make use of scattering theory. Rather, it uses the Kubo formula and considers a very long piece of disordered material. This approach leads directly to the impurity diagram expansion. Following early work on coherence properties of wave functions scattered in a random potential, it was this approach which led to a quantitative understanding of conductance fluctuations [21, 22, 23, 24, 25, 26]. (A method not mentioned in section 3, the method of the transfer matrix specific to solid-state problems, was employed in ref. [23] and, in conjunction with a maximum entropy approach kin to stochastic modelling, was developed further in ref. [26]. Lack of space does not allow me to go into the details of this interesting approach.) This standard approach does not usually pay attention to the coupling to external leads save for the use of boundary conditions. It is the merit of the model of section 2 that this coupling is explicitly included and the connection to Landauer's formula is established. An additional strength exploited recently in connection with localization [27] is the high degree of invariance under symmetry transformations of the model of section 2.

For sufficiently large lengths $L_x \gg l$, both $\overline{g}$ and var(g) coincide with the results of earlier calculations, indicating that the random-matrix model of section 2 which maps a three-dimensional problem onto a many-channel one-dimensional model is succesful. Differences arise for smaller values of L_x/l; these are quantified below. This shows that coupling to the leads does have an influence on the conductance and its fluctuations. A particular strength of the approach lies in the ease with which one can deal with geometries that are more complex than the one shown in figure 2.1.

Figure 5.1 displays the significance of the three energy scales d, E_c, and Γ on the behaviour of $\overline{g}$ which for $\alpha = 1$ is explicitly given by

$$\overline{g} = \frac{2\Lambda}{2 + (K-1)\frac{\pi^2}{2}\frac{\Gamma}{KE_c}} + \dots \tag{5.4}$$

The dots indicate the term of next order in the asymptotic expansion (the weak localization correction to $\overline{g}$). (Note that $\Gamma/(KE_c)$ is independent of length L_x).

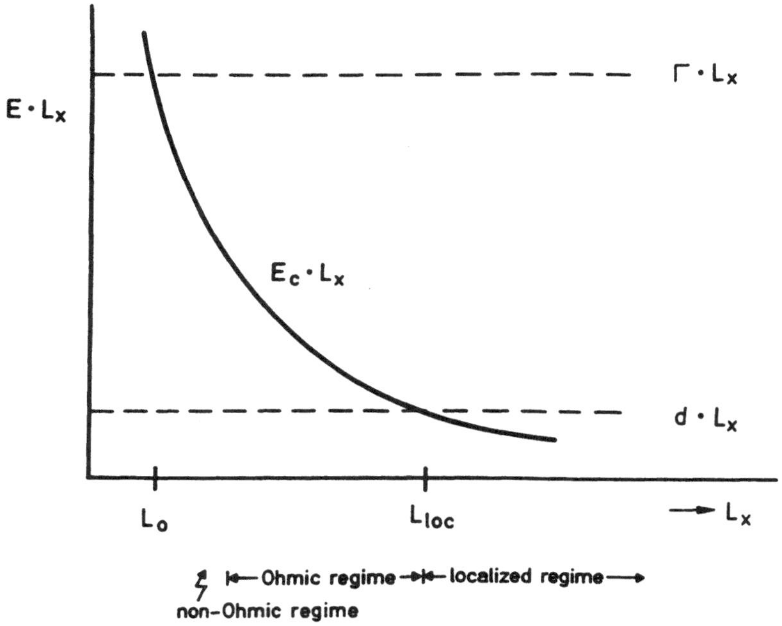

Fig. 5.1. The three characteristic energy scales of the model, multiplied by L_x, versus L_x (schematic).

In figure 5.1, we plot schematically energy $\cdot\, L_x$ versus length L_x. The dashed straight lines give $d \cdot L_x$, and $\Gamma \cdot L_x$ (we recall $\Gamma \gg d$), and the full curve, $E_c \cdot L_x$ (which falls off with L_x^{-1}). The curves for $E_c \cdot L_x$ and $d \cdot L_x$ intersect at length L_{loc} determined by $E_c = d$ or by the condition that spreading width and mean level spacing are equal. For $L_x > L_{loc}$, the mixing of wave functions seizes to be effective over the entire sample, the eigenfunctions fall off exponentially at either end, and so does $\bar{g}$: We have reached the localization transition with L_{loc} the localization length. Inspection of the terms indicated by dots shows that the asymptotic expansion (5.4) is valid for L_x-values less than L_{loc}. – The curves for $E_c \cdot L_x$ and $\Gamma \cdot L_x$ intersect at another critical length, L_0. Eq. (5.4) shows that for $L_0 < L_x < L_{loc}$, i. e. for $T_{diff} > T_{dec}$, $\bar{g}$ is nearly Ohmic, i. e. $\bar{g} \sim K^{-1}$. This is because the diffusion time through the sample determines the behaviour of the system. For $L_x < L_0$, on the other hand, we have $T_{diff} < T_{dec}$, so that internal diffusion is more rapid than the decay into the open channels. The electron density fills the sample uniformly before decay sets in. Therefore, $\bar{g}$ is (nearly) independent of length L_x and (nearly) equal to half its maximum possible value $2\,\Lambda$ because electron emission into both leads is equally probable.

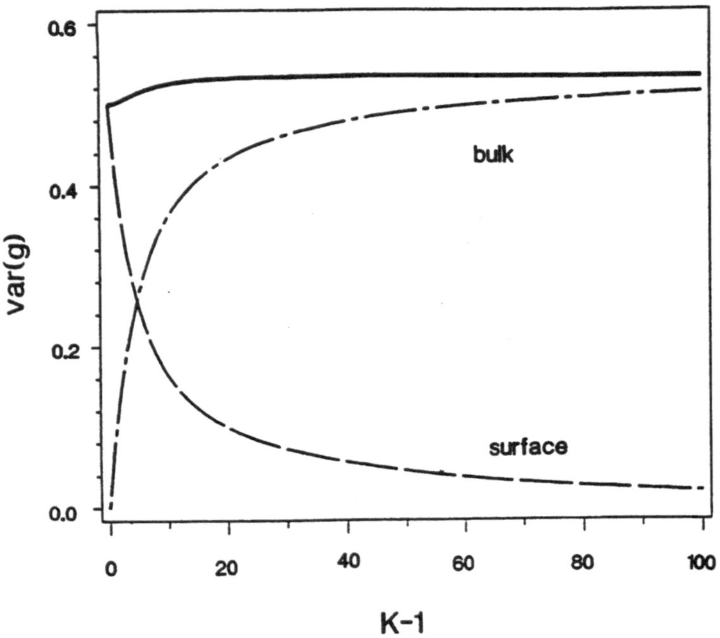

Fig. 5.2. The dependence of var(g) on $K = L_x/l$ for $\alpha = 1$ (see text). Taken from ref. [10].

We now turn to the fluctuations of g, expressed by the variance and the autocorrelation function. An early result from disorder perturbation theory was that var(g) is indeed of order unity. For the geometry of figure 2.1 (quasi one-dimensional problem), var(g) $= \frac{8}{15}$. This value is also obtained in the approach of section 2 in the Ohmic regime, more precisely in the limit $L_x \gg l$ but $L_x \ll L_{loc}$. The behaviour of g near the localization transition has been a topic of intense study and debate. It is studied with the help of renormalization group techniques which go beyond the scope of these lectures. Suffice it to say that for $L_x \simeq L_{loc}$, the distribution of g is not normal but logarithmically normal.

For values of L_x not too large in comparison with l, we expect to see an influence of the coupling to the channels. Fig. 5.2 shows var(g) for $\alpha = 1$ versus $K = L_x/l$. The dashed and dotted curve gives the bulk contribution to var(g). Asymptotically ($L_x \to L_{loc}$) it approaches the value 8/15 known from previous studies of the problem. For small values of L_x/l, it is dwarfed by the surface contribution (proportional to the coupling to the channels, dashed curve). The two curves add to a nearly constant curve (full line) for var(g). This is fortuitous, as shown by figure 5.3 where var(g) is plotted versus $K = L_x/l$ for three values of α. We observe that var(g) is influenced by the surface terms, and that this influence extends over dozens of mean free paths.

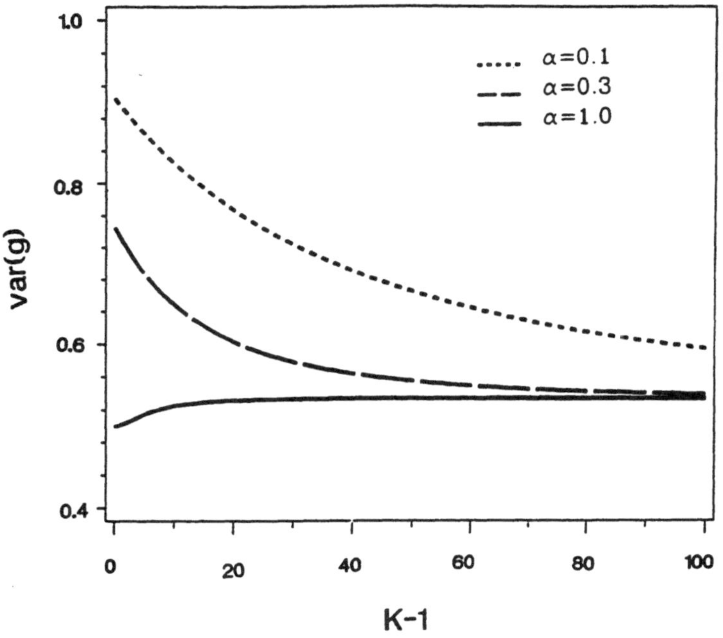

Fig. 5.3. The dependence of var(g) on $K = L_x/l$ for three values of α (see text). Taken from ref. [10].

The influence of the coupling to the open leads is even more pronounced for the autocorrelation function. This function falls off almost like a Lorentzian. Fitting it to a Lorentzian, one obtains a value for the correlation width (an energy). Figure 5.4 shows values of the width obtained in this way (dots) in comparison with both the decay width Γ and the Thouless energy E_c. The transition from one regime to the other is seen very clearly.

For a quantitative comparison with experiment, it is necessary to calculate the autocorrelation function not as a function of a difference in Fermi energies (this was done to obtain figure 5.4) but rather as a function of magnetic field strength. Here, it is necessary to account for the breaking of GOE symmetry by the magnetic field. Moreover, it is crucial for quantitative (rather than qualitative) comparison with experiment to take account of the exact geometry of the probe. This obviously applies to ring-shaped probes with external leads where the random fluctuations are superposed by Aharanov-Bohm oscillations, cf. figure 1.8. But it also applies to geometries involving a sizeable number of external leads attached to a simple wire because each new lead affects the coherence properties of the wave function. To the best of my knowledge, such a quantitative comparison has not yet been published. For the case of rings, new results [28] do indicate quantitative agreement with experiment.

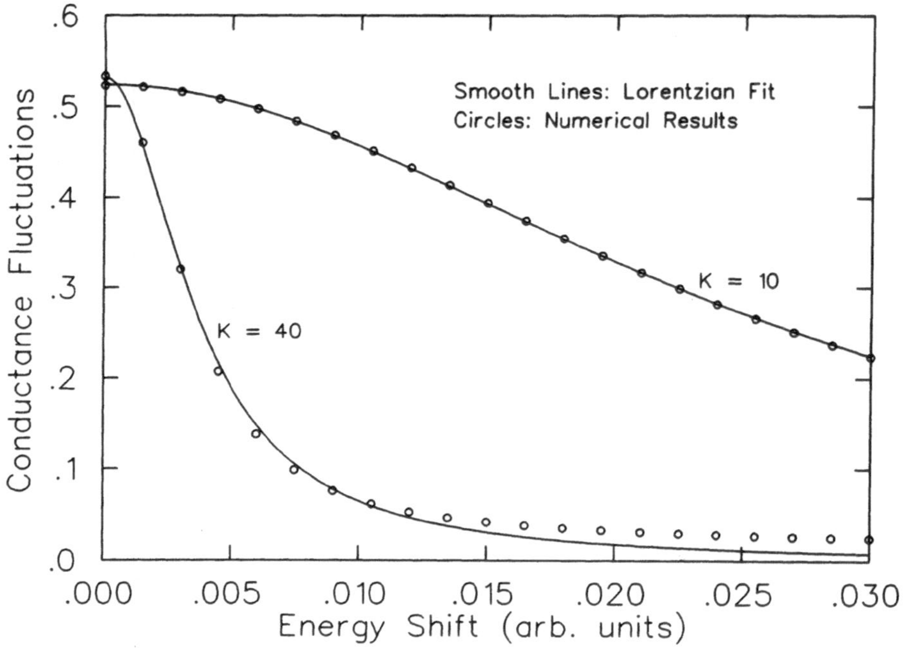

Fig. 5.4. The energy correlation width of the autocorrelation function of the conductance g versus sample length. For details, see text. Taken from ref. [11].

In concluding this secion, I wish to draw a comparison between the stochastic fluctuations of nuclear cross-sections for the case of strongly overlapping resonances ("Ericson fluctuations": figure 1.3) and the conductance fluctuations. Such a comparison deepens the understanding of both phenomena. I recall that in both cases, we deal with strongly overlapping resonances, $d/\Gamma \ll 1$, and use an asymptotic expansion in this parameter. I also recall that in the first case, the observable is a cross-section proportional to $|S_{ab}|^2$ whereas in the second, the observable is g, a sum of many such terms as given in eq. (2.5). This fact alone explains qualitatively why $\mathrm{var}(\sigma) \sim \sigma$ in fig. 1.3 while $\overline{g} \gg \mathrm{var}(g)$ for the conductance fluctuations. I finally recall that in the nuclear case, S_{ab} is modelled in terms of a single GOE (eqs. (2.1, 2.2)) while in the case of conductance fluctuations, we need a number K of GOE's with K given by the length of the disordered sample (eqs. (2.3, 2.4)).

More quantitatively, it might be expected that the elements S_{ab} of the S-matrix for compound-nucleus (CN) scattering have a Gaussian distribution, and this is indeed what theory predicts when many channels are open, and most $T_a \simeq 1$. More precisely: $S_{ab}(E)$ is a complex Gaussian random process, uncorrelated with matrix elements carrying channel indices (a', b') of which at least one differs from both a and b. In the case that $\overline{S_{ab}} = 0$ for $a \neq b$, this implies that $|S_{ab}|^2$ has an exponential distribution. This theoretical expectation is borne out by the data: A plot of the frequency of occurrence of the cross-section values in fig. 1.1, taken at fixed energy intervals, yields an exponential. It follows immediately that

$$\overline{(|S_{ab}|^2)^2} - \left(\overline{|S_{ab}|^2}\right)^2 = \left(\overline{|S_{ab}|^2}\right)^2 \quad . \tag{5.5}$$

This shows that the fluctuation $\left(\overline{\sigma_{ab}^2} - \overline{\sigma_{ab}}^2\right)^{1/2}$ equals its mean value $\overline{\sigma_{ab}}$. Modifications of this result are due to two causes. Often more than one channel contributes to the cross section at fixed angle, and direct reactions cause $\overline{S_{ab}}$ to differ from zero for $a \neq b$. The influence of both modifications on the cross section fluctuations is well understood.

These statements apply to leading order in the asymptotic expansion, i. e. whenever $\sum_a T_a \gg 1$. In cases where this condition is violated, and where terms of higher order in the asymptotic expansion must be included, there have never been enough data on fluctuation properties of CN reactions to call for an extension of the theoretical analysis, except for the case $\Gamma \ll d$ where, however, interest was focussed on the spacing distribution of CN resonances.

Turning to conductance fluctuations, we naturally expect the elements S_{ab}^{LR} of the S-matrix in eq. (2.3) also to be complex Gaussian random processes, and to leading order in the asymptotic expansion, this is indeed correct. We use this and the fact that S_{ab}^{LR} and S_{cd}^{LR} are uncorrelated (to the same order) for $(a,b) \neq (c,d)$ to work out the conductance fluctuations:

$$4\sum_{abcd} \overline{|S_{ab}^{LR}|^2\,|S_{cd}^{LR}|^2} - 4\left(\sum_{ab} \overline{|S_{ab}^{LR}|^2}\right)^2 = 4\sum_{ab}\left(\overline{|S_{ab}^{LR}|^2}\right)^2 \quad . \tag{5.6}$$

But $\overline{g} = 2\sum_{ab} \overline{|S_{ab}^{LR}|^2}$. Assuming that the contributions to this sum from all channels is about the same, we have $\overline{g} = 2\Lambda^2\overline{s^2}$. Putting this back into the right-hand side of eq. (5.6), we obtain

$$4\sum_{ab}\left(\overline{|S_{ab}^{LR}|^2}\right)^2 \simeq 4\Lambda^2\overline{s^2}^2 \simeq \frac{\overline{g}^2}{\Lambda^2} \quad . \tag{5.7}$$

Since $\overline{g} \simeq 2\Lambda$ for $\Gamma \leq E_c$, we find that the fluctuations are indeed of order unity. We also note, however, that the fluctuations given by the expression (5.7) decrease rapidly with length since $\overline{g} \sim L_x^{-1}$ in the Ohmic region, thus destroying both theoretical universality and, more importantly, the agreement with experiment.

What went wrong in the calculation of the expression (5.7)? Büttiker et al., who used the reasoning just described, gave the answer. The S-matrix elements are not Gaussian, and S_{ab} is correlated with S_{cd} for $(a,b) \neq (c,d)$. Indeed, such correlations are found when the asymptotic expansion in powers of $(\sum_{a,c} T_a^c)^{-1}$ is carried to higher orders than the first. And terms of higher order are not negligible even though $\sum_{a_c} T_a^c \gg 1$ because of the fourfold summation in eq. (5.6):

Each summation essentially contributes a term $\sum_a T_a^c \gg 1$. And it is the contribution of such higher-order terms which accounts for the universality of the conductance fluctuations in the Ohmic regime $E_c < \Gamma$. This is clearly seen in figure 5.2 where the surface terms derive partly from contributions like (5.7) while the "bulk" terms are due to the higher-order correlations.

We conclude that Ericson and conductance fluctuations, although both caused by a stochastic Hamiltonian, are manifestations of quantum chaotic scattering in very different regimes. Ericson fluctuations occur in cross section data involving few channels. They are direct manifestations of the Gaussian distribution of the S-matrix elements. Conductance fluctuations, on the other hand, are usually measured in the regime $E_c < \Gamma$ (the Ohmic regime). In this regime, the Gaussian term contributes little to the fluctuations. The universality of the fluctuations is due to higher-order correlations between S-matrix elements. They dominate the fluctuations as L_x increases and approaches the localization length yielding a universal value for the fluctuations.

The chaotic dynamics of the CN and the dynamics of disordered conductors are fundamentally similar. In particular, they give rise to similar fluctuation properties. This is the reason for the fundamental similarity between Ericson fluctuations and conductance fluctuations. Both phenomena can be viewed as random or chaotic scattering processes, governed by a scattering matrix in the domain of strongly overlapping resonances. The elements of this matrix are, to lowest order in an asymptotic expansion involving the inverse of the number of open channels, uncorrelated complex random variables with Gaussian probability distribution.

Differences arise because Ericson fluctuations are most conspicuous when only few channels contribute to the reaction, whereas the conductance is necessarily a sum over many channels. Therefore, the r. m. s. deviation of the nuclear cross section is of the order of the average cross section, while conductance fluctuations are only a small ripple on a large background.

A further difference arises because the CN is an equlibrated system while in typical mesoscopic systems, the diffusion time through the disordered sample is large compared to the emission time of the electron into the leads. Moreover, in an Ohmic situation, the fluctuations would be strongly suppressed if the elements of the S-matrix were indeed strictly Gaussian random variables. It is only through deviations from Gaussian statistics, manifest in correlations terms of higher order in the asymptotic expansion, that the fluctuations survive and, for large length scales of the mesoscopic sample, attain their universal value.

A final difference is manifest in the autocorrelation functions of nuclear cross sections, and of the conductance. In the CN case of an equilibrated system, the correlation width is given by the inverse lifetime of the system for particle emission, while in typical mesoscopic systems, it is close to the inverse diffusion time through the system.

A topic not addressed in this section but of considerable significance for the understanding of experimental data is the role of dephasing by inelastic scattering of the electrons. The inelastic mean free path L_ϕ increases with decreasing temperature. At temperatures around 100 mK or so where experiments on mesoscopic systems are typically performed, L_ϕ is usually larger than the biggest linear dimension of the sample. This is why we have modelled electron transport entirely in terms of elastic scattering. It has to be borne in mind, however, that actual data are taken on systems for which figure 2.1 is only a poor model: The ideal external leads do not in fact exist, and the coherence length of the electron's wave function is not infinite (as supposed in this model) but determined by L_ϕ. Moreover, the conductance is often investigated as a function of temperature, and here L_ϕ plays a decisive rôle. Analytical procedures for including L_ϕ in the calculations have only been developed in the last few years.

6. Persistent Currents in Mesoscopic Rings.

Novel aspects of mesoscopic physics were unravelled in an experiment by Levy et al. [29] and the theoretical discussion that came with it. I devote a lecture to this subject even though it does not fit into the frame of stochastic *scattering* theory. I do so because of two reasons. First, at zero temperature a theoretical understanding of the experiments transcends the frame of disorder perturbation theory, and requires the full machinery of Grassmann integration. Second, the experiment is not really understood even after this tour de force. This fact casts a shadow of doubt on our present understanding of mesoscopic phenomena altogether.

In the experiment, electron lithography was used to produce 10^7 isolated copper "rings" on a silicon surface of about 7 mm^2 size. Each "ring" has a circumference of about 2.2 μm, and an area of about 0.3 μm^2. (The "rings" were actually squares with a square-shaped hole in the center. This fact is immaterial for what follows; we shall omit the quotation marks in the sequel.) Each ring carries about 10^{10} electrons. Data are taken at temperatures in the range of 100 mK. It is known that for the material at hand, the inelastic mean free path $L_\phi > L$ at $T = 1.5$ K, so that dephasing is of no concern at temperatures of a fraction of a K. The elastic mean free path l due to impurities, on the other hand, is ≈ 200 Å, so that $L/l \simeq 10^2 \gg 1$. A slowly time-dependent magnetic field perpendicular to the surface and with amplitude ≈ 100 G induces in each ring a persistent current. The sum of these currents is measured in terms of the induced magnetic dipole moment of the sample. It yields a dipole moment per ring which is about 15 Bohr magnetons. In this lecture, I address the question: Can this experiment be understood on the basis of the standard model for mesoscopic systems, i. e. in terms of the model of electrons moving independently in a white-noise potential? (Or in terms of a number of Gaussian ensembles with nearest-neighbour coupling as modelled in section 2). To approach the answer and attain an understanding of the mechanism that causes persistent currents, I begin with ideal rings (no impurities).

An ideal normal metal ring threaded by a magnetic flux carries an equilibrium current as long as the phase coherence of the electron's wave function is preserved. This fact, emphasized as early as 1983 by Büttiker, Imry and Landauer [30], was investigated in detail by Gefen, Riedel and collaborators [31]. I consider for simplicity here the case of zero temperature, and the case of a one-dimensional ideal ring.

A magnetic flux ϕ threading the ring imposes on the single-particle wave function $\chi_j(x)$ (with x a length along the cricumference) the condition

$$\chi_j(L) = \chi_j(0)\,\exp(2i\pi\phi/\phi_0) \tag{6.1}$$

where $\phi_0 = hc/e$ is the elementary flux quantum. Eq. (6.1) follows easily from gauge invariance. With $\chi_j(x)$, all observables of the system are periodic functions of ϕ with period ϕ_0.

At zero temperature, the induced persistent current $I(\phi)$ is simply obtained by summing the contributions from the N (lowest) occupied states. With $\epsilon_j(\phi)$ the single-particle energy in level j, we have

$$I(\phi) = \frac{\text{charge}}{\text{time}} = -2c\sum_{j=1}^{N}\frac{\partial\epsilon_j(\phi)}{\partial\phi} \quad . \tag{6.2}$$

The factor 2 is due to spin. Eqs. (6.1) and (6.2) show that every single-particle state carries a persistent current. The magnitude of the total current can be estimated from the first eq. (6.2). With a time for the electron to once pass around the ring given by L/v_F (where v_F is the Fermi velocity), we arrive at

$$I \simeq 2ev_F/L \quad . \tag{6.3}$$

This estimate is in keeping with detailed calculations. For the latter, it is important to realize that $\phi = 0$ and $\phi = \phi_0/2$ are special values of the magnetic flux, for the following reasons. For $\phi = 0$, the condition (6.1) means periodicity. The ground state is non-degenerate (χ_0 = constant), all excited states are twofold degenerate (both sin and cos are independent solutions). Near $\phi = 0$, the spectrum has a shape as indicated on the left-hand side of figure 6.1. We conclude that when the number N of electrons on the ring is odd, the contributions to the sum in eq. (6.2) near $\phi = 0$ nearly cancel, and I is small. If, on the other hand, N is even, the contribution from the last occupied state jumps discontinuously at $\phi = 0$, and this contribution is big in comparison to all the others. For ϕ near $\phi_0/2$, the argument is the same but for an interchange of even and odd. This is because the condition $\chi_j(L) = -\chi_j(0)$ cannot be met by a constant and the lowest state is twofold degenerate like all the rest (right-hand side of figure 6.1).

It is not known how many electrons are located on any one ring. Therefore, it is meaningful to average the theoretical calculations for ideal rings over N even and N odd. The result is a sawtooth curve which is periodic in $\phi_0/2$ and jumps discontinuously at $\phi = 0, \pm\frac{1}{2}\phi_0$. Except for a factor $\frac{1}{2}$ due to averaging, the amplitude of the jump is given by the estimate (6.3). This result carries

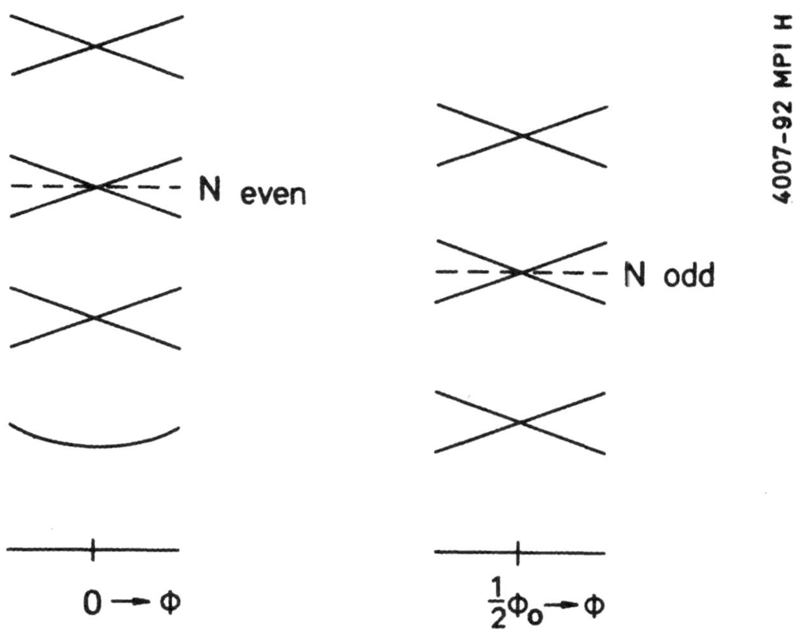

Fig. 6.1. Eigenvalues of an ideal ring versus flux ϕ (schematic).

over to ideal two-dimensional rings (in the calculation, one disregards the fact that the flux penetrates not only the interior but also the body of the ring).

How does random impurity scattering modify the result for ideal rings? This question was investigated perturbatively in a series of papers [32]. Before giving a quantitative answer, I describe some qualitative arguments, and computer simulations.

We expect an impurity potential to lift the degeneracies of the eigenvalues shown in fig. 6.1, and thereby to smooth the discontinuity of the sawtooth-shaped curve for the current which results from the degeneracy. Moreover, we expect impurity scattering to weaken the coherence properties of the wave function which, by virtue of the boundary condition (6.1), are necessary to produce a persistent current. A qualitative estimate for the current in the presence of impurity scattering relies on the argument leading to eq. (6.3) but replaces [29] the time L/v_F by the diffusion time L^2/D with D the diffusion constant, $D = \frac{1}{3}v_F L$ for one-dimensional systems. This yields

$$I \approx (2ev_F/L) \cdot \left(\frac{1}{3}\frac{l}{L}\right) \quad . \tag{6.4}$$

This estimate is smaller than (6.3) by a factor $\frac{1}{3}l/L \simeq (300)^{-1}$ and yields about 50 Bohr magnetons per ring, roughly a factor 3 more than the measured value [29].

Unfortunately, a more quantitative calculation yields a significantly smaller value and disagrees with experiment [32]. It is illuminating for the special features of such mesoscopic systems to describe the problems of the calculation in some detail.

Going back to the right-hand side of eq. (6.2), I write the current as

$$I = -2c \int_0^{\mu} \mathrm{d}E' \left[\sum_{j=1}^{\infty} \frac{\partial \epsilon_j}{\partial \phi} \delta(E' - \epsilon_j) \right] \quad . \tag{6.5}$$

Here, μ is the chemical potential. The form (6.5) is useful because the content of the angular bracket can be expressed as the ϕ-derivative of a Green function. In this way, the connection to disorder perturbation theory or the method of the generating functional is established, and the ensemble average of I can be evaluated. The result is

$$\overline{I} \simeq \exp(-L/l) \tag{6.6}$$

which is negligbly small [33]. The reason is interesting and instructive: It lies in the use of the grand canonical ensemble, or in the use of a fixed chemical potential μ. As ϕ varies, the single-particle energies $\epsilon_j(\phi)$ change, and it will happen that the largest $\epsilon_j(\phi)$ which for $\phi = 0$ lies below μ crosses μ from below as ϕ increases, or conversely that the smallest $\epsilon_j(\phi)$ which lies above μ for $\phi = 0$ crosses μ from above as ϕ increases. In either case, we effectively change the number of electrons on the ring, from N to $N-1$ and from N to $N+1$, respectively. And we remember that for ideal rings, the contribution of the last occupied orbital is decisive for the value of the induced current. As we average the expression (6.2) over the ensemble, such a crossing of the $E = \mu$ line will happen at virtually every value of ϕ, and the result is an exponentially small average current.

This difficulty was overcome in refs. [32] allwing μ to depend on ϕ and by treating this ϕ-dependent term in μ as a small perturbation. An equivalent expression was developed later [34] using a different approach. I prefer to describe the latter because I find it more intuitive, and because it allows to identify the small parameter with respect to which the expansion in refs. [32] is made.

We start with the observation that for fixed impurity realization, the number of levels $\epsilon_j(\phi)$ which intersect the line $E = \mu$ for $0 \le \phi \le \phi_0$ is of order unity. This follows from numerical simulations [35]. It also follows from theoretical arguments. Indeed, changing ϕ is analogous to changing the deformation of a nucleus. In the latter case, such a change produces "shell effects", i. e. systematic oscillations of the Fermi surface. The size and shape of such oscillations can be calculated semiclassically. Applying the same idea to the present problem, one finds that the shell effect is of the order of the mean level spacing.

The experiment uses $\approx 10^7$ rings. Although produced under identical circumstances, these rings are, of course, not completely identical in size and shape.

It seems reasonable to guess that they differ on the level of one part in 10^3 or so. This implies that not all rings carry the same number of electrons (about 10^{10}), and that this number also varies in the per mille region, i. e. by about 10^4. To calculate the observed current, it is reasonable to *average* over this number. In other words, we average the expression (6.2) over N, allowing N to range from $N_0 + 1 = 10^{10}$ to $N_0 + K$ with $K \simeq 10^4$.

It is useful to realize that in this way, we perform a double average: An ensemble average over the distribution of impurities for fixed N involving about 10^6 rings, and an average over N involving about 10^4 rings. Using the steps which lead from eq. (6.2) to eq. (6.5), we arrive at

$$\overline{I}(\phi) = -2c\overline{\frac{1}{K}\sum_{N=N_0+1}^{N_0+K}\sum_{j=1}^{N}\frac{\partial\epsilon_j}{\partial\phi}} = \tag{6.7}$$

$$-2cK^{-1}\lim_{\delta\to 0^+}\overline{\int_S \mathrm{d}E\int_0^E \mathrm{d}E'\left[\left\{\sum_{j=1}^{\infty}\frac{\partial\epsilon_j}{\partial\phi}\delta(E'-\epsilon_j)\right\}\left\{\sum_{N=1}^{\infty}\delta(E-\epsilon_N-\delta)\right\}\right]}\,.$$

The energy interval S is chosen in such a way that it contains the K consecutive eigenvalues $\epsilon_{N_0+1},\ldots,\epsilon_{N_0+K}$ but no other ones. As it stands, we still cannot perform the ensemble average over the r. h. s. of eq. (6.7) because the position and length of the interval S change with both ϕ and the impurity realization. But now we use the fact that for fixed impurity realization, S changes only by one or two mean level spacings d as ϕ varies from 0 to ϕ_0. Neglecting this change induces an error of order $K^{-1} \simeq 10^{-4}$ and is permissible. And choosing the *same* interval S for different impurity realizations means that we choose different N_0 values. But we have every reason to expect that the average current does not depend on N_0. In summary, we take S fixed and approximate eq. (6.7) by

$$\overline{I}(\phi) = \tag{6.8}$$

$$-2cK^{-1}\lim_{\delta\to 0^+}\int_S \mathrm{d}E\int_0^E \mathrm{d}E'\left[\overline{\left\{\sum_{j=1}^{\infty}\frac{\partial\epsilon_j}{\partial\phi}\delta(E'-\epsilon_j)\right\}\left\{\sum_{N=1}^{\infty}\delta(E-\epsilon_N-\delta)\right\}}\right]\,.$$

The content of the square bracket can be written as the product of a Green function and of the derivative of a Green function; this allows the calculation of the ensemble average in the usual way without yielding an exponentially small result like (6.3). Moreover, the average is independent of $E + E'$, dependent only on $E - E'$, and falls off exponentially for large $E - E'$, so that eq. (6.8) takes the form

$$\overline{I}(\phi) \simeq -2cd\int_0^{\infty}\mathrm{d}\epsilon\,\overline{\left\{\sum_{j=1}^{\infty}\frac{\partial\epsilon_j}{\partial\phi}\delta(E'-\epsilon_j)\right\}\left\{\sum_{N=1}^{\infty}\delta(E-\epsilon_N)\right\}}\Bigg|_{\epsilon=E-E'}\,. \tag{6.9}$$

This form of $\overline{I}$ is equivalent to the starting point of refs. [32]. We see that K^{-1} plays the role of the small parameter.

Using disorder perturbation theory to calculate the expression (6.9), the authors of refs. [32] arrived at a disappointing result: After introducing a cutoff to cure the divergence of the perturbation expression (this cutoff is justified because the experiment is done at finite temperature $T > d$), the authors find an average current which is 1.5 to 2 orders of magnitude smaller than the experimental value.

A systematic investigation of the model of independent electrons in a random white-noise potential, moving at zero temperature in a ring threaded by a flux ϕ corroborates this result [34]. In this case, perturbation theory is inadequate but a (nearly) exact calculation is possible. This calculation yields an average current which is antisymmetric about $\phi = 0$ and periodic in ϕ with period $\phi_0/2$. $\overline{I}$ vanishes at $\phi = 0$ and at $\phi = \frac{1}{4}\phi_0$, rises steeply near $\phi = 0$ with slope $4E_c/\phi_0^2$, and reaches a maximum value given by $\overline{I}_{max} \simeq 0.3\,(E_c d)^{1/2}/\phi_0$ at $\phi \simeq 0.16\left(\frac{d}{E_c}\right)^{1/2}\phi_0$, and then falls off gently to its zero at $\phi_0/4$. The maximum value is again too small by 1.5 to 2 orders of magnitude. At nonzero temperature, the value of $\overline{I}$ is expected to be decreased further.

What is the cause of this failure of theory? There has been much speculation that the model of *independent* electrons must be replaced by a model of *interacting* electrons, but no results are available. Looking back at the calculations reported in section 5, we are led to wonder why in the case of conductance fluctuations the model of independent electrons seems to work, while it fails for isolated rings. In any case, many questions are open in this exciting field of physics.

7. Conclusions

We have identified two causes which justify a stochastic modelling of physical systems: Chaos and disorder. In chaotic systems, stochastic modelling is expected to yield the generic features which relate to sufficiently long time scales, and uses input information (average S-matrix elements etc.) which relates to short time scales. The separation between the two kinds of scales is expected to become ever wider as the number of degrees of freedom of the system increases, and the usefulness of other approaches (semiclassical theory) is expected to decrease correspondingly. For disordered systems, stochastic modelling is anyway the only theoretical tool available.

In classical physics, stochastic fluctuations are usually associated with the coupling to a heat bath. In these lectures, we have considered small and isolated systems. Here, the fluctuations are due to the intrinsic stochasticity of the systems. In principle they are deterministic and reproducible, but from a practical point of view they are unpredictable and random. Stochastic modelling reproduces the characteristic features of fluctuations with a small number of input parameters. The latter are found semiclassically from the dynamics of the system.

The theme dominating the entire discussion has been: Wave coherence survives random elastic scattering. How and to what degree it does so accounts for the phenomena observed in such diverse fields as microwave chaotic scattering, compound nucleus scattering, resonance fluorescence scattering of light by complex molecules, passage of light through media with a random index of refraction, and conductance properties of mesoscopic systems. All these phenomena are but different aspects of an emerging new discipline: The statistical mechanics of small systems. Section 6 has shown one example (of many) for the statement that this field contains important open questions.

Acknowledgements

I am grateful to Prof. D. Heiss for inviting me to the Bleydepoort Summer School and for the opportunity to deliver these lectures in a congenial atmosphere and pleasant surroundings.

References

[1]H. J. Stöckmann and J. Stein, Phys. Rev. Lett. **64** (1990) 2215
[2]E. Doron, U. Smilansky, A. Frenkel, Phys. Rev. Lett. **65** (1990) 3072
[3]A. Richter, private communication
[4]Aa. Bohr and B. R. Mottelson, *Nuclear Structure II*, W. A. Benjamin, Reading (1975)
[5]T. Ericson and T. Mayer-Kuckuk, Ann. Rev. Nucl. Sci. **16** (1966) 183
[6]R. Jost and M. Lombardi, Lecture Notes in Physics **263** (1986) 133
[7]T. Guhr and H. A. Weidenmüller, Chem. Phys. **146** (1990) 21
[8]J. Freund et al., Phys. Rev. Lett. **61** (1988) 2328
[9]S. Washburn and R. A. Webb, Adv. Phys. **35** (1986) 375
[10]S. Iida, H. A. Weidenmüller and J. Zuk, Ann. Phys. **200** (1990) 219
[11]A. Altland, Z. Phys. B (in press)
[12]H. Nishioka. J. J. M. Verbaarschot, H. A. Weidenmüller and S. Yoshida, Ann. Phys. (N. Y.)**172** (1986) 67
[13]J. D. Bowman et al., Phys. Rev. Lett. **65** (1990) 1192; C. M. Frankle et al., ibid **67** (1991) 564.
[14]K. Binder and D. W. Heermann, *Monte Carlo Simulation in Statistical Physics. An Introduction*. Springer Series in Solid State Sciences Vol. 80. Springer, Berlin 1988
[15]A. A. Abrikosov and L. P. Gor'kov, Sov. Phys. JETP (engl. transl.) **8** (1958) 1090
[16]K. B. Efetov, Adv. in Phys. **32** (1983) 53
[17]J. J. M. Verbaarschot. H. A. Weidenmüller and M. R. Zirnbauer, Phys. Rep. **129** (1985) 367
[18]C. Mahaux and H. A. Weidenmüller, Ann. Rev. Nucl. Part. Science **29** (1979) 1
[19]U. Smilansky, in *Chaos and Quantum Physics* (M.-J. Giannoni, A. Voros and J. Zinn-Justin, editors), Elsevier Science, New York 1990
[20]C. Lewenkopf and H. A. Weidenmüller, Ann. Phys. (N. Y.) **212** (1991) 53

[21]B. L. Altshuler and B. I. Shklovskii, Sov. Phys. JETP (engl. transl.) **64** (1986) 127
[22]B. L. Altshuler, V. E. Kravtsov and I. Lerner, Sov. Phys. JETP (engl. transl.) **64** (1986) 1352
[23]Y. Imry, Europhys. Lett. 1 (1986) 249
[24]P. A. Lee, A. D. Stone and H. Fukuyama, Phys. Rev. B **35** (1987) 1039
[25]K. A. Muttalib, J.-L. Pichard and A. D. Stone, Phys. Rev. Lett. **59** (1987) 2475
[26]P. A. Mello, Phys. Rev. Lett. **60** (1988) 1089
[27]M. R. Zirnbauer, University of Köln preprint (1991)
[28]A. Müller-Groeling, Ph. D. Thesis, Heidelberg (1992) and to be published
[29]L. P. Levy et al., Phys. Rev. Lett. **64** (1990) 2074
[30]M. Büttiker, Y. Imry, and R. Landauer, Phys. Lett. **96** A (1983) 365
[31]H. F. Cheung et al., Phys. Rev. B **37** (1988) 6050
[32]B. L. Altshuler, Y. Gefen and Y. Imry, Phys. Rev. Lett. **66** (1991) 88; A. Schmid, ibid. 80; F. von Oppen, E. K. Riedel, ibid. 84
[33]H. F. Cheung, E. Riedel and Y. Gefen, Phys. Rev. Lett. **62** (1989) 587
[34]A. Altland, S. Iida, A. Müller-Groeling and H. A. Weidenmüller, to be published
[35]H. Bouchiat and B. Montambaux, J. Phys. (Paris) **50** (1989) 2695

Atomic and Molecular Physics Experiments in Quantum Chaology

by
Peter M. Koch
Physics Department
State University of New York at Stony Brook
Stony Brook, NY 11794-3800 USA
internet: pkoch@ccmail.sunysb.edu

To be published in the proceedings of Eighth South African Summer School in Theoretical Physics: **Chaos and Quantum Chaos**
(13-24 January 1992, Blydepoort, Eastern Transvaal, Republic of South Africa)
in the Springer-Verlag *Lecture Notes in Physics* series

Abstract

Several systems open to experimental study in atomic and molecular physics are furnishing important information about the behavior of a quantal system whose classical counterpart exhibits a transition to chaos. Some of these systems are autonomous (time-independent), and one gains information through detailed studies of their photoabsorption or stimulated-emission spectra. Others of these systems are time-dependent, being strongly driven by an external driving field. The aim of this paper is to review from the point of view of an actively involved experimenter our present understanding of the time-dependent process of the excitation and ionization of excited hydrogen atoms by a strong, linearly polarized microwave electric field. Many comparisons of experimental data with classical and quantal calculations have revealed six different regimes of dynamical behavior in this periodically driven system. Certain scaling relationships present in the classical Hamiltonian dynamics for the driven Kepler system are found to be at least approximately obeyed in the corresponding quantal system, sometimes in surprising ways. Exploiting these scaling relationships has greatly facilitated the interpretation of experimental data, which reveal a number of fascinating phenomena.

1 Introduction

This paper contains some of what was covered by the author in lectures at the Eighth South African Summer School in Theoretical Physics during 13–24 January 1992, at the FH Odendaal Resort at Blydepoort in the Eastern Transvaal. With the conference/resort site overlooking the Blyde River canyon, this was a stirring venue for a school devoted to *Chaos and Quantum Chaos*. The author is indebted to the Organizing Committee of the summer school, in particular its Chairman Professor W.D. Heiss, for their organization of the school, warm hospitality, and opportunity to visit some of their country during its present period of change and anticipation.

The author's experience at this summer school was similar that at to other conferences and workshops on the subject of chaos and quantum chaos in the following way. Such meetings are typically dominated by theorists, with usually only a small percentage of experimenters. If his count was correct, the author was one of only two or three experimenters present out of all the lecturers, organizers, students, or other participants. The organizers requested that he speak on *Observational Evidence of Chaos in Atomic and Molecular Systems*. Let us consider the challenges this poses. Even though most physicists, the author included, will admit to not understanding it deeply, nearly all of us believe in (nonrelativistic) quantum mechanics as the guiding theory for (nonrelativistic) physical phenomena in the microscopic world. (You know the old joke: "Only seven people in the world understand quantum mechanics, and I'm not one of them!") Maybe Professor Joe Ford has a different point of view, thinking he understands it well enough to say repeatedly in talks and in print that conventional quantum mechanics is flawed. Boiling down his arguments presented in Ref. [48], in which algorithmic complexity theory applied to a certain model system figures prominently, it goes something like this: *(i)* generic classical systems (such as the Arnol'd cat map model system in the cited paper) are chaotic; *(ii)* the quantal treatment of such a system is not chaotic; *(iii)* comparing *(i)* and *(ii)* gives us a result at odds with what we expect from the correspondence principle.

The present author was asked to give "observational evidence" for chaos in sytems from the microscopic world that we all know to be quantal. Others assert, even prove subject to certain conditions, that chaos is impossible in quantum systems. Presumably the reader is now recognizing the problem facing a lecturer on this subject. This lecturer will emphasize, moreover, that nowhere attached to his atomic physics apparatus will one find a "chaos meter". Lots of power supplies, vacuum pumps, lasers, all sorts of electronics, but NO chaos meter. So discussions in his laboratory certainly don't sound like, "Well, today's been a good day: we measured $34 \pm 5\%$ chaos. Let's write it up and send it in!"

A subject that is constantly discussed in his Stony Brook laboratory is how to understand and interpret a long succession of experimental results on a simple atomic system, excited hydrogen atoms exposed to a strong oscillatory electric field (microwaves). Stong here means that the external force on an electron is comparable to the Coulomb force binding it to its partner proton, indeed, strong enough for their bound system to be ripped apart. This *microwave ionization* phenomenon [7,8] was an accidental experimental discovery [83] during the author's Ph.D. dissertation research with Jim Bayfield nearly two decades ago, and many of its details are still not understood.

However, thanks to lots of experimental work, to theoretical work both classical and quantal and both numerical and analytical, and to comparisons between them, enormous progress progress has been made. A classical treatment of the hydrogen atom strongly driven periodically reveals chaotic dynamics at play, if only transiently [127], though there is no classical analytical solution to the problem. Rather, the insights come from approximate treatments, including iterated maps, and from numerical integration of the classical (Newton's or Hamilton's, say) equations of motion. All these give insights into the complicated dynamics of this driven Kepler system in a phase space divided into regular and irregular regions.

Nor can the nonrelativistic quantal treatment, so easily formulated with the Schrödinger equation, be solved analytically. Some approximate treatments have been worked out, including iterated quantum maps, but much effort has been spent on large numerical calculations. In all these quantal treatments, there has been nary a sign of any chaos. Thus, if there isn't any, one must be careful when talking about *quantum chaos*. Most practicioners in the field that quantum chaos is about the study of quantal systems whose classical counterparts are chaotic. It goes without saying that this involves a deep understanding of both the classical and quantal systems. In the author's opinion, the great theoretical insights in such problems come when one can mount a successful semiclassical attack on the problem. This inevitably involves using the classical dynamics as a foundation upon which one builds an approximate version of the quantal dynamics.

Berry's opinion on this subject was even presented in the title of his paper [13]: *Quantum Chaology, Not Quantum Chaos.* His presented definition of the subject is "Quantum chaology is the study of semiclassical, but nonclassical, phenomena characteristic of systems whose classical counterparts exhibit chaos." This will do just fine for all that follows in this paper. In the problem of microwave ionization of excited hydrogen atoms, the reader will find, as has the author, his research group, his theoretical collaborators, and other interested parties, that this particular system is replete with semiclassical phenomena. The key word here is semiclassical, meaning that classical actions (assuming for now that they can be computed) are large compared to Planck's quantum of action h, but not infinitely large. Inverting the comparison and perverting the mathematical language, letting h approach zero (Remember: it's a constant. One really means letting actions get large compared to it.) probes the semiclassical limit of a problem, which can be entirely different from its classical limit, when h=0. Going from one case to the other involves nonalyticity, so hiding in this not-understood limiting process, and in the other limit that must be taken, $t \to \infty$, is all the interesting physics. As is emphasized in [15], these two limits do not commute for classically chaotic systems.

In atomic and molecular physics (AMP), one has systems of two or more particles bound by Coulomb forces. These forces are important precisely because they are well understood *and* can be written down. (One cannot do this exactly in nuclear physics, where strong forces are involved.) Moreover, one has lots of experience with high-resolution spectroscopy in AMP, particularly since the advent of the laser as a routine laboratory tool. Its high spectral brightness, narrow linewidth, and, in some cases, continuous or step-tunability, make it the ideal spectrscopic tool for probing quantum systems with a high density of states or for producing atoms or molecules excited into such states for use in further experimentation.

Spectroscopic studies of autonomous (time-independent) systems are part of the research on quantum chaology in AMP. For a useful, recent compendium of articles on these autonomous systems, on some time-dependent systems, and on related model systems, the reader should consult *Irregular Atomic Systems and Quantum Chaos* [53], which is an important collection of articles edited by J.-C. Gay.

Before we briefly discuss these spectroscopic studies, however, it is important to understand a feature they all share, which has both advantages and disadvantages. When one uses a laser to drive transitions out of a given initial state to a dense region of final states, all of the energetically accessible final states are not sampled. Only those whose electric dipole transition matrix element with the initial state is large enough will be observed. If the initial state is tightly bound and, therefore, localized near the nucleus, only final states having non-negligible wavefunction density near the nucleus will be sampled. In such a situation, a spectrum would contain no (direct) information about final states having no density near the nucleus. Semiclassically, one would be sensitive only to orbits that pass near the nucleus and insensitive to orbits that never do so.

1.1 The Diamagnetic Kepler Problem

In atomic physics, much attention has and is being devoted to the diamagnetic Kepler problem, also called the quadratic Zeeman effect, which involves the hydrogen atom in a strong static magnetic field $\vec{B}$. With use of laser excitation into high-lying states only weakly bound by the Coulomb force, one may readily study quantal states for which the diamagnetic potential, proportional to B^2, is comparable to Coulomb potential. In this regime, a hydrogen atom treated classically is a two degree-of-freedom system exhibiting a transition to chaos. A few experimental groups, most notably that of Professor Karl Welge in Bielefeld and that of Professor Dan Kleppner at MIT, have been mining the wealth of this spectroscopic problem, but one should point out that their experiments, as well as those in other groups of which this author is aware, have used tightly bound initial states. Therefore, the comments in the preceeding paragraph apply: These experiments have not yet probed the full (enormous!) density of states near the ionization limit.

The former group has faced the challenge of developing ultraviolet pulsed laser techniques for studies with real hydrogen atoms. The latter group decided to work with what they call "poor man's hydrogen", that is, lithium atoms in highly excited, or Rydberg, states. The other two electrons in a Li Rydberg atom make up a spherically symmetric, spinless "core" that perturbs the outer Rydberg electron in a relatively benign way that theory (quantum defect theory) is reasonably well able to handle. Use of Li atoms permitted Kleppner's group to use tunable, continous-wave dye lasers, the resolution of which is much higher than in the pulsed tunable lasers used in the Bielefeld experiments.

Kleppner's group has obtained well-resolved negative and positive energy photoabsorption spectra of Li Rydberg atoms in a 6.113 Tesla field. (E=0 is the one-electron ionization limit.) These are conditions for which the corresponding classical hydrogen atom is well above the transition to chaos. How well does quantum mechanics solve the spectral problem? The interested reader should compare the experimental spectrum (for Li) and computed spectrum (for hydrogen) presented in Fig. 1 of Ref. [65] to see that quantum mechanics does very well, indeed. Delande, Bommier, and Gay carried out state-of-the-art numerical computations in Paris, and the agreement between their calculations and the MIT experiment of Iu *et al.* can only be described as excellent.

Though the photoabsorption spectrum has been measured and calculated numerically however, it has not yet been understood in depth. It is full of both broad and sharp features, including Rydberg-like series of lines associated with the combined effect of the Coulomb electrostatic and external magnetic fields, and even including sharp features above $E=0$ that signal the presence of long-lived (*i.e.*, slowly decaying) states well into the region where the classical phase space is nearly entirely chaotic. This author believes that the diamagnetic Kepler problem will not be understood until one has a clear semiclassical picture of what produces the rich array of spectral features.

Welge's group has emphasized the power of Fourier transforming their swept-laser-frequency photoabsorption spectra out of the 2p state of hydrogen in a strong magnetic field. The transforms produced peaks at certain times, which they and others have showed to be associated with classical periodic orbits or even just closed orbits that return near the nucleus. This group later exploited the use of previously found classical scaling relations for the diamagnetic Kepler problem and carried out scaled-energy spectroscopic experiments for this system. Fourier transforming these scaled spectra produced an even simpler array of peaks in the conjugate variable. Many, but not all, of these peaks could be associated with families of classical periodic orbits that they classfied into primary vibrators, primary rotators, or "exotics". A recent short review article [100] by Welge's group covers their work and gives many useful references.

We might mention that their use of scaled variables to motivate and interpret experiments resonates strongly with our long use of scaled variables in experiments on another strongly perturbed Kepler problem, the microwave ionization of hydrogen atoms, see Sec. 3. Use of scaling relations in another Kepler problem, the helium atom, has been crucial to enormous recent progress in the semiclassical treatment of this three-body problem, see Sec. 1.3 below.

For the diamagnetic Kepler problem, one may find discussions of the above-mentioned scaling relations in two recent reviews [49,61], and a book [59]. All three references contain some discussion of quantal solutions for "scarred" and other wavefunctions, and the first two discuss statistical properties of spectra and their relationship with random matrix theory. The latter subject is treated generally in some depth in articles elsewhere in this volume.

1.2 Spectroscopy of Highly Excited Polyatomic Molecules

In molecular physics studies related to quantum chaology, much the effort has been directed toward understanding the spectroscopy of dense, high-lying vibrational levels of polyatomic molecules such as acetylene (C_2H_2), the sodium-trimer (Na_3), and others. Most of the spectroscopic experiments have exploited the powerful technique of stimulated-emission pumping (SEP) by tunable lasers. The articles in a special topical issue *Molecular Spectroscopy and Dyamics by Stimulated-Emission Pumping* of the J. Opt. Soc. Am. 7, September 1990, are a useful collected source of information on this subject. One usually wants to know how many (approximately) good quantum numbers there are in a given system and what they are. This is usually done by trying to "classify" measured spectra to see if they fit *e.g.*, rotational sequences, vibrational sequences, *etc.* If reasonably pure sequences of levels can be found in portions of spectra, then one reckons that the different modes of motion (in a classical sense) are not strongly coupled, for example, different vibrations and rotations.

If one cannot find pure sequences in well-resolved spectra belonging to a known symmetry class, then one usually reckons that two or more modes of motion are strongly coupled and good quantum numbers are not to be found associated with them. Some molecular spectroscopy groups have become adept at applying statistical measures to their dense, but resolved, SEP spectra in order to see if techniques developed in random matrix theory can be helpful. Some times they are useful, but other instances have led to puzzling conclusions, with apparent lack of pure sequences in some spectra indicating strong coupling of vibrational modes, but the statistical properties of the same spectra tending to indicate the opposite behavior. For an example of this puzzling behavior in the spectroscopy of high-lying vibrational levels in acetylene, see Ref. [41].

Recent progress has also been made in the production [34] and computation using classical methods [55] of SEP spectra for the sodium-trimer.

Another useful article concerned with how to extract dynamical information from spectra of molecules that are classically chaotic, including use of Fourier-transform and other techniques, is Ref. [98].

1.3 The Helium Atom

A celebrated failure of the old quantum theory, and one that spurred the devlopment of the "new" quantum mechanics in the mid 1920's, was the quantization of the helium atom. This is a three-body problem bound by Coulomb forces, so from the classical orbital mechanics point of view, it certainly suggests chaos. A classical treatment of the allowed motions in this system is a challenging task, because it lives in a high-dimensional phase space divided into irregular regions of fully chaotic motion, regions of regular motion, and mixtures of the two.

Particularly challenging for both experimental and quantal treatment are doubly-excited states lying energetically above the first ionization limit. These decay by autoionization, or one electron taking most of the internal energy and whizzing off to infinity, at the expense of the other one that drops into a lower-lying state to conserve total energy. Much progress has been made during the past few decades in the ability to calculate the positions and decay widths for these levels with use of "conventional" quantal methods. However, there has been continuous evolution in the debate about how these levels are organized into classes and families and how different "motions", in a classical sense, are responsible for this organization. For example, one popular view has been that the so-called Wannier-ridge mode (or motion on the Wannier saddle potential energy surface), in which the two electrons undergo an in-phase symmetric stretch motion, was responsible for organizing certain families of resonances seen experimentally and found in large quantal calculations.

Thus, it is coming as a bit of a shock to some, but as an impressive piece of work to all, when during the last year or two the semiclassical quantization of the doubly-excited helium atom was accomplished. Ref. [131] gives an account of the recent work of Wintgen and co-workers in this regard, as well as references to the contributions of other workers.

What has this work accoamplished thus far? For one, it was shown, though not rigorously proved, that the symmetric stretch, or Wannier-ridge, mode is fully chaotic classically and, hence, does not support the families of quasi-bound resonant states that others had previously associated with this mode. Rather, the near-collinear intra-shell resonances were shown to be associated with the (fundamental) asymmetric stretch motion of the electron pair. However, it was possible to apply the formidable tools of semiclassical quantization of "hard chaos" systems to find quantal resonances associated with the Wannier mode. These tools included symbolic dynamics, the Gutzwiller trace formula, and cycle expansions.

For another, this work showed the existence of fully stable classical motion around the collinear mode with both electrons on the *same* side of the nucleus. Note that this configuration *maximizes* rather than minimizes the potential energy of repulsion between the electrons! Moreover, approximate torus quantization on the island structures in the classical phase space was carried out and accurate predictions made for the position of the associated quantum mechanical resonances. Such exotic resonant states have not (yet) been observed experimentally, but this semiclassical theoretical work is certainly stimulating an experimental search.

What is important about this new theoretical work is the semiclassical approach giving keen insight into the dynamical organization of whole families of quantal resonances. When one is able to visualize the classical motions, one can begin to understand the structure of wavefunctions and gain intuition about the quantum dynamics. This is a point that was already made earlier in this introduction and that will continue to apply throughout this paper.

1.4 Swift Ions Traversing Foils

One branch of atomic physics (that began with Rutherford) uses accelerated ions and studies their interactions with gas targets and solid targets such as foils. If the projectile ions are not fully stripped, they carry their own electrons into the target, and having kinetic energies in the many keV or MeV range, far above the binding energy of electrons in the target constituent(s), they also liberate electrons from them. Though the foils are mechanically very thin, to the ions and electrons penetrating them, they are thick. That is, the swift ions and electrons undergo many collisions (plural scattering) inside the foil and can achieve therein what is called charge-state equilibrium: The charge state (number of attached electrons) of the ion emerging from the foil is independent of the incident charge state.

Because gas targets can trivially be made so thin that the probability is high only for single collisions, it should not be surprising that there are important differences between, say, electron emission spectra for ions traversing foils *versus* the same ions traversing a thin gas target. For example, it has been observed [133] that the angular momentum or ℓ-distribution of highly excited projectile states emerging from a foil is much flatter with increasing ℓ than the distribution obtained with a thin gas target.

This observation was first clearly explained [36] with use of calculations based on classical transport theory. The assumed model is relatively simple. A swift ion with net positive charge enters a solid, amorphous (so channeling may be neglected) foil and while traveling through it distorts the medium. Inside the medium nearby particles see not only the radially symmetric Coulomb electrostatic field of the ion, but also a so-called wake potential. Electrons can be carried along with the ion, either those brought into the foil with it or those captured from within the foil. In the moving projectile frame, the electrons and ion are constantly being pelted with particles (ions and electrons) from the foil medium, much as one running through a falling rain is pelted with raindrops. Some of the collisions are elastic, and others are elastic. For all these collisons, there is a distribtution of impact parameters and momentum transfers.

The idea of the model is to take these collisions to be occurring randomly, subject to distribution functions that model important properties the foil medium, such as its density. One then solves the classical transport problem by Monte Carlo sampling of a discrete (but large) ensemble of initial phase-space coordinates. These are propagated according to a Langevin equation that models the stochastic collisions. Because of these random forces, this problem is unlike the other systems discussed in the present article, all of which are fully deterministic Hamiltonian systems. However, it has been shown [36] that the motion of the electron in the combined Coulomb + wake field of the swift ion is chaotic.

This classical transport theory has been able to reproduce and give insight into the reasons behind the ℓ-distribution observed in [133]. It was also subsequently applied to other phenomena observed when swift ions or neutral particles are incident on and penetrate solid targets. A recent paper [114], for example, uses this theory to simulate so-called convoy-electron emission. References to other applications of the classical transport theory may be found in this paper.

1.5 What This Paper Covers and Does Not Cover

The earlier parts of this introduction have given a brief indication of the range of quantal systems in atomic and molecular physics being affected by insights gained from the study of their chaotic classical counterparts. The rest of this paper will be concerned with one subject, but it is a broad one whose experimental study is a major part of the work in the laboratory of the author, the microwave ionization of excited hydrogen atoms. All our experiments thus far have involved linearly polarized fields, so we shall limit our discussion to them. This does not mean that the cases of circularly and elliptically polarized fields in uninteresting. On the contrary, they too are attracting experimental attention with alkali Rydberg atoms [50] as well as classical theory for hydrogen atoms [107,64]. The last two sources give references to earlier classical theoretical work on hydrogen atoms in a circularly polarized field. Experiments are now being mounted in our laboratory at Stony Brook that will subject excited hydrogen atoms to circularly and elliptically polarized fields.

Other experiments that we [103,104,106] have carried out with excited hydrogen atoms and others have carried out with sodium Rydberg atoms [119] or with rubidium Rydberg atoms [35] have exposed these atoms to a microwave field consisting of two frequencies. This is an interesting subject whose experimental study is continuing in our laboratory, as least partly motivated by new classical calculations [63] for hydrogen driven by two frequencies, but we shall not consider futher in this paper the subject of two-frequencies. For a critical comparison of the two-frequency microwave ionization experiments that have been published, see Ref. [106].

Another class of experiments that we shall not cover here involve hydrogen or Rydberg atoms being driven by microwave fields whose coherence has been decreased by added shot-noise fields. The most experimental and theoretical work thus far carried out has been on rubidium Rydberg atoms and is reported in [19,4]. Some experimental results have also been reported for hydrogen atoms and on the use of a classical model to explain them. Ref. [11] presents some of this work and gives references to earlier work.

Some of the following sections of this paper have been adapted from a recent review article [106] written by L. Moorman and this author. The interested reader should consult that article for some of the subjects in microwave ionization, *e.g.,* two-frequency driving and the microwave ionization of non-hydrogen Rydberg atoms, that are not covered herein.

2 Apparatus and Experimental Method

2.1 Apparatus

Fig. 1 shows a simplified picture of a typical layout of the apparatus used in our experiments at Stony Brook. [91,52,77,80,82,116,87]. A fast beam of hydrogen atoms, with typical kinetic energy near 15 keV and uniform to about 0.1%, was produced by electron-transfer collisions of protons with Xe gas. In separate regions of static field called F_1 and F_3, atomic transitions were driven sequentially with a double resonance excitation technique introduced in Ref. [74], but now using two $^{12}C^{16}O_2$ lasers neither of whose beams entered the microwave interaction region. Typically, F_1 was about 30 kV/cm, and F_3 was between 1–500 V/cm. With each laser linearly polarized parallel to the static field direction, only $\Delta m = 0$ transitions were driven. A typical scheme involved excitation of the extremal Stark state with parabolic quantum numbers $(n_0, n_1, |m|) = (7,0,0)$ to the state $(10,0,0)$ with the F_1-laser, and from there to $(n_0, 0, 0)$ with the F_3-laser. The energy splitting between the pair of states for each transition was Stark-tuned into resonance with the Doppler-shifted laser photon energy. Spectroscopic scans [77,82] showed that available laser lines and experimental spectroscopic resolution allowed population and resolution of states of the hydrogen atom with each principal quantum number in the range $24 \leq n_0 \leq 90$. That one may populate such a wide range of n_0 has been of key importance in our experiments.

Though a unique substate was populated in F_3, that was not what entered the microwave cavity interaction region. When the atoms flew through a nominally zero field region after laser excitation in F_3 and before the cavity, where 'zero' means dominated by stray fields, the distribution of atomic substates was altered. Diagnosis [91,87] by static electric field ionization, supported by calculations, indicated that a a statistical distribution of all substates of the F_3-laser-excited n_0-value entered the cavity; classically this corresponds [111,93] to an ensemble of orbits with a microcanonical distribution of orbital planes and eccentricities $\epsilon \leq 1$. We shall refer to such an ensemble or distribution as 3d, meaning that

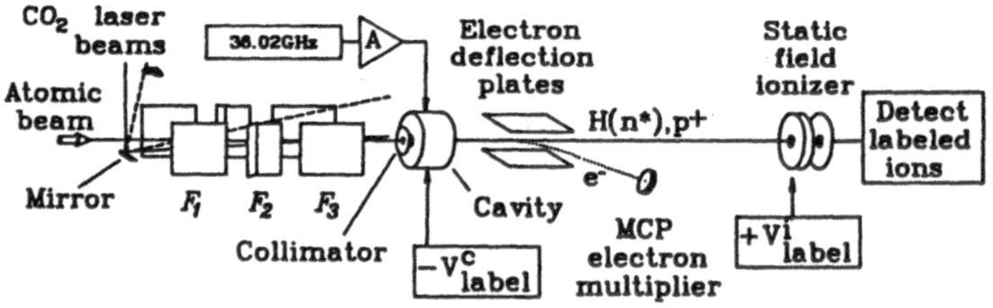

Fig. 1. *A schematic view, not to scale, of the apparatus used in Stony Brook for studying microwave "ionization" and excitation of excited atoms. Atoms in a fast beam enter from the left, are laser-excited into a state with principal quantum number n_0, and are further (de-)excited or "ionized" when interacting with the microwaves in the resonant cavity. The end products of this strong field interaction, either surviving atoms, protons, or electrons, may be detected using various schemes described in the text.*

it fills all three spatial dimensions. We shall later refer to 1d models, in which the electron bounces back and forth on an axis connecting the nucleus and its classical turning point in the Coulomb potential; this roughly corresponds to the quantal case an extremal parabolic state [125,73,16,89,54,9]. This state of the 3d atom was the initial state populated by the F_3-laser.

A collimator limited the diameter of the atomic beam entering the right-circular-cylindrical copper cavity with holes machined through the center of the upstream and downstream endcaps for beam entrance and exit, respectively. We have used each of several cavities resonantly driven in a mode TM_{0pq} with p=2, 3, or 4 and q=0 or 2. The number of field oscillations experienced by each atom as it flew threw the cavity, as well as the overall shape of the pulse, was determined jointly by the beam speed (*e.g.*, for 14.6 keV H atoms, $v = 1.67 \times 10^6$ m/s), the microwave frequency $\omega/2\pi$ (typically between 7.6 and 36 GHz), the length L of the cavity (typically a few to 10 cm), and how the spatial distribution of the electric amplitude of the chosen resonant mode was perturbed by the endcap holes.

The perturbed spatial field distribution in each cavity was determined by numerical modeling. A public-domain software package called 'Superfish' [60], which solves the Helmholtz equation for a cavity resonator whose fields vary over only two dimensions, was applied to each relevant axially symmetric mode of the cavity. Fig. 2 shows for three different modes in one of our cavities the calculated electric field amplitude E_z along the beam direction

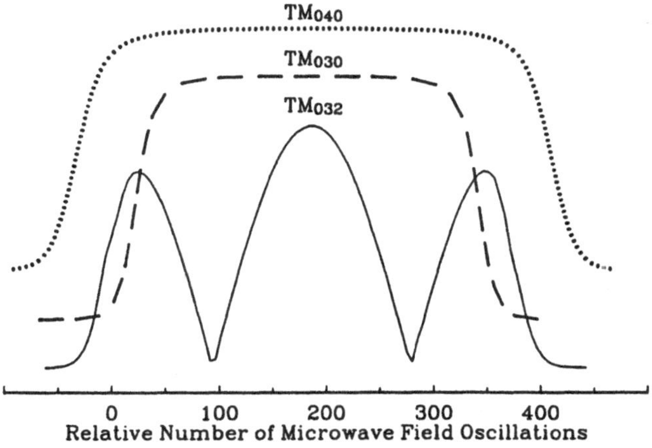

Fig. 2. *Displaced vertically for clarity, the curves show the absolute value $|A(t)|$ of pulse shapes used in microwave ionization experiments with 14.6 keV hydrogen atoms, for each of three microwave resonant frequencies in the same cylindrical cavity: Top, 36.02 GHz; middle, 26.43 GHz; bottom, 30.36 GHz. The cavity was 2.007 cm long, 3.130 cm in diameter, and had 0.64 cm thick endcaps with 0.258 cm diameter holes on the cavity axis for beam entrance and exit. The leakage of the microwave field out the endcap holes caused the initial rise and final fall of each numerically calculated curve.*

that was seen by the atoms in the beam at a fixed radial distance ρ very close to the cavity symmetry axis, which was also the atomic beam axis. For each of these three modes, the atoms on the axis saw the largest maximal field amplitude.

We refer to the E_z curves in Fig. 2, whose values are normalized to that on the cavity axis, as the envelope function $A(z)$ (for a given ρ-value). Seen in the rest frame of the atoms, they give the temporal pulse-shape of the field, $A(at)$, where $a^{-1} = L/v$ is an important time scale: how long it takes an atom to traverse the cavity. Because v was uniform to better than 0.1%, each atom saw the same pulse shape. The amplitude inside the endcap holes fell off rapidly because, compared to the field wavelength, the holes had small radii and the endcaps were thick: the tubular openings through the endcaps acted as cylindrical waveguides beyond cutoff. The amplitude already dropped to 50% where the cavity axis crossed the inside plane of each cavity endcap, and it was negligible outside the beam holes [105,106]. By symmetry, the radial component E_ρ of the field amplitude vanished at ρ=0. At finite ρ, it was negligible inside the cavity, *i.e.*, more than a hole-diameter in from the endcaps. To a good approximation the microwave electric field vector remained longitudinal in all regions where its amplitude was near maximal [105,106], *i.e.* where one expects microwave ionization of the atoms to be most probable.

We have carried out explicit tests of the accuracy of these calculations of the amplitude envelope. This included *in situ* measurements [123] carried out by passing a small dielectric 'bead' through several of the cavities and determining the relative spatial variation of the square of the electric field amplitudes in each cavity from measured resonance frequency shifts. These tests demonstrated that for the cylindrically symmetric modes used in the experiments described herein, the pulse shape was known precisely, at the percent level, and is thus available for use in numerical calculations of the atomic dynamics.

For a typical collimator radius of 0.7 mm, atoms on beam trajectories with different allowed values of ρ experienced slightly different maximal field amplitudes F_{max} [105,106]. This produced a spread, or 'radial droop' ΔF_{max}, that was largely determined by a J_0 Bessel function, which gives a quadratic falloff near ρ=0 of E_z for the TM_{0p0} modes. For example, for the TM_{030} and TM_{040} modes that resonated at 26.4 GHz and 36.0 GHz, the radial droop was 4% and 7%, respectively. (In other experiments carried out at Stony Brook [106] but not discussed here, the beam was so tightly collimated that ΔF_{max} was only 0.2%.)

We made a number of comparisons between amplitude functions calculated by Superfish, with the endcap tubes included, to analytic expressions that would hold for the case of an ideal mode in a cavity without endcap tubes. For example, in the cavity midplane, we found the numerical solution to match well the radial dependence of E_z given by a J_0 Bessel function for an ideal cavity without endcap tubes [105,106]. Moreover, we were able to determine maximal cavity field amplitudes F_0 absolutely to an estimated accuracy of 5% with use of a special calibration technique described in detail in [121].

Important note: All microwave amplitudes given in this paper for data obtained at Stony Brook correspond to 'median' values evaluated at a distance $\rho_c/\sqrt{2}$ from the cavity(beam) axis, where ρ_c is the collimator(beam) radius used in each experiment. It is called the median value because half the atoms in the beam experienced a value smaller than this, and the other half of the atoms experienced a value larger than this. For careful comparisons with these experimental data, one must average theoretical results over the cross section of the beam, *i.e.*, over the radial droop.

2.2 Experimental Methods

To detect the "ionization" produced by the microwave field, we used three methods with somewhat differing objectives. Their names are meant to indicate which particle we intended to detect that emerged from the cavity: neutral atom, proton, or electron, respectively; see Fig. 1.

A) ***Atom Detection*** (AD) method. Atoms emerging from the cavity still in a highly excited state are ionized in a longitudinal static field ionizer located downstream from the cavity. The ions produced there are detected in the same particle multiplier used in the PD method. Because this method is sensitive to the atoms that survive, *i.e.*, are not *quenched*, by the microwave field, we have also called it the quench method. Defined below is the concept of an n-cutoff, which for the AD method we use the symbol n_c^q.

B) ***Proton Detection*** (PD) method. Protons produced in the cavity by the interaction of the atoms with the microwave field and with any fringe or stray static electric fields are detected in a particle multiplier downstream from the cavity. We have also earlier called the PD method the proton-"ionization" method. We use the symbol n_c^p for the n-cutoff in the PD method.

C) ***Electron Detection*** (ED) method. Electrons produced in the cavity by the interaction of the atoms with the microwave field and with any fringe or stray static electric fields are electrostatically deflected after the cavity for detection with a microchannelplate electron multiplier. We have also earlier called the ED method the electron-"ionization" method. We use the symbol n_c^i for the n-cutoff in the ED method.

For discrimination against background signals, all three detection methods used a label voltage [77] on that part of the apparatus where the charged particles of interest were produced. Charged particles produced in the label-voltage region were, upon leaving it, accelerated toward ground potential. This boosted their speed above that of the original beam and gave them a 'labeled kinetic energy'. For the PD method, the cavity body was typically held about 100 V above ground potential; for the ED method, the label voltage was near -6 V; for the AD method, the cavity body was grounded and the the first electrode in the static field ionizer assembly [116,87] was held between 70–140 V. This energy label allowed for charged-particle kinetic-energy selection before the particle detector in an electrostatic filter-lens system, not shown in Fig. 1. For additional background suppression and signal definition, charged particle measurements were performed by phase sensitive detection synchronized to the mechanical chopping of the F_1-laser beam in the range 0.33–1 kHz.

All three methods were affected by electrostatic fields produced by the applied label potential or by other fields after the cavity that were used for charged particle deflection. For example, in the PD and ED methods the label voltage on the cavity produced electrostatic fields just outside the endcap tubes. Similar statements apply to the AD method, concerning electrostatic fields after the cavity and before the longitudinal field ionizer. These fields were intentionally kept too weak to affect, *e.g.*, ionize, atoms entering the cavity in the initially prepared excited state with principal quantum number n_0. However, the microwaves drove transitions from n_0 not only to the continuum, true ionization, but also to higher (or lower) bound states. For a given array of electrostatic fields outside and beyond the exit endcap tube of the cavity, no matter how weak, there is always a n-cutoff value $n_c > n_0$ above which these electrostatic fields can cause ionization and produce a signal that is detected in the same way as the true microwave ionization signal. Therefore, an important part of specifying the conditions for each method involves knowing or estimating the maximal electrostatic field strength and the corresponding value of n_c. For the experiments discussed in this paper, values of n_c were determined only by static fields outside the microwave interaction region; consequently, the atom-microwave interaction dynamics inside the cavity was free from static field interaction effects, at least at the level of 10 mV/cm or so. This is different from other experiments done both in our laboratory and elsewhere [9,17,91,17,10,87], in which static electric fields in the range 1–10 V/cm were intentionally applied in the microwave interaction region.

IMPORTANT: we use the term "ionization" in quotation marks to refer always to experimental signals registering the sum of true ionization plus excitation to bound states $n \geq n_c$. The reader should be aware of "ionization", as opposed to ionization.

Experimental data for the PD and ED methods were usually collected by repetetive, slow sweeps of the microwave power dissipated in the cavity, and consequently F_0 inside the cavity, between empirically set low and high values. The low value was usually below that at which the "ionization" signal began to increase, and the lack of variation here was used to set the zero level for the "ionization" probability $P_{\text{"ion"}}$. If this was intentionally not the case, then the $P_{\text{"ion"}}=0$ level was determined separately with even lower power levels. At some high power level, the "ionization" signal finally leveled off and no longer increased with higher power. This meant that all atoms prepared in the initial n_0-level were being "ionized", and this signal level was used to set where $P_{\text{"ion"}}=1$. Determination of values of $P_{\text{"ion"}}$ between these asymptotes followed from the linearity of the detection system.

Experimental data for the AD method were collected similarly, but because atoms were detected that had *not* ionized, *i.e.*, that had survived exposure to the microwave field, the signal decreased from $P_{surv}=1$ to $P_{surv}=0$ with increasing microwave power. We call a graph of P_{surv} *vs.* F_0 a quench curve. Inverting the quench curve, *i.e.*, $(1 - P_{surv})$ *vs.* F_0, gives a curve showing "ionization" *vs.* F_0. Understandably such inverted AD curves are very similar to curves obtained with the PD and ED methods if the n-cutoff in all three cases is either similar or far above n_0.

Another difference for the AD method is its being sensitive to microwave de-excitation to n-levels far below n_0. How far below is determined by the minimum value of n that will be ionized in the static field ionizer and subsequently detected. This is easily diagnosed, however, by increasing the field in the ionizer to a higher value, which lowers this value of n even further. We kept the value of the static field in the ionizer high enough to be insensitive to this effect. Hence, the AD-method data discussed in this article are actually "ionization" data.

A way to extract useful information from complete experimental curves is to determine the electric field amplitude at which $P_{\text{"ion"}}$ or $(1 - P_{surv})$ first reaches X%. We shall call these amplitudes for each n_0-value the X%-"ionization" threshold amplitude, $F_0(X\%)$, or in shorthand, the X%-threshold. Because they are, respectively, near the onset of "ionization" and near complete "ionization", frequently used values of $X\%$ are 10% and 90%. Also useful are 50%-thresholds. Many comparisons [116,87] of threshold amplitudes carried out for a 9.92 GHz field driving hydrogen atoms with n_0 ranging from 32 to approximately 70, showed that PD and (inverted) AD curves gave nearly the same experimental information, even when they contained structure(s). However, when n_0 approached the value of n_c for one of the methods but not for the other(s), differences in the measured curves became apparent. This is to be expected. For example, experiments carried out at 36 GHz with the PD and ED methods produced data that differed when n_0 was no longer far below both values of n_c, which were usually quite different for the two methods. We shall return to this point later.

3 The Hamiltonian and Scaled Variables

To obtain atomic units appropriate for an actual hydrogen atom, with non-infinite nuclear mass, set the square e^2 of the the electronic charge, the reduced electronic mass μ (where $\mu/m_e = 0.9994557$), and $\hbar$ all equal to one. The resulting reduced-mass atomic units for various physical quantities are those for the ground state of the Bohr atom having the correct reduced mass. To six significant figures: the distance unit is $a_1 = a_0(\mu/m_e)^{-1} = 0.529465 \times 10^{-10}$ m, where $a_0 = 0.529177 \times 10^{-10}$ m is the Bohr radius for an infinite nuclear mass; the speed unit $v_1 = \alpha c = 2.18769 \times 10^8$ m/s, where $\alpha = (137.036)^{-1}$ is Sommerfeld's fine-structure constant, and c is the speed of light; the angular frequency unit is $\omega_1 = v_1/a_1 = 4.13188 \times 10^{16}$ rad/s; the unit of energy is $2E_0(\mu/m_e) = 27.1966$ eV, where E_0 is the binding energy of the ground state of the Bohr atom with infinite nuclear mass; the electric field unit is $F_1 = e/(4\pi\epsilon_0 a_1^2) = 5.13660 \times 10^{11}$ V/m. (It appears that the curious *sub-zero* in the well-known symbol a_0, rather than a sub-one that would be more suitable for the $n = 1$ ground state, comes from Bohr in 1913 having begun with zero his enumeration of quantum numbers in his model hydrogen atom, with a "+1" tacked onto it in various formulae. By the time the old quantum theory gave way to quantum mechanics in the 1920's, the enumeration had begun in the ground state with what every student learns now: one.)

With all quantitities measured in reduced-mass atomic units, the correct non-relativistic Hamiltonian for describing the microwave ionization of hydrogen atoms moving through a TM_{0pq}-mode cavity whose symmetry axis is along the beam axis was shown in [96] to be, subject to approximations valid for experimental conditions,

$$\mathcal{H} = \frac{1}{2\mu}\left(\mathbf{P} - \frac{eF}{\omega}A(t)\hat{z}\cos(\omega t + \delta)\right)^2 - \frac{e^2}{r}, \tag{1}$$

$$\mathbf{P} = \mathbf{p} + \frac{e}{c}\mathbf{A} \approx \mathbf{p} + \frac{eF}{\omega}A(t)\hat{z}\cos(\omega t + \delta), \tag{2}$$

where μ is the reduced mass of the electron, $(\mathbf{P}, \mathbf{r})$ are conjugate momentum and coordinate variables, F and ω are the microwave electric amplitude and frequency, respectively, $\mathbf{A}$ is the vector potential of the electromagnetic field, δ is an initial phase, and $A(t)$ is the envelope function or pulse shape that turns on, maintains, and turns off the microwave electric amplitude.

With use of a canonical transformation $(\mathbf{P}, \mathbf{r}) \to (\mathbf{p}, \mathbf{r})$, one obtains the dipole-gauge Hamiltonian

$$\mathcal{H}_d = \frac{1}{2\mu}p^2 - \frac{e^2}{r} - \left(\frac{eF}{\omega}\right) z \frac{d}{dt}\left[A(t)\cos(\omega t + \delta)\right]. \tag{3}$$

As was emphasized in [96], but which appears not to be so commonly known, only if $A(t)$ varies little during one field period, *i.e.*, if $\dot{A}/A \ll \omega$, then a good approximation to Eq. 3 is the much more commonly used form of the dipole-gauge Hamiltonian (in atomic units)

$$\mathcal{H}(t) = \frac{p^2}{2} - \frac{1}{r} + zA(t)F\sin(\omega t + \varphi), \tag{4}$$

where we are now labeling the initial phase φ. Thus far, all our experiments have averaged over this phase and have used a slowly varying $A(t)$. However, that neither of these conditions is forced upon the experiments is motivating new experiments at Stony Brook.

Let us investigate the scaling properties [92] of Eq. 4 with use of action-scaled variables denoted with a tilde. Using where convenient n_0 in place of the initial classical principal action I_0 [1] (semi-classical quantization yields $I_0 = n_0\hbar$),

$$\tilde{\mathbf{p}} = n_0\mathbf{p}, \qquad \tilde{\mathbf{r}} = \mathbf{r}/n_0^2, \qquad \tilde{t} = t/n_0^3, \qquad \tilde{\varphi} = \varphi,$$

Eq. 4 may then be re-expressed as

$$\tilde{\mathcal{H}}(\tilde{t}) = n_0^2\mathcal{H}(t) = \tilde{p}^2/2 - \tilde{r}^{-1} + \tilde{z}A(n_0^3\tilde{t})\tilde{F}\sin(\tilde{\omega}\tilde{t} + \tilde{\varphi}), \tag{5}$$

where $\tilde{\omega} = n_0^3\omega$ is the *scaled frequency*, the ratio of the driving frequency and the initial Kepler frequency, and $\tilde{F} = n_0^4F$ is the *scaled amplitude*, the ratio of the driving electric amplitude and the mean Coulomb field binding the initial Kepler orbit. When $n_0^3\omega = 1/1$, the driving frequency is equal to the classical Kepler frequency, or quantally, the photon energy is very near the mean of the $\Delta n = \pm 1$ energy splittings for the free atom. This we know from Bohr's correspondence principle.

For the present experiments with "3d atoms", whose atomic (parabolic) substate distribution is uniform across a given n_0-shell — classically, the microcanonical distribution [92] — when one takes either a constant initial phase φ or averages over it, the *classical* dynamics for Eq. 5 does not depend separately on the three quantities n_0, ω, and F; for $A(t)$=*const.*, it depends only on the two scaled quantities [92] $n_0^3\omega$ and n_0^4F. If there is a real pulse-shape, the exact *classical* scaling can still be maintained. Experimentally this would be done by varying the atomic flight time through a given, spatially fixed microwave field in such a way that the respective number of Kepler periods (for the initial value of I_0) during (initial) rise, middle, and (final) fall of $A(t)$ are kept constant.

However, quantum mechanics does not obey the classical scaling. Evaluating the commutator of the quantal operators associated with the classically scaled variables leads to $\tilde{\hbar}$, an effective-$\hbar$; *e.g.*, $[\tilde{x}, \tilde{p}_x] = i\tilde{\hbar}$, where $\tilde{\hbar} = \hbar/n_0$. Thus, with fixed $n_0^3\omega$ and n_0^4F (and correctly scaled pulse-shape), increasing [decreasing] n_0 affords the interesting possibility of experimentally being able to decrease [increase] the effective-$\hbar$ to move closer to [farther away from] the semiclassical limit. We shall return to this point later.

Let us whet the reader's appetite for most of what will follow by showing first an example of experimental data that nicely demonstrate classical scaling and second another example of experimental data whose details do not obey classical scaling.

Fig. 3a shows five "ionization" curves obtained from measurements on hydrogen atoms with three different pairs of initially prepared n_0-level and driving frequency: n_0=67 (two curves obtained for $\omega_\alpha/2\pi$=9.9 GHz); n_0=48 (one curve obtained for $\omega_\beta/2\pi$=26.4 GHz); and n_0=43 (two curves obtained for $\omega_\gamma/2\pi$=36.0 GHz). All curves were obtained with the AD-method described in Sec. 2.2, for which the n-cutoff n_c^q was near 90. The electric field amplitudes in V/cm necessary for "ionization" increase by about a factor 6 from the highest to the lowest n_0-value in the above trio. From a quantal point of view, one might not expect these three cases to have much in common because each requires a significantly different number of photon energies to reach the free-atom continuum: the binding energy of the respective initial states is $73\hbar\omega_\alpha$, $53\hbar\omega_\beta$, and $47\hbar\omega_\gamma$. Thus, it may come as a surprise that these three curves may be made to lie nearly on top of each other with a simple rescaling of the electric field amplitude F: multiply by n_0^4 and divide by the reduced-mass atomic unit F_1 given above. Use of this rescaling produces the curves shown in Fig. 3b. The physical significance of the rescaling is its expressing F in units of the mean Coulomb electrostatic field binding the 3d hydrogen atom in its initial n_0-manifold.

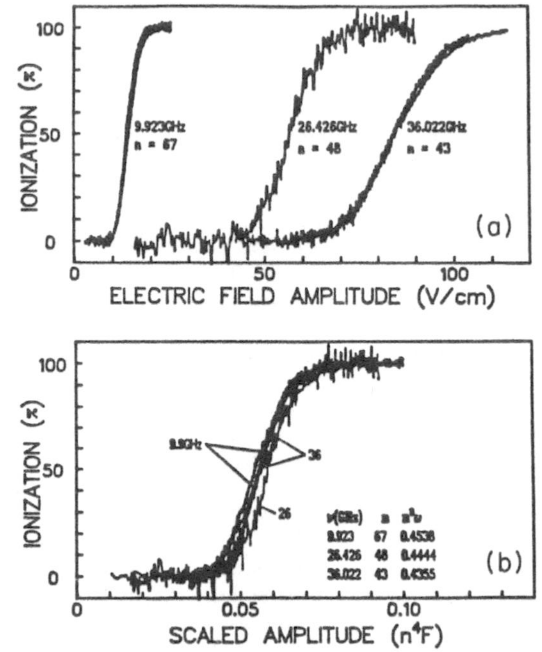

Fig. 3. *Example of "ionization" curves in the semiclassical Regime–III obtained with three different values of* n_0, *each for a different microwave frequency* $\omega/2\pi$. *The* (n_0, ω) *pairs were chosen because all three cases correspond to very nearly the same scaled frequency,* $n_0^3\omega = 0.44 \pm 0.01$. *Frame (a) shows the "ionization" probability plotted vs. microwave electric field amplitude in laboratory units V/cm. Notice the wide separation of the curves. Frame (b) shows the same curves plotted vs. scaled microwave electric amplitude* n_0^4F. *That the curves in frame (b) now closely coincide is a demonstration of classical scaling of the periodically driven, quantal hydrogen atom.*

One needs to realize that the three curves in Fig. 3a were not chosen for n_0-values taken at random. On the contrary, all three values of $n_0^3\omega$ are very nearly the same. For ω measured in reduced-mass atomic units of ω_1 introduced in the beginning of this section, $n_0^3\omega$ expresses the driving frequency in units of the Kepler orbital frequency for a classical hydrogen atom with initial principal action [1] $I_0 = n_0\hbar$. By Bohr's correspondence principal, the Kepler frequency for $n_0 \gg 1$ is very nearly equal to the mean of the $n_0 \rightarrow (n_0 \pm 1)$ frequency splittings in the quantal (Bohr) atom. For the cases shown in Fig. 3a, $n_0^3\omega_\alpha$=0.45 for n_0=67; $n_0^3\omega_\beta$=0.44 for n_0=48; and $n_0^3\omega_\gamma$=0.44 for n_0=43. That is, they are all equal to better than few percent.

Though the pulse *shape* $A(t)$ for all three (n_0, ω) pairs in Fig. 3 was nearly the same, their durations in Kepler periods for the initial n_0-value were comparable but not the same. The "flat-top" $A(t)$ functions for 26.43 GHz and 36.02 GHz are shown in Fig. 2 in units of field oscillations. We shall measure the rise as the time between 5% and 95% of the peak, similarly for the fall, and the top as the time between 95% crossings. For $(n_0, \omega_\beta/2\pi) = (48, 26.43 \text{ GHz})$, the flat top [rise and fall] lasted 551 [135] Kepler periods. For $(n_0, \omega_\gamma/2\pi) = (43, 36.02 \text{ GHz})$, the flat top [rise and fall] lasted 767 [188] Kepler periods. For $(n_0, \omega_\alpha/2\pi) = (67, 9.923 \text{ GHz})$, the same pulse shape was used, whose flat top [rise and fall] lasted 509 [174] Kepler periods.

The near coincidence of the scaled curves in Fig. 3b is the previously promised example of classical scaling, which we see here to be true even for unequal durations in Kepler periods of the three $A(t)$ functions. We have observed many other examples of classical scaling in microwave "ionization" data.

However, also previously promised was an example of experimental data whose details do not obey classical scaling. Figs. 4a,b show 3d experimental "ionization" curves for two different $(n_0, \omega/2\pi)$ combinations such that $n_0^3\omega = 0.166 \pm 0.001$, along with 1d- and 3d-classical and 1d-quantal theoretical curves; see Ref. [116]. As is expected from classical scaling, notice the agreement (within Monte Carlo calculation statistical errors) between the classical 1d [classical 3d] calculations in panels (a) and (b), though the 1d results rise more quickly and at lower n_0^4F-values than the 3d results. This behavior with dimensionality is also to be expected. Notice, however, that the quantal 1d curves here *do not* obey classical scaling. The two scaled ($P_{\text{"ion"}}$ *vs.* n_0^4F) experimental curves for 3d atoms are nearly the same at relatively strong driving, *i.e.*, where $P_{\text{"ion"}} \gtrsim 0.8$, and where each agrees with the 3d classical Monte-Carlo calculations [52,92,116,113]. At lower driving amplitudes, particularly near the onset of "ionization", each experimental curve departs significantly from the classical 3d results. In (a) "ionization" begins experimentally at lower n_0^4F-values than 3d-classically. This "subthreshold ionization" effect [16,20,21,116] is also shown by the quantal 1d curve compared to the classical 1d results in (a). In (b), however, "ionization" begins experimentally at higher n_0^4F-values than 3d-classically. This "quantal suppression" [116] of the onset of ionization is also shown in the quantal 1d curve compared to the classical 1d results in (b).

As subsequent sections of this paper will show, at Stony Brook we have thus far varied [83,84,85,52,103,82,116,91,87,123,122] $n_0^3\omega$ between 0.021–2.8, a wide dynamic range of over 130, with use of cavity frequencies between 7.6 and 36 GHz and n_0-values between 24 and 90. Though the variation of $n_0^3\omega$ has been in discrete steps of a few percent, such a large coverage has allowed us to investigate the first five of the six regimes discussed below. Indeed, combined experimental–theoretical investigation of this wide range of scaled frequencies, and particularly the comparison between experimental data and calculations that closely model the experiments, has led to our recognition of these different dynamical regimes. (Even more may be waiting to be discovered.)

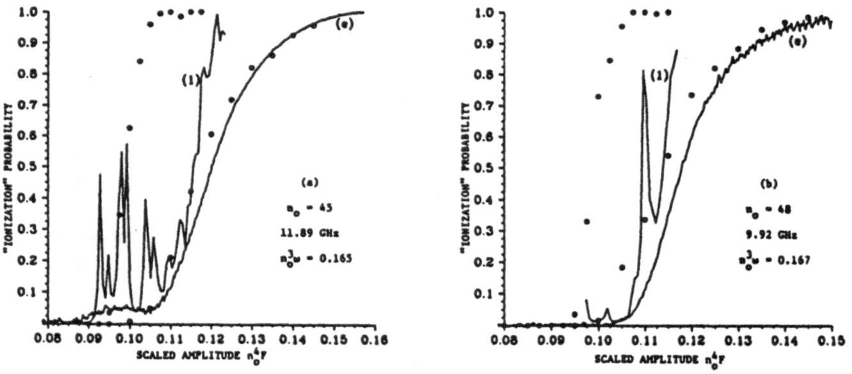

Fig. 4. *Comparisons taken from Ref. 116. Curve (e): measured probability for "ionization" (ionization plus excitation to above an n-cutoff $n_c^i \simeq 75$) of "3d" hydrogen atoms; (∘): 3d classical Monte-Carlo simulation; (•): 1d classical Monte-Carlo simulation; curve (1): 1d quantal calculation using the adiabatic basis of states. Though the values of n_0 and ω differ in frames (a) and (b), the scaled frequencies are very nearly the same, $n_0^3\omega = 0.166 \pm 0.001$. Because of classical scaling the results (∘) in frames (a) and (b) agree, as do the results (•). Quantal resonance effects in this scaled frequency regime, called Regime–II in the text, cause the quantal results, (e) and (1), not to scale classically from frame (a) to (b). Notice in frame (a): curve (e) begins to rise before (∘), an example of quantally enhanced "ionization" near threshold; see this also for the 1d theoretical results, curve (1) and (•). Notice, conversely, in frame (b): curve (e) begins to rise after (∘), an example of quantally suppressed "ionization"; see this also for the 1d theoretical results, curve (1) and (•).*

4 Regimes of Behavior

Experiments and classical and quantal theoretical studies, and particularly comparisons among all these, have made it clear that in a strong field, which means that n_0^4F is strong enough for "ionization" to be possible, the driven-atom dynamics depends strongly on the value of the scaled frequency $n_0^3\omega$. As previously mentioned, we have thus far found different dynamical behavior for six ranges of non-zero $n_0^3\omega$, but we emphasize that the boundaries between adjacent regimes are fuzzy: as $n_0^3\omega$ increases the behavior in one regime slowly passes to that in the next higher regime.

We shall focus in this paper on what is known about the behavior in each of the first five regimes. We note that the highest scaled frequency achieved in published microwave "ionization" experiments is a bit below 3, and for the conditions of those experiments the energy of about 15 microwave photons was needed to reach the free-atom continuum. In the sub-discipline of multiphoton ionization in atomic physics, whose language is based on perturbation theory, this would be called 15-photon ionization. When there is absorption of more than the minimum number of photons that is energetically required for ionization, atomic physicists speak of above-threshold ionization (ATI), although some prefer the name excess-photon ionization.

Unlike the other regimes, where one speaks of scaled-amplitude thresholds for the "ionization" to occur, Regime–VI behavior for weak driving fields, the well-known textbook problem of the photoelectric-ionization effect, is characterized by a frequency threshold: for one-photon ionization to be allowed energetically, the photon energy $\hbar\omega$ must be greater than the binding energy of the initial state, $(2n_0^2)^{-1}$ for hydrogen atoms.

For superstrong high-frequency driving fields, however, there are theoretical predictions that more exotic phenomena, such as so-called "stabilization" of the atom, will take place: As the driving electric amplitude increases, the ionization rate would decrease rather than increase. The recent flurry of theoretical activity in this area is a renascent development of ideas first advanced in the early 1970's. Several experimental groups are probing the validity or lack thereof of the recent theories, which have led to both quantal and classical calculations and not a small amout of lively debate at workshops and in the literature. The interested reader is advised to keep abreast of developments in this active area. A useful source of literature references on this problem is one recently published paper [51]; it emphasizes the results of 3d classical Monte Carlo calculations using the same methods that other authors earlier used for microwave ionization calculations that are presented later in this paper.

4.1 "Ionization" Curves

Our typical raw experimental data are complete "ionization" curves, such as those shown in Figs. 3,4. These record the "ionization" probability $P_{\text{"ion"}}$ as a function of F, where the reader should recall from Sec. 2.2 the importance of the quotation marks around ionization. Contributing to the signal is an ensemble of atoms having an initial distribution of substates belonging to a fixed value of principal quantum number n_0. Obviously, the substates that ionize more easily will contribute to the initial rise of the curves from $P_{\text{"ion"}}=0$, whereas those that ionize less easily will contribute to the final approach of the curves to $P_{\text{"ion"}}=1$. In an external static or linearly polarized oscillatory electric field, or in a parallel combination of the two, the absolute value $|m|$ of the magnetic quantum number is a good quantum number, but the angular momentum quantum number ℓ is not. A suitable quantal basis is the parabolic basis with good "weak field" (Stark) quantum numbers $(n_0, n_1, n_2, |m|)$, which is a usefully descriptive but redundant set because $n_0 = n_1 + n_2 + |m| + 1$. The situation is obviously more complicated for oscillatory fields.

We now make a crude semiclassical argument that suggests why low-$|m|$ substates from a given n_0-shell "ionize" at lower microwave amplitudes than high-$|m|$ substates, meaning that for an "ionization" curve measured with a statistical (microcanonical) distribution of substates, the low-$|m|$ substates determine the lowest driving amplitudes at which "ionization" begins. First consider frequencies ω much less than the $\Delta n = \pm 1$ splittings to the nearest n-shells above and below n_0; this means $n_0^3\omega << 1$. A quasi-static approximation to the dynamics will be reasonable, with the system moving on a slowly oscillating Coulomb-Stark potential $V = -r^{-1} + zF\sin(\omega t)$. At each instant in time, one may separate the time-independent Schrödinger equation in parabolic coordinates (ξ, η, ϕ) into a free-particle equation in ϕ and two coupled, 1d 'radial' equations in ξ and η. The effective potential energy functions for the latter two are [90]

$$V(\xi) = \frac{m^2 - 1}{8\xi^2} - \frac{\beta_1}{2\xi} + \frac{F\xi}{8} \qquad V(\eta) = \frac{m^2 - 1}{8\eta^2} - \frac{\beta_1}{2\eta} - \frac{F\eta}{8}, \tag{6}$$

with $\beta_1 + \beta_2 = 1$ for the hydrogen atom. $V(\eta)$ is unbounded from below for large η, and ionization (escape from small to large η) takes place when F is large enough for escape over or wave-mechanical tunneling through the barrier in $V(\eta)$. Notice, however, that the first term in each effective potential is a repulsive barrier for $|m| > 1$ that keeps the electron from sampling small values of ξ or η. Therefore, large-$|m|$ substates have very small wavefunction density near the field axis and, therefore, near the nucleus; classically, large-$|m|$ orbits don't pass near the nucleus as these represent almost circular orbits in the plane whose normal is the field axis. Because an oscillatory field does not increase the average energy of a *free* electron, one expects that the energy and momentum will be transferred from the field to the electron, which is necessary for ionization, only when the electron passes close to the nucleus. Hence, substates with larger $|m|$-values will require larger field amplitudes to be driven close enough to the nucleus for ionization to occur.

These quasi-static arguments are very crude, and the whole quasi-static picture is invalid when $n_0^3\omega$ is not low. However Richards [117] has performed some numerical and analytical studies of 1d, 2d, and 3d classical orbits for the sinusoidally driven hydrogen atom for some scaled frequencies above 1. His results show that large-$|m|$ orbits can have very large "ionization" thresholds; for example, for $n_0^3\omega$=5.6, the circular orbit with $m = \ell$ is calculated to have a scaled "ionization" threshold near 0.29, which is roughly an order of magnitude higher than that needed to "ionize" m=0 orbits at this scaled frequency. It is also interesting to remark that the calculations showed that 3d classical 10%-thresholds are reasonably well approximated by 1d 10%-thresholds only for scaled frequencies up to about 3, but even above there they are reasonably well approximated by the 2d 10%-thresholds. 2d orbits lie in a plane containing the oscillatory field vector and thus have m=0.

The importance of these results is that 10%-"ionization" thresholds measured in our experiments with 3d atoms have often been compared to classical or quantal calculations performed for 1d atoms. Published experimental data have thus far been limited to $n_0^3\omega \leq 2.8$. If Richards's results for 3d classical calculations are also true for 3d quantal calculations, then the good, often excellent, agreement (*e.g.*, see [91,120,72]) between the 10%-thresholds measured for 3d atoms and the quantal 10%-thresholds calculated for 1d atoms will no longer continue when $n_0^3\omega$ rises above 3. This should be tested experimentally.

Several different theories have been used to model our experiments. Although summarizing them is beyond the scope of the present paper, we may sense the broad range of theoretical difficulties by quoting from a recent paper [118] of two active theorists:

> The theory for the microwave "ionisation" of excited hydrogen atoms is difficult. The field strengths of interest are sufficiently strong to ionise the atom, the density of states is very large so there are very many coupled states, and the continuum needs to be included, although paradoxically it does not necessarily significantly affect the bound state dynamics. For these reasons most quantal calculations have dealt with a 1d model ...; a full quantal 3d treatment is probably not possible on present computers. The numerical integration of Hamilton's equations is, however, relatively straight forward if regularisation is used to remove the Coulomb singularity.

Some solace may come from realizing that a "full quantal 3d treatment" is not quite as daunting as it sounds, at least for linear polarization. Because $|m|$ is conserved in this polarization, one needs 'only' to do 2d quantal calculations [37,39,128] for each allowed $|m|$-value. The general case of an elliptically polarized field, however, would necessitate 'really full' quantal 3d calculations. With the necessity of including the continuum in some way or other, this latter case seems to be beyond the present numerical state-of-the-art for microwave ionization of Rydberg atoms.

5 Static Field Ionization

Before discussing ionization of excited hydrogen atoms by a static electric field F_S, we need to understand what this really means experimentally. Static means unchanging in time. Given the small amounts of ripple, *etc.*, in modern dc power supplies, this idealization is reasonably well met in the laboratory. In an apparatus, however, one must get the atom into the static field in order to study it, and there are a few possibilities. One might prepare the atom in a zero-field eigenstate and then slowly bring up the value of F_S. Because of the zero-field n^2-degeneracy of excited non-relativistic hydrogen atoms, particularly for large n it is not at all easy to keep pure a zero-field eigenstate. Therefore, it is preferable when studying excited hydrogen atoms to prepare them in the presence of a static field in a parabolic (or Stark) eigenstate with some kind of laser excitation out of a lower-lying state. Then, if the value of F_S is changed slowly enough compared relevant atomic time scale(s), the atomic evolution is adiabatic and good parabolic quantum numbers (n_1 and $|m|$, but not n_0 unless F_S is weak) are preserved.

If highly excited hydrogen atoms are allowed to spend time in a zero-field region, *i.e.*, one dominated by small stray fields whose strength *and* direction is uncontrolled, the good parabolic quantum numbers are *inevitably* not preserved.

In the static field ionization experiments described in [75,77], hydrogen atoms were prepared in each of a few parabolic substates of n_0=30 and 40. The good quantum numbers were preserved, and measurements of the F_S-dependence of the static field ionization rate $\Gamma_S(F_S)$ were measured for rates between about 10^5 to 10^8 s^{-1}. The data were found to be in excellent agreement with the results of quantal numerical calculations carried out by Damburg and Kolosov (see [75] for references), and in less good but still reasonable agreement with rates obtained from a semi-empirical formula developed by these authors.

We use static field ionization as a backdrop to what follows next, because quantum mechanics systematically affects static field ionization compared to classical mechanics. As follows from Oppenheimer's 1928 theoretical treatment [109] for the hydrogen ground state, the wave-mechanical effect of barrier penetration (tunneling) systematically allows ionization to proceed at values of F_S below that at which classical escape of the electron from the Coulomb force field of the proton becomes possible. For a 1d hydrogenic atom with nuclear charge Ze, or in the so-called surface-state-electron (SSE) model, the classical (scaled) static threshold field is known analtyically [67] to be $(I_0^4/Z^3)F = (4\sqrt{2}/3\pi)^4 \simeq 0.130$, where I_0 is the classical (principal) action. For a real 3d hydrogen atom, the classical (scaled) static threshold fields have been calculated numerically in [5] for a wide range of allowed classical orbits (neglecting exceptional orbits) and summarized graphically. The most easily ionized classical orbits begin to ionize at a scaled threshold field very close to the classical 1d result just given, whereas the least easily ionized classical orbits don't ionize until scaled threshold fields exceeding $(I_0^4/Z^3)F \simeq 0.35$.

6 Regime–I: The Dynamic Tunneling Regime

Ref. [122] presents microwave "ionization" data obtained for the lowest range of scaled frequencies yet covered for excited hydrogen atoms. With a 9.908 GHz electric field in a TM_{020}-mode cavity and values of n_0=24,. . . ,32, the experiment covered the range $n_0^3\omega$=0.021–0.050. The AD-method described in Sec. 2.2 was used, and the n-cutoff $n_c^?$ was above 75.

Fig. 5 compares "ionization" curves for three different n_0-values with the results of classical and quantal calculations. These were carried out for 3d hydrogen atoms because the experiments were performed with a uniform distribution of substates across each initial n-shell, classically the microcanonical ensemble. Compare first the experimental data (dots)

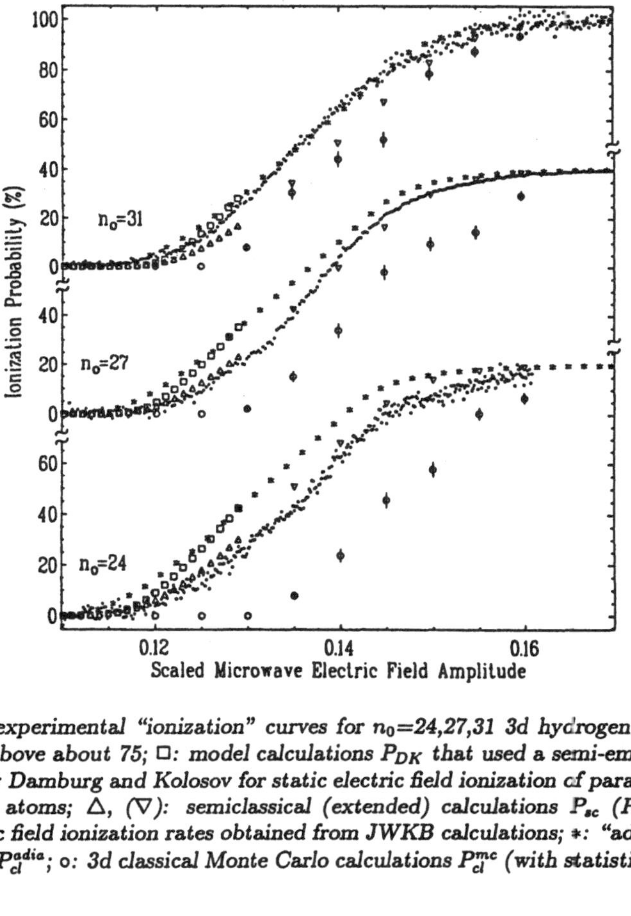

Fig. 5. •: *experimental "ionization" curves for* n_0*=24,27,31 3d hydrogen atoms and an* n*-cutoff* $n_c^?$ *above about 75;* □: *model calculations* P_{DK} *that used a semi-empirical formula developed by Damburg and Kolosov for static electric field ionization of parabolic substates of hydrogen atoms;* △, (▽): *semiclassical (extended) calculations* P_{sc} (P_{sc}^{ext}) *that used static electric field ionization rates obtained from JWKB calculations;* ∗: *"adiabatic" model calculations* P_{cl}^{adia}; o: *3d classical Monte Carlo calculations* P_{cl}^{mc} *(with statistical error bars).*

to the 3d classical Monte Carlo calculations (open circles with Monte Carlo statistical error bars). The shape of a smooth curve through each set of calculated points is similar to that for the corresponding measured data points for $P_{\text{"ion"}}$ *vs.* n_0^4F, but in each case there is a systematic shift: The experimental value of n_0^4F needed to reach each fixed value of $P_{\text{"ion"}}$ between 0 and 100% is systematically lower than the classical 3d calculated value. The amount of the shift in Fig. 5, however, decreases with n_0, being largest for n_0=24 and smallest for n_0=31.

One correctly surmises that the systematic shift is at least partly the result of quantal tunneling through the potential barrier that oscillates with the sinusoidal field. The simplest quantal model for calculating ionization is to assume that at each instant along the oscillatory variation of the electric field, the ionization rate for each parabolic substate is given by its instantaneous static value. For the lowest in the range of n_0-values considered here, this turns out to be an excellent approximation, and the most accurate calculations for the static field rates here are those obtained with uniform semiclassical (JWKB) methods (upright and inverted triangles). These give the best agreement for n_0=24 and 27, showing that the tunneling static field ionization mechanism may be used to obtain quantitative agreement with very low scaled-frequency "ionization" data. Results (open squares) obtained using a semi-empirical formula for the static field ionization rate, see Sec. 5, are somewhat less accurate quantitatively.

For n_0=31 the agreement with the JWKB calculations is no longer good. Compared to these calculations, the experimental data are systematically shifted towards lower n_0^4F. (Why the apparent agreement with the open squares is accidental is explained in [122].) By the time n_0 has risen to 31, the ionization mechanism in the 9.908 GHz field is no longer just tunneling. Notice, however, that the asterisks are in good agreement with the n_0=31 data, though they do not agree for n_0=24 and 27.

The assumptions leading to the "adiabatic" theoretical model producing the asterisks give some insight into the ionization mechanism that has come into play for n_0=31: *(i)* $F_0 \equiv n^4F$ is large enough to mix strongly the initial n_0 and adjacent (n_0+1) manifolds; *(ii)* F_0 exceeds the classical critical (ionization threshold) field F_{cl}^{crit} for some (n_0+1) substates; *(iii)* these ionize completely. The adiabatic ionization probability P_{cl}^{adia} is obtained from an average of the classical critical [5] field $F_{cl}^{crit}(\mathbf{n})$ over the (n_0+1) manifold. In this limit P_{cl}^{adia} depends only upon F_0 and can be computed by Monte Carlo integration over the microcanonical substate distribution, if one allows for the influence of the alternating sign of the oscillatory field on $F_{cl}^{crit}(\mathbf{n})$. This approximation, which assumes field-driven n-state changing transitions, is complementary to tunneling, which assumes no transitions between adiabatic basis states.

For all the 9.908 GHz experimental data between n_0=24,...,32, and the corresponding classical and quantal calculations, Fig. 6 shows the scaled amplitudes $F_0(10\%)$ needed to produce $P_{\text{"ion"}}$ = 10%. These comparisons summarize the statements made above. The classical 3d Monte Carlo calculations are all systematically high. The JWKB quantal calculations are systematically high for the upper three or four n_0-values but are in good quantitative agreement for the rest at lower n_0. Notice that this agreement begins to occur just where the experimental values go through a mild peak at n_0=28. This mild peak is a signature for the changeover from one ionization mechanism for the higher n_0-values, involving n-state changing transitions, to another ionization mechanism for the lower n_0-values, dynamic tunneling.

Several tunneling-based formulae, whose validity are said in the literature to depend on various (double) inequalities being satisfied, have often been used to give a dynamic tunneling interpretation to "low-freqency" laser-multiphoton experimental data for atoms and ions of noble gas and other species. The accuracy of the microwave ionization data presented in [122], however, was good enough to show that these formulae are quantitatively inaccurate for excited hydrogen atoms. Some of the formulae underestimate the ionization threshold amplitude F by a factor of two. This is a huge error for a tunneling ionization process, in which the ionization rate grows exponentially with the field amplitude. Moreover, some of the inequality conditions previously claimed in the literature for these formulae to be valid were also shown in [122] not to be accurate.

At the low scaled frequencies for which dynamic tunneling comes into play as the dominant ionization mechansism in an oscillatory field, chaotic classical dynamics really has little to do with the physics. Classical chaos, of the transient type for orbits that become irregular [127] for awhile and eventually escape and ionize, requires the electron to be strongly externally driven but still to stay in the vicinity of the proton long enough for the Kepler orbital motion and back-and-forth jiggling motion induced by the external field to become strongly mixed. When the external field oscillates very slowly on the time scale of the initial Kepler orbit(s), the electron on an orbit that is going to ionize receives such a large energy and momentum boost from the external field that it just whizzes off to infinity. One may readily see by following 2d classical orbits in configuration space on the graphics screen of a personal computer that ionizing orbits generally do not look chaotic at all for very low scaled frequencies.

The following sections explore how high the scaled frequency must be for transient chaos [127] to play a significant role in the classical ionization in an oscillatory field.

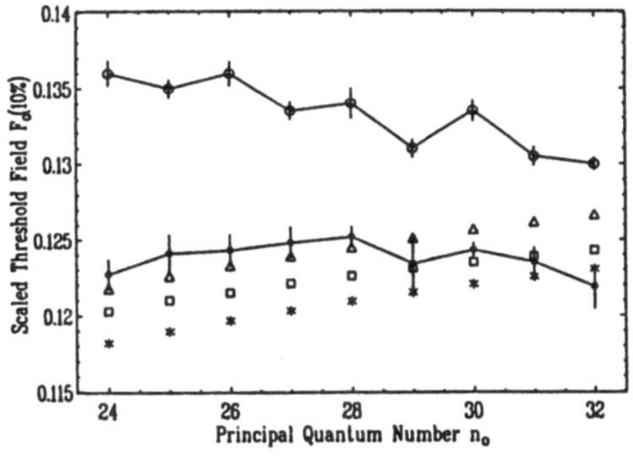

Fig. 6. *Scaled thresholds $n_0^4F(10\%)$ for 10% "ionization" probability as a function of n_0. The symbols are the same as those in Fig. 5. The experimental error bars show the reproducibility for each n_0-value. Not shown is an overall 5% uncertainty in the determination of the absolute experimental field strength.*

7 Regime–II: The Low Frequency Regime

For $0.05 \lesssim n_0^3\omega \lesssim 0.3$, some experimental "ionization" curves have shown structures [76,78,79, 80,104,103,116] including non-monotonic bumps, steps, or changes in slope. Here, unlike in Regime–I, quantal excitation or de-excitation resonances, respectively, decrease (Fig. 4a) or increase (Fig. 4b) the stability of the real atom compared to its classical counterpart. Both frames in Fig. 4 are taken from the large number of graphical comparisons in [116] between experimental data with 3d atoms and theoretical results including classical 1d and 3d calculations and calculations based on a quantal 1d theory [115] that exploits the low scaled frequency via use of an efficient, adiabatic basis of quantal states. This theory is said to be applicable to $n_0^3\omega \lesssim 0.4$. Whereas this 1d quantal theory and the two mentioned three and four paragraphs below correctly explain and even predict the n_0-values and field amplitudes F at which such structures appear for a given driving frequency, none reproduces the complete F-dependence of $P_{\text{"ion"}}$ for the experiments with 3d atoms.

To make "apples *vs.* apples" comparisons in Fig. 4, one should directly compare each classical 3d curve with the corresponding experimental curve (*i.e.*, the quantal analog computation!) obtained with 3d atoms, and then compare each 1d classical curve with the corresponding 1d quantal curve.

The 1d quantal theory [115,116] used in Fig. 4 gives insight into the switchover discussed in Sec. 6, see also Figs. 5,6, from the time-integrated static-tunneling model to the ***ad hoc*** but now physically motivated "adiabatic" model, also discussed there, that assumed strongly mixed n_0 and (n_0+1) states. The coupling constant [115] between 1d adiabatic states n_0 and (n_0+1) is $C_{n_0} = 1.5 n_0 (n_0^4 F)(n_0^3 \omega)$, where F and ω are in unscaled atomic units. For fixed n_0 and ω, C_{n_0} varies as n_0^4. In particular, the ratio $C_{24}/C_{31} \simeq 1/3$. Thus, the $n_0 = 31 \rightarrow n = 32$ excitation amplitudes are three times greater than those for $n_0 = 24 \rightarrow n = 25$. For similar values of $n_0^4 F$, as $n_0^3\omega$ grows even higher, couplings from the initial adiabatic state to even higher adiabatic states will obviously become appreciable. This supplies a suggestive mechanism for many strongly coupled states coming into play for scaled frequencies above those in Regime–II, and the resultant switchover to more classical behavior in Regime–III discussed in the following section.

However, the above adiabatic picture is based on a basis of quantal states that ride up and down on the oscillating Coulomb-Stark potential energy function. For high enough scaled frequencies, this adiabatic picture will break down, so one does not expect to follow it to arbitrarily high values of $n_0^3\omega$. We shall see in our discussion below of the high-scaled-frequency behavior in Regime–V that other simplifying approximations will give insight into at least some of the experimentally observed "ionization" behavior.

Another 1d quantal theoretical treatment of the low-scaled-frequency structures that characterize Regime–II was presented in [16,18,20,21] and included discussion of the quantal de-stabilization, or "subthreshold ionization" effect. These authors based their numerical calculations on zero-field hydrogenic basis states, including bound-continuum couplings. With this choice of basis states, many bound states were strongly coupled, even at low scaled freuqencies. (This shows how calculational details may vary with choice of basis, though in a fully converged calcuation, the results should be independent of this choice. The rub is what happens when the size of the basis is truncated for calculations.) Physical interpretation of the results of these calculations were, in part, based on state-mixing effects near avoided crossings of Floquet states of the [atom + field] system. These calculations led to a prediction that a "spiky" substructure should be present in $P_{\text{"ion"}}$ *vs.* F, at least for 1d atoms. Each "ionization" spike would be caused by upward transitions to so-called window states, with the transitions mediated by interactions at avoided-crossings of Floquet states. "Ionization" was explained theoretically as occurring from the window states, those that are coupled strongly to continuum states. Thus far, such spikes have not been observed in our experiments [106], but this may be a consequence of experimental conditions not allowed for in the theory, such as use of 3d atoms and actual pulse shapes. Rigorous application of the Floquet method requires the driving field amplitude, period, and polarization to be constant for all time, whereas in real experiments the amplitude varies in time according to the pulse-shape, $A(t)$ in Eq. 4.

Subsequent quantal 1d calculations [26,25], see also [23,24,27,28,29,30,31,32], based on Floquet methods have assessed how slowly $A(t)$ must vary in order that pulse-shape-dependent effects are relatively negligible — for $\varphi = 0$ or π in Eq. 4, only a few field oscillations for low scaled frequencies. Though these calculations have not yet included continuum states, the endpoint for true ionization, they give important insights into the contributions to "ionization" of de-excitation and excitation processes among the bound states. In particular, they have shown how the time scale associated realistic pulse shapes can greatly truncate the effective number of Floquet states that must be included in an approximate treatment of the strong-field 1d dynamics.

"Ionization" by very short pulses (or, for that matter, predicted atomic "stabilization" against ionization in such short pulses, see the text just above Sec. 4.1) will likely require very strong field amplitudes that will involve non-negligible couplings to continuum states. If Floquet methods are to be of use in this case, it will be necessary for the Floquet theory to treat the continuum in some way; see *e.g.*, [44,6].

8 Regime–III: The Semiclassical Regime

Fig. 7 shows and compares with 3d classical calculations [113] some 9.9 GHz "ionization" data [87,83,84] obtained with the PD and AD methods discussed in Sec. 2.2 for hydrogen atoms with each n_0-value in the range 32–90. The open circles are scaled 10%-thresholds $n_0^4F(10\%)$, taken over the average of several experimental runs; the open squares show average 90%-thresholds, $n_0^4F(90\%)$. These "ionization" thresholds are plotted *vs.* scaled frequency $n_0^3\omega$ over the range 0.05–1.1. Straight lines have been drawn between neighboring 10% data points and between neighboring 90% data points. The two vertically displaced curves produced in this way are similar: on the average they decrease with increasing $n_0^3\omega$ until they start to rise again for $n_0^3\omega$ above about 0.8. However, each curve also displays a number of structures that appear as series of local maxima or minima. One easily notices that most local maxima occur at values near $n_0^3\omega = 1/p$, where p is an integer value in the range 1–6, but these are not the only maxima. Compared to the peak at 1/2, the comparable, neighboring peak at $n_0^3\omega$ near 0.57 in the 10% data becomes the dominant peak in the 90% data. The 10% curve has a noticeable peak at $n_0^3\omega$ near 0.4, or 2/5, and there are also other "shoulders" in the data, such as for $n_0^3\omega$ near 2/3.

The dashed curves in Fig. 7 are the results of 3d classical Monte Carlo calculations [116,113] that modeled important features of the experiments [91,116,82,87], such as the

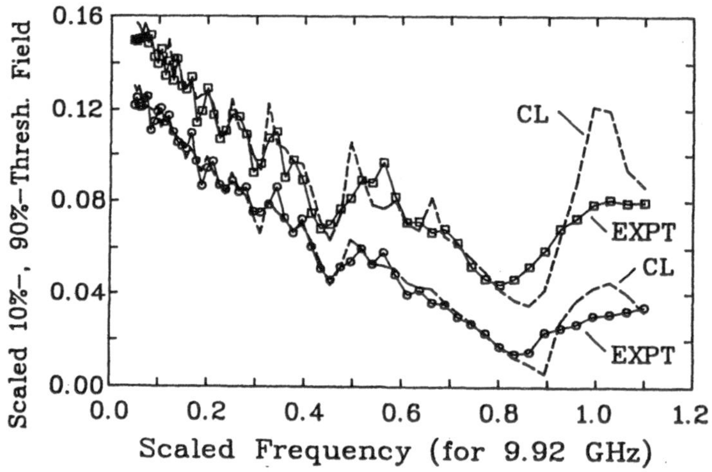

Fig. 7. *Classically scaled threshold amplitudes* n_0^4F vs. *classically scaled frequency* $n_0^3\omega$ *that were obtained from 9.92 GHz measurements of the 10% (○) and 90% (□) "ionization" probability. 3d hydrogen atoms prepared in each of the principal quantum numbers* n_0*=32,...,90 were used in the measurements. The dashed lines show the results of 3d classical Monte Carlo calculations that modeled the experiments.*

amplitude envelope function $A(t)$ in Eq. 4 and the n-cutoff n_c, see Sec. 2. There were no adjustable parameters in the calculations that used a microcanonical distribution of initial orbits having fixed initial principal action I_0, which corresponds to the uniform distribution of quantal substates of a given n_0-manifold [111] that entered the microwave cavity in the experiment .

The classical results reproduce well the general trends in the experimental data. Below $n_0^3\omega$=0.8, with exceptions near some local maxima in the data, the agreement can be at the few percent level. We emphasize the quality of this agreement by noting that the calculations were carried out with no adjustable parameters, and the ***absolute***, not relative, comparison between experiment and theory using scaled variables n_0^4F *vs.* $n_0^3\omega$, is presented on a ***linear plot***, not on a logarithmic one. Moreover, the classical calculations also approximately reproduce the experimentally observed local maxima and minima.

However, at the low end of scaled frequencies for which data are shown in Fig. 7, one may discern that the measured 10%-, and particularly the 90%-threshold amplitudes lie systematically below the classical calculations. Indeed, it was this systematic difference that motivated the dynamic tunneling experiments that were discussed in Sec. 6 for even lower scaled frequencies. The reader may glance again at Fig. 6 to see a continuation down to n_0=24 of the 9.9 GHz 10%-data of Fig. 7, for which the lowest n_0-value was 32. In Fig. 6 one clearly sees the systematic lowering of measured (quantal) 10%-thresholds below classical values.

Above $n_0^3\omega$=0.8 in Fig. 7, the agreement between experiment and classical calculations is only qualitative. A likely contribution to the lack of quantitative agreement is as follows. Since the experiment was carried out at fixed driving frequency ω, higher values of scaled frequency were achieved by increasing n_0. As n_0 approached the n-cutoff n_c, the amount $(n_c - n_0)$ of upward excitation in n that was needed to produce the excitation-contribution to the "ionization" signal decreases. The highest value of n_0 used for the data in Fig. 7 was 90, and the value of n_c was estimated to lie between 90–95. Because the data for $n_0^3\omega$=0.8–1.1 were obtained for value of n_0 that approached n_c ever more closely, the effect of the uncertainty in the precise value of n_c would become increasingly magnified.

However, one might expect this uncertainty to produce a systematic shift, which is not the case for $n_0^3\omega$ between 0.8–1.1. Note that the experimental 10% and 90% data lie ***above*** the classical calculations for $n_0^3\omega$ between about 0.8–0.92 but ***below*** them between about 0.92–1.1. Assuming that excitation upward in n was dominating the experimental "ionization" signal for $n_0^3\omega$=0.8–1.1, one concludes that the classical model ***overestimates*** the amount of upward excitation for n_0=0.8–0.92 but ***underestimates*** it for 0.92–1.1. Note that the underestimation occurs right at the position of a large local maximum in the classical calculations, which, as is discussed next, is the result of a nonlinear trapping resonance near the scaled frequency 1/1. Based on the comparisons shown in Fig. 7, one concludes that for the range 0.05–1.1 of scaled frequency, the classical model is quantitatively least accurate near the resonances causing the local maxima in the data.

A simple and direct interpretation for these local maxima comes from a $(1+1)$-dimensional (space + time) treatment of the classical dynamics: the electron moves in a Coulomb potential $-x^{-1}$ for $x \geq 0$ and reflects from a hard wall at x=0. This is the so-called SSE model [66] named after a real physical system, surface-state-electrons. An electron *in vacuo* above the planar surface of liquid helium sees a weak attractive Coulomb potential between it and its image charge below the surface, but Pauli exclusion produces a repulsive potential of about 1 eV (huge compared to the SSE binding energy in the meV range) that prevents it from going into the liquid. Such surface-state-electrons have been produced experimentally [58], and when the electron areal density is low enough to avoid interactions between different SSE's, they exhibit a hydrogenic energy spectrum. Moreover, because of the lack of reflection symmetry for the potential and, hence, for the classical orbits or quantal wavefunctions, the SSE have large permanent dipole moments and exhibit large Stark-energy-shifts in static electric fields applied normal to the helium surface [58].

Though this actual physical system has not yet been used for microwave "ionization" experiments, Jensen [66] began the use of the SSE *model* for a classical interpretation of microwave "ionization" experiments with real 3d hydrogen atoms. Quantal applications of this 1d model followed soon after. (See [3] for a report of static field ionization experiments carried out with surface-state-electrons at electron areal densities high enough for electron-electron correlation effects to be important. These were modeled theoretically with use of an effective potential.)

Poincaré sections in an action-angle (I, θ) phase space [67] of the bound classical motion in the driven SSE model can reveal around elliptic fixed points nonlinear resonance ***islands of stability*** embedded in a chaotic sea. This is the divided phase space well known from Hamiltonian theory [42,97]. The overlap of adjacent chains of resonant islands as a simple mechanism for the onset of global chaos in driven, low-dimensional model systems was first emphasized by Chirikov [42]. Meerson, Oks and Sasorov [101] first realized that this stochastic instability of a driven nonlinear oscillator would supply a classical mechanism for explaining the onset of microwave ionization of hydrogen atoms. Using the SSE model, Jensen [66] showed the first Poincaré sections that graphically illustrated this classical mechanism for the driven, 1d Kepler problem, and he made simple Chirikov-resonance-overlap estimates for the microwave amplitudes at which global chaos and, hence, ionization, would begin classically.

The stabilizing influence of the resonance islands supply a classical explanation for the experimentally observed local stability around certain values of the scaled frequency, *i.e.*, the local maxima in Fig. 7. Using the classical SSE model Sanders *et al.* [120] carried out numerical calculations for the onset of chaos and compared their results to 10% thresholds measured in our laboratory for 3d hydrogen atoms. They found distinct peaks centered at scaled frequencies $1/p$, $p = 1, 2, 3, 4, 5$, just as in the experiments, Fig. 7, and they also found local stability near $n_0^3\omega = 2/3$.

In their classical SSE-model calculations, Blümel and Smilansky [20] estimated thresholds for the onset of global chaos, which permits ionization, by determining when the action-half-width of the $1/p$ primary islands grew enough for these resonances to touch the higher-order $2/(2p-1)$ resonances. Their estimates generally reproduced the measured 10%-thresholds presented in Fig. 7 to better than 30%. They also found that an interpolation based on higher order classical perturbation theory, which necessitated solving an implicit equation, was reasonably successful for calculating the relative variation of the thresholds around the classical resonances. Moreover, they showed that this result can also be obtained from quantal perturbation theory.

A different approach using the SSE model is the 1d adiabatic low frequency quantal theory of Richards [115], further developed in [116]. As was discussed in Secs. 6,7, it uses a quantal basis of adiabatic states that one may truncate at sufficiently low frequency to only two adiabatic states that are strongly coupled, which allows analytic forms to be derived. One finds resonances when $n_0^3\omega$ is near $[n_0(n_0+1/2)]/[(n_0+1)^2 p]$, where p is a small integer; this result also agrees well with that of the 1d classical theory.

8.1 Classical Kepler Maps for 1d Motion

For bounded motion the 1d classical Hamiltonian in Eq. 4 can be rewritten [101] in terms of the action variable $I = (2\pi)^{-1} \oint pdq$ and conjugate angle variable θ

$$\begin{aligned} H(I,\theta,t) &= -\frac{1}{2I^2} + FI^2 \sum_{m=0}^{\infty} V_m \cos(m\theta - \omega t) \qquad (7) \\ V_0 &= \frac{3}{2}; \\ V_m &= -\frac{J'_m(m)}{m} \approx -0.411 m^{-5/3} \end{aligned}$$

[67,125,16]. A Poincaré section in phase space, or phase-space portrait, is a useful way to show the dynamics of such a system. For time-dependent systems some authors plot the (I,θ) points once per forcing period [67,20,63], but we shall do it once per period $T(I)$ of the unperturbed Kepler motion, which depends on I (nonlinear system). The resulting equations define a 'Kepler Map' (KM) that projects the (I,θ) plane onto itself, giving points at time $t_0+T(I)$ from those at t_0 [56]. An initial condition $(I_0,\theta_0;t_0)$ for each orbit produces a set of points $(I_1,\theta_1;t_0+T(I_1));(I_2,\theta_2;t_0+\sum_{i=1,2}T(I_i));\ldots;$ *etc.*. There have been several derivations of Kepler maps [38,39,56,57,107].

Gontis and Kaulakys [56] derived a Kepler map (GK-map) using a rescaled energy $\epsilon = s^{-2/3} = (-2E)/(\omega Z)^{2/3}$ and rescaled amplitude $\Phi_0 = FZ/4E_0^2$. The initial total energy $E_0 = -Z^2/2n_0^2$, where $I_0 = n_0$ defines an initial principal quantum number (in atomic units, *au*) for the 1d atom and Z is the nuclear charge in units of $|e|$. The GK map is

$$\begin{aligned} \epsilon_{k+1} &= \epsilon_k - \pi\epsilon_0^2\Phi_0 f(\epsilon_{k+1},\theta_k) \qquad (8) \\ \theta_{k+1} &= \theta_k + 2\pi/\epsilon_{k+1}^{3/2} - \pi\epsilon_0^2\Phi_0 g(\epsilon_{k+1},\theta_k), \qquad (9) \end{aligned}$$

where

$$f(\epsilon_{k+1},\theta_k) = 4\epsilon_{k+1}^{-1}[\mathbf{J'}_{s_{k+1}}(s_{k+1})]\sin(\theta_k), \qquad (10)$$

and the function g follows from Liouville's theorem and the requirement that the map be area preserving. Note that this constraint produces an implicit mapping requiring that a set of equations be solved at each iteration. With use of Newton's method and analytical expansions given in [56], this is numerically straightforward. The minus signs in Eqs. 8,9 are not present in [56], but they are consistent with [39] and place the stable, elliptic fixed points of the primary islands at an angle $\theta = \pi$; in [56] the stable fixed points were at $\theta = 0$. In Eq. 10 J' is the derivative of the Anger function, for which Ref. [56] gives useful asymptotic expressions. However, note that the $\sin(\theta_k)$ factor in Eq. 10 was inadvertently omitted in Eq. 15 of [56].

L. Moorman, A. Haffmans, and the present author have used the GK-map to investigate

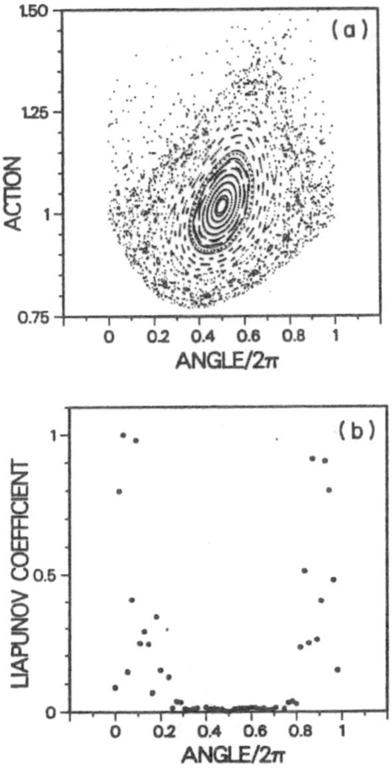

Fig. 8. *Frame (a): A phase-space portrait for 52 orbits, obtained with an area-preserving classical 2d map that models the 1d hydrogen atom driven by a microwave electric field. The initial scaled action for all orbits was 1.00, and initial angle variables were uniformly distributed betwen 0 and* 2π. *The scaled frequency was* $n_0^3\omega=1.00$, *and the scaled amplitude was* $n_0^4F=0.02$. *For each orbit the map was iterated 150 times. Frame (b): the result of calculating the Lyapunov exponent after 500 map iterations for each of the 52 orbits used for frame (a). Orbits with small Lyapunov coefficients, which are associated with those in the nonlinear trapping resonance clearly observable in frame (a), did not "ionize" in the calculation. (Most of) the orbits with significantly larger positive Lyapunov coefficients, which are clearly associated with the 'chaotic sea' in frame (a), did "ionize" in the calculation.*

the 1d classical dynamics. Our study of this and other Kepler maps was greatly facilitated by a menu-driven, graphics-intensive software package begun in TurboPascal© for MS-DOS© computers by a graduate student in our laboratory, Mr. S. Knop, and developed by Mr. Haffmans. We are grateful to Mr. Haffmans for preparation of Fig. 8. For a scaled frequency $n_0^3\omega$=1.00 and a microwave field suddenly switched on to an amplitude n_0^4F=0.02, Fig. 8a shows a phase-space portrait for 52 orbits all having initial action (initial scaled action of 1.00) corresponding to n_0=69, with angle variables uniformly distributed between 0 and 2π, and with 150 iterations of the map. The smallest 'bubble' near the center of the frame is a quasiperiodic orbit associated with the main nonlinear trapping resonance (island of stability) at $n_0^3\omega = 1/1$. Near its center is a stable, elliptic fixed point corresponding to an isolated periodic driven orbit, but the initial conditions used to generate this plot did not include this orbit. Surrounding the bubble are other quasiperiodic orbits and a few 'Birkhoff chains' of secondary islands [2]. Farther away is the 'chaotic sea' of points belonging to trajectories that will eventually rise to very large actions and "ionize". The abrupt scarcity of points below the chaotic sea near the bottom of the frame is caused by a confining 'KAM (Kolmogorov-Arnold-Moser) surface' (torus) [88,42,97].

Fig. 8b shows the results of using the program to run the same 52 orbits, but this time the vertical axis gives the Lyapunov exponent computed for each orbit after 500 map iterations with use of the method of Wolf *et al.* [132]. It is clear that the maximal width of the island in the angle direction, when sampled at fixed initial action of n_0=69, can be read immediately from the graph: all orbits inside the island have small Lyapunov exponents, on the order of 0.01. Orbits in the chaotic sea, however, have Lyapunov coefficients that are about an order of magnitude larger and fluctuate greatly with initial angle. The calculation of these positive Lyapunov exponents is the most direct demonstration that the orbits outside the island have an exponentially unstable character and may, therefore, be called transiently chaotic [127]. This supports Meerson *et al.* [101] having ascribed the classical ionization mechanism to the stochastic instability of a nonlinear oscillator and Jensen having interpreted phase-space portraits for the 1d classical Hamiltonian flow as exhibiting stable islands inside a sea of chaotic orbits [67].

The above examples were done with a sudden turn-on of the field with constant amplitude. We also used our graphics program to study the effect of different pulse shapes $A(t)$. Jensen [68] first brought clearly into focus how a slowly rising and falling pulse shape may enhance the stabilizing effect of the nonlinear resonance islands. While rigorous adiabatic theory on this problem is difficult, because orbits on or crossing seperatrix layers involve zero frequency and are thus never truly adiabatic, watching a phase-space portrait develop on a computer screen is highly instructive. If the amplitude F turns on slowly enough, the area inside a resonance island is an approximate adiabatic invariant. As the vertical extent (in action I here, usually called the island 'width' in the literature, so the reader should avoid confusion) grows, approximate conservation of phase-space area causes its horizontal extent (in angle) to decrease. It is *not* an attractor such as one has in dissipative systems, but it is as if the island sucks in orbits. If the amplitude is turned on suddenly, orbits outside the island (in angle) lay in the chaotic sea and eventually "ionize". When it is turned on slowly, however, many of these orbits become part of the island. If during the finite interaction time (see the discussion of transient chaos in Ref. [127]), they are not "ionized", one may say that they have been stabilized by the *combined* effect of the island and the slow turn-on. Conversely, we have seen occasional numerical counter-examples, particularly for relatively large amplitudes: there are orbits that would not "ionize" for a sudden turn-on that did "ionize" for a slow one. These crude statements call for further serious theoretical study.

Given the approximations used in deriving it, one expects the GK-map to follow reasonably well the dynamics of the 1d classical Hamiltonian for scaled frequencies $\gtrsim 1$. We have found the GK-map to compare favorably with the results of 1d classical Monte Carlo calculations for scaled frequencies as low as 0.5, or into the middle of what we label Regime–III. This is consistent with findings of Casati *et al.* [39], who calculated the error per map iteration. Though the GK-map may not approximate well the evolution of individual orbits at, say, $n_0^3\omega=0.5$, we have found it to reproduce reasonably well the averaged quantity of 'what percentage of orbits ionizes' for scaled frequencies at least this low.

9 Regime–IV: The Transition Regime

Fig. 9 shows in scaled units the same 9.9 GHz experimental 10%-thresholds and 3d classically calculated 10%-thresholds presented in Fig. 7; also shown are 36.02 experimental 10%-thresholds and the results of corresponding 3d classical calculations. (The 9.9 GHz and 36.02 GHz "ionization" curves in Fig. 3 give the respective 10%-thresholds shown in Figs. 7,9.) All these data were obtained with the AD or PD methods discussed in Sec. 2.2, and the respective n-cutoffs were $n_c^q \simeq 90$ and $n_c^p \simeq 75$. (Only AD data are plotted for $n_0 \gtrsim 70$.) The agreement for $n_0 \lesssim 0.8$ among the two sets of experimental and corresponding 3d classical 10%-thresholds, including the local maxima and minima, is a demonstration both of the accuracy of 3d classical calculations and of classical scaling being obeyed in this part of Regime–III. However, the agreement between the 9.9 GHz and 36.02 GHz 10%-thresholds for $n_0^3\omega$ between about 0.8–1.1 is not as good. This is, at least in part, an n-cutoff effect.

For each given value of $n_0^3\omega$ the initial n_0-value in the 9.9 GHz experiments were much closer to the n_c^q than they were in the 36.02 GHz experiments. Therefore, it is not surprising that for $n_0^3\omega$ between 0.8–1.1 in Fig. 9, the scaled 10%-threshold amplitudes for 9.9 GHz are systematically lower than those for 36.02 GHz. This also suggests that there were non-negligible amounts of microwave excitation probability to final bound states between about 80–90 for the $n_0^3\omega$=0.8–1.1 data taken at 9.9 GHz, compared to that for 36.02 GHz.

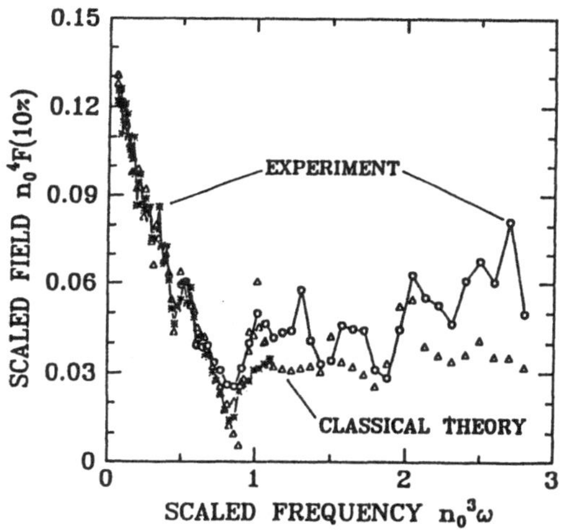

Fig. 9. *Scaled amplitudes for 10% "ionization" probability for 3d hydrogen atoms. (∗): 9.92 GHz experiments [82,87,91,116] with each n_0=32,...,90 and an n-cutoff $n_c \simeq 92$; (o): 36.02 GHz experiment [52] with each n_0=45,...,80 and an n-cutoff $n_c \simeq 89$; (△): 3d classical Monte-Carlo simulations [52,113,116] of the experiments.*

For $1 \lesssim n_0^3\omega \lesssim 2$, Fig. 9 show the classical dynamics beginning to break down. For example, the experimental bumps near $n_0^3\omega$=1 and 2 are partially reproduced by classical nonlinear trapping resonances around scaled frequencies 1/1 and 2/1, but the experimental dip near $n_0^3\omega$=1.5 and rise just beyond anti-correlates with the classical local stability near 3/2. Notice that the large measured peak near $n_0^3\omega$=1.3 is entirely absent in the classical calculations. (But see the next paragraph.) In Regime–IV, as in Regime–II, one sees that quantal effects may either lower experimental threshold fields below or raise them above classical values.

Fig. 10 shows more recent 10%-threshold data [123] in Regime–IV, obtained using the ED method discussed in Sec. 2.2 and with an n-cutoff $n_c^i \simeq 114$, about 15 n-units higher than for the data shown in Figs. 7,9. The data in Fig. 10 were obtained with each of three different driving frequencies, 26.43, 30.36, and 36.02 GHz and ranges of n_0-values; the respective pulse shapes for each data set are those shown in Fig. 2. Also shown are the results of 3d classical calculations that modeled the 36.02 GHz experiment, including the value of n_c^i.

Because they were obtained with the same flat-top pulse shapes (though not exactly the same durations in initial Kepler periods, see Secs. 2,3), let us focus first on the 26.43 GHz and 36.02 GHz data in Fig. 10. Each agrees quite closely for $n_0^3\omega$ below about 1.05 with the 3d classical calculations for 36.02 GHz, yet again a demonstration of the accuracy of 3d

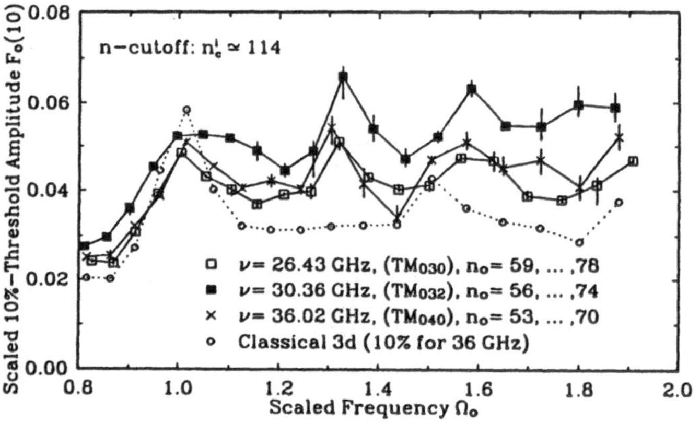

Fig. 10. *Scaled 10%-threshold amplitudes $n_0^4F(10\%)$ observed to produce "ionization" probability $P_{\text{"ion"}}$=10%* vs. *scaled frequency $n_0^3\omega$, for three different experimental data sets. All cases had the n-cutoff $n_c^i \simeq 114$ that was determined by a 3.4 V/cm static field outside the microwave cavity. Also shown are 10%-thresholds obtained from 3d classical Monte Carlo calculations carried out by Richards for the conditions of the 36.02 GHz experiment. Straight lines join each set of plotted points.*

classical calculations and of classical scaling being obeyed in Regime–III. For higher values of $n_0^3\omega$ in Regime–IV, however, even though the classical calculations no longer reproduce the experimental data, one notes that these two sets of data continue rather closely to obey classical scaling. There are certainly local exceptions, *e.g.*, for $n_0^3\omega$ near 1.4, 1.7, and just below 1.9, but, overall, the degree to which these data follow classical scaling is remarkable.

In particular, notice the significant peaks for $n_0^3\omega$ near 1.3 and just below 1.6. The former is entirely absent classsically; the latter is displaced towards higher $n_0^3\omega$ from the classical peak that is caused by the nonlinear trapping resonance at 3/2.

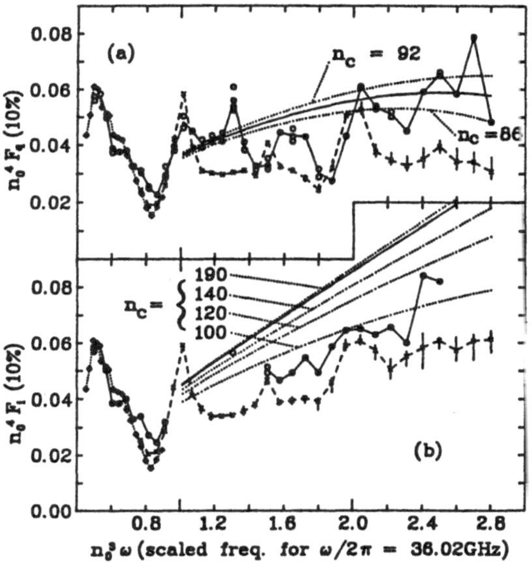

Fig. 11. *36.02 GHz experimental data for the scaled amplitudes n_0^4F for (a): 10%-quenching probability, obtained with the AD method, see Sec. 2.2, and (b): 10%-"ionization" probability, obtained with the ED method, see Sec. 2.2, for 3d hydrogen atoms with principal quantum numbers n_0 in the range 45-80, compared with theoretical results. (∘): experimental data from Ref. [52] for an n-cutoff n_c^q in the range 86-92 in frame (a) and n_c^i in the range 160-190 in frame (b). (×): 3d classical Monte-Carlo simulations from Ref. [52] of the 36.02 GHz experiments, including the experimental pulse shape and the effect of the different n_c-values in (a) and (b). (⋄): 3d classical Monte Carlo simulations from Ref. [113] of the 9.92 GHz experiments with $n_c \simeq 92$. Solid lines: the n_c-corrected 1d quantal delocalization border (ncqdb) for 10%-excitation probability above $\overline{n_c} = 89$ in frame (a) and above $\overline{n_c} = 175$ in frame (b). Eqs. (a1-a3) of Ref. [40] were used to evaluate these borders. The dotted lines, show how sensitive the calculated ncqdb is to variations in n_c.*

These features were also present in the first experimental penetration in 1988 [52] into what we now label Regime–V, that for $n_0^3\omega \gtrsim 2$, which the next subsection covers. Fig. 11 shows two sets of 36.02 GHz experimental 10%-thresholds for n_c^q between 86–92 for frame (a) and n_c^i between 160–190 for frame (b), and with use of the pulse shape shown for this frequency in Fig. 2. (The data in frame (a) are those shown for 36.02 GHz in Fig. 9.) Also shown are 3d classical calculations that modeled each set of data.

9.1 Nonclassical Local Stability and "Scars"

The comparisons so far between experimental 10%-thresholds and 3d classical theory displayed several examples of local stability, which means "ionization" threshold amplitudes that are higher for some n_0- or $n_0^3\omega$-values than those for neighboring n_0- or $n_0^3\omega$-values. We shall call classical local stability (CLS) those cases for which the effect is echoed by classical theory, where it is caused by the locally stabilizing influence of nonlinear trapping resonances at scaled frequencies such as the fundamental resonances at 1/1 and 2/1, the subharmonic resonances at $1/p$, p=2,3,..., and, possibly, higher-order classical resonances at q/p with $q, p > 1$. However, we recall the remark from Sec. 8 that even in the semiclassical Regime–III, right on these classical resonances the agreement between experiment and 3d classical theory is less good than off them. A prominent example in Fig. 7 is the double bump in the experimental 10%- and 90%-threshold data for $n_0^3\omega$ near 0.5: The classical theory exhibits only a single bump associated with the 1/2 trapping resonance. Moreover, notice that the second bump, the nonclassical one at $n_0^3\omega$ near 0.57, becomes the dominant one in the 90%-threshold data.

We shall call nonclassical local stability (non-CLS) those cases for which the effect is not found in classical calculations. At least three examples have just been given: $n_0^3\omega$ near 0.57, 1.3, and just below 1.6. (The perceptive reader will note that 1.3 is very near 4/3, where a higher-order classical nonlinear trapping resonance is located that might be expected to produce CLS. Indeed, there is a resonance island centered there in the classical Poincaré sections, but it encloses significantly less area than the lower-order islands and, therefore, does not play [72] an important stabilizing role.)

Nonclassical implies a quantal (or at least semiclassical) origin for the non-CLS, which is confirmed by Fig. 12. It shows a composite of 9.92 GHz [91,79,80,82,81,87] and 36.02 GHz [52] 10%-threshold data sets, each obtained with n_c^q near 90, along with the results of two independent quantal 1d-numerical calculations that modeled important experimental conditions, such as the pulse-shape, $A(t)$ in Eq. 4, and the value of n_c^q. That of [69,70] was a supercomputer calculation. That of [94] used a basis of states in the coomputationally-efficient "compensated-energy representation", which is physically justifiable [94] for $n_0^3\omega \gtrsim 2.5$, *i.e.*,in Regime–V; see Sec. 10. Notice that these quantal 1d numerical calculations approximately reproduce nearly all of the important resonant features of our 3d-experiments, in particular, the non-CLS at $n_0^3\omega$ near 1.3. One obviously wants to know the physical origin of the non-CLS, not just that it can be reproduced in large numerical calculations.

In 1989 [70] Jensen *et al.* used theory for the 1d model of the sinusoidally driven hydrogen atom to give a non-perturbative quantal explanation for at least some of the examples of non-CLS that were previously observed with 3d hydrogen atoms and were discussed above.

See also the review [72]. These authors ascribed observed examples of non-CLS to the stabilizing influence of "scarred" wavefunctions, which are somehow tied to the tangle of unstable periodic orbits around classical resonances and possibly to other features embedded in the the chaotic sea, all of which can be visualized in Poincaré projections of the action-angle phase-space.

The traditional textbook examples of semiclassical Bohr-Sommerfeld-Wilson [Einstein-Brillouin-Keller (EBK)] and Jeffreys-Wentzel-Kramers-Brillouin (JWKB) quantization procedures for single [multiple] degree of freedom systems are well known for classically regular motions. Here the central role played by classical periodic orbits is clear, introductory textbooks emphasizing that it is along these orbits that wave-mechanical constructive interference builds up wavefunction density. What is less obvious but is at the heart of the considerable activity in this area (for recent monographs, see [59,110]) is that in semiclassical treatments of classically chaotic systems, ***unstable periodic orbits*** play a key role. In a system with a phase space divided into regular and irregular regions, both stable and unstable periodic orbits exist. In a hyperbolic (fully chaotic) system there are only unstable periodic orbits. Though the periodic orbits proliferate exponentially with increasing energy in classically chaotic systems such as billiards, they are statistically negligible ('of zero measure') compared to the chaotic orbits. Nevertheless, the periodic orbits are the backbone of the semiclassical dynamics [59]. For a useful collection of recent articles on and

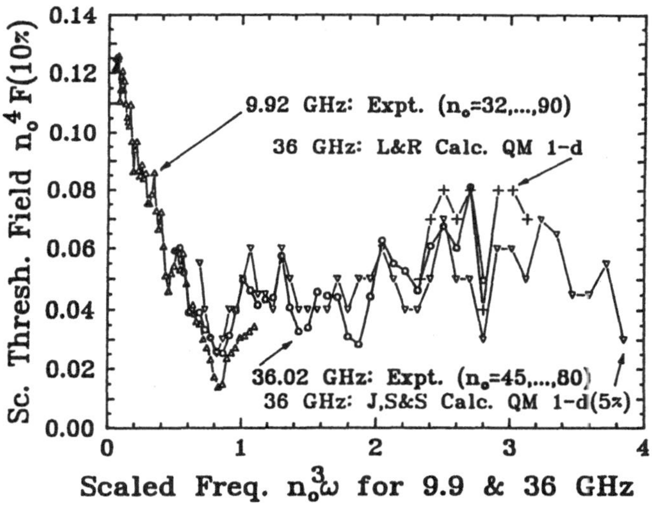

Fig. 12. *Experimental 10%-threshold data obtained with 3d hydrogen atoms for the frequencies and ranges of n_0 shown, compared to 1d quantal calculations for (∇): 5%-thresholds calculated in Ref. [69], and (+): 10%-thresholds calculated in Ref. [94] with use of the quasi-resonant-compensated-energy basis of states.*

many references to so-called periodic orbit theories (*e.g.*, Gutzwiller- , Selberg- , and other trace formulae, van Vleck propagators, *etc.*) now being investigated for use in semiclassical quantization of chaotic systems, mostly for model systems exhibiting "hard chaos", the interested reader should peruse Vol. 2, No. 1 (1992) of the AIP journal *CHAOS*; see also Smilansky's article elsewhere in this volume.

From numerical studies of the chaotic Bunimovich stadium billiards, Heller first demonstrated that the plotted density of configuration space wavefunctions for many excited states could be seen by eye to peak along one or more *unstable* classical periodic orbits. Heller said that such a quantal state is *scarred by the unstable periodic orbit(s)*, or more briefly, to exhibit a *scar* [62,108]. Subsequent numerical calculations on the stadium emphasized a phase-space approach [47]. Indeed, it has been conjectured [14] that a phase-space approach offers the most fruitful approach for developing a general theory of scars, which is currently lacking. Scars have been found theoretically in other time-independent problems, such as the hydrogen atom in a uniform magnetic field [130,46] and two coupled anharmonic oscillators [129]. Recent calculations on the quantum kicked-rotor [112] and the hydrogen atom in strong oscillatory fields [69,72,71] have exhibited scars, so it is clear there is an important role for them to play in time-dependent problems, too.

Now we return to the explanation for non-CLS observed in [52,86,84,123]. Jensen *et al.* (JS^3) [70], see also [72], decomposed each calculated strong-field 1d wavefunctions $\Psi(t)$ on a basis of quasienergy states (QES) and projected coarse-grained-Wigner (or Husimi) transformations of each QES onto a classical phase-space portrait calculated in action-angle (I,θ) variables for the same parameters as $\Psi(t)$. Relevant earlier work had been performed by Stevens and Sundaram (SS) [126], who used coherent states on a basis of unperturbed states and plotted results in an (x,p_x) phase-space representation. SS found that for $n_0^3\omega\gtrsim 1$ and at $n_0^4F=0.04$, the quantal phase-space evolution could remain localized while the classical evolution was diffusive. In this regime, SS commented that the quantal evolution seemed to be better characterized as tunneling rather than diffusion. They thought this behavior was consistent with the quantal transport being inhibited by cantorus-barrier(s), see [99,102].

JS^3 concluded from their calculations [70] that the inhibition of quantal transport upward in action (suppression of "ionization") could be due to excitation of individual QES whose Husimi transformation was highly localized near classical unstable periodic orbit(s).

With the Hamiltonian expressed in action-angle variables, JS^3 constructed Poisson-like coherent states on a basis of unperturbed states

$$|I,\theta>=\sum_{n=0}^{\infty}[A_n(\alpha)I^{\alpha n}e^{-\alpha I}]^{1/2}e^{i2\pi n\theta}|n>, \tag{11}$$

where, consistent with Heisenberg's uncertainty relation, the squeezing parameter α determines the relative θ- and I-widths of the wavepacket. These can be viewed as the coarse-grained probe functions in the procedure of JS^3. A state $\rho_n(I,\theta)=<n|I,\theta>$ would be peaked at action $I=n$ as a squeezed Poisson distribution whose I-width is $\Delta I=\sqrt{(\alpha n+1)/\alpha^2}$ and whose θ-width is maximal, *i.e.*, uniformly distributed in θ between 0 and 1 in units of 2π. The Husimi distribution of an arbitrary coherent state is defined by a projection of the wavefunction on the probe functions

$$\rho(I,\theta)=|<I,\theta|\Psi>|^2, \tag{12}$$

where

$$\Psi = \sum_{n=0}^{\infty} a_n |n> . \tag{13}$$

Using this prescription JS[3] carried out numerical experiments for a relatively slow turn-on of the amplitude, 20 field oscillations, which is comparable to the experiments reported in [52,123] with 3d atoms, see Fig. 2. The typical calculated situation in the strong field after the turn-on was a wavefunction made up of several QES. The Husimi distribution for one or more of these QES could be concentrated in certain regions of the classical phase space, whereas others extended to actions far above the initial value n_0 and were obviously the ones that were contributing to the "ionization". An example of this is n_0=61 in a 36.02 GHz field, for which $n_0^3\omega$=1.24; at n_0^4F=0.05, about 6 QES were appreciably excited, some of which projected onto very high actions.

A quite different situation was found for certain initial conditions, the most prominent of which was n_0=62 in a 36.02 GHz field, which corresponds to $n_0^3\omega$=1.30. After the 20-oscillation turn-on to a final scaled amplitude n_0^4F=0.05, JS[3] found the 1d wavefunction to be 97.8% a *single* QES. (This degree of wavefunction purity was quantitatively confirmed in other 1d quantal Floquet calculations by Breuer *et al.* [30], see Secs. 7,10.) Fig. 13 shows the Husimi distribution of that calculated QES projected onto the classical phase-space portrait calculated for the same conditions. The closed solid curves from the quantal calculations are contour lines of equal probability density, much as in a topographic map. The heart-shaped structures from the classical calculations surround the main classical resonance centered at an action giving the scaled frequency very near $I_0^3\omega = 1/1$; the six small blobs outside the hearts are a Birkhoff chain of secondary islands [2], and the sprinkling of dots is the 'chaotic sea', sharply delineated from below by a confining KAM-curve. Note the area enclosed by the box in the upper right-hand corner of Fig. 13: it approximately shows the quantal cell in phase space for the calculated parameters. (It is actually shown slightly smaller than it should be.) Crudely, it is over this amount of area that quantum mechanics washes out any structure in the classical phase plane shown in the figure.

The Husimi distribution in Fig. 13 shows no contour lines above about n=72 nor below about n=52, which means that the support of the wavefunction is restricted to a band of n-values (action) between these values. However, notice in Fig. 13 that the Husimi distribution is not uniformly spread in θ: it is concentrated in the chaotic sea ***outside*** the main classical nonlinear resonance but near an unstable (hyperbolic) periodic orbit at the 'X-point' near n=56 and θ=0,1 and also near its stable and unstable manifolds, which are shown by the (homoclinic) oscillating curves of nested chevrons.

A glance back at Figs. 9–12 will remind the reader that n_0=62 exhibited a non-classical local stability in the 36.02 GHz driving field: its 10%-"ionization" threshold was significantly higher than those of its neighbors, n_0=61,63, and this local stability was not found in the 3d classical calculations that simulated the experiment [52]. According to JS[3], the quantal reason for this local stability is shown by Fig. 13 and the discussion above. We recall further from Figs. 7,9–12 that the 10%-threshold for n_0=62, approximately 0.055 for the 36.02 GHz data (and just below this for the 26.43 GHz data for n_0=69), was insensitive to the n-cutoff being raised from n_c^q=86–92 to n_c^i=114 [123], or even 160–190 [52]. This is also neatly explained by Fig. 13. Both n-cutoffs lie well below the range of action over which the Husimi distribution of the experimentally excited QES is concentrated; hence, even the lower n-cutoff does not 'chop off' the top of the Husimi distribution.

The intriguing aspect of the semiclassical explanation of JS[3], which was based on 1d calculations for the non-classical local stability that was observed with 3d atoms, is its intimate connection with strong-field, semiclassical wavefunction quantization on invariant structures in the classical phase space. (Of course, one's attention need not be restricted to unstable periodic orbits. Any invariant phase-space structure is a likely candidate, including cantori; see Sec. 10, Ref. [112], and Prange's article elsewhere in this volume.)

An obvious experimental test of the idea advanced by JS[3] to explain our 1988 36.02 GHz data [52] was to see if the non-classical local stability scales classically: if one varies n_0

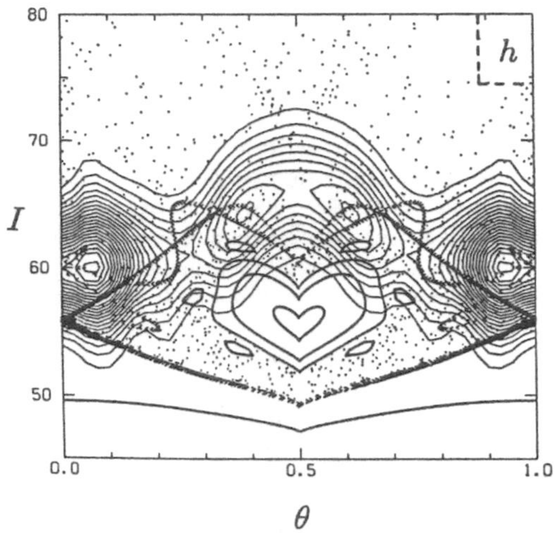

Fig. 13. *The contours show a coarse-grained Floquet wavefunction (Husimi distribution) obtained from a 1d quantal calculation for an n_0=62 1d hydrogen atom in a 36.02 GHz field with scaled amplitude n_0^4F=0.05 after a slow (20 field oscillation) turn-on. It has been projected onto a phase-space portrait of the classical motion for the 1d atom driven by the same field. The closed, heart-shaped curves are associated with the nonlinear trapping resonance island at scaled frequency $n_0^3\omega = 1/1$, which is surrounded by a Birkhoff chain of secondary islands and a 'chaotic sea'. The chevrons show the stable and unstable manifolds of orbits associated with the unstable, hyperbolic fixed point near the left/right edge of the figure. Note that the projected 1d 'scarred' wavefunction is concentrated in the chaotic sea near the hyperbolic fixed point and the stable and unstable manifolds. The phase-space area associated with Planck's quantum of action h is shown (slightly too small) in the upper right-hand corner. Reproduced from Ref. [70] with permission; see also Ref. [72].*

and ω, does the stability remain tied to certain values of the scaled frequency $n_0^3\omega$? That this is indeed the case was discussed earlier and is shown in Fig. 10.

However, what has not yet been mentioned is that one of the three cases in the experiments in [123] involved a rather different pulse shape from the other two: the 30.36 GHz data were obtained using the three-lobed pulse shape shown in Fig. 2, which should be contrasted with the other two, flat-topped ones in that figure. In Fig. 10 one readily certain notices differences between the scaled 10%-threshold data sets: *(i)* at each value of $n_0^3\omega$, the $n_0^4F(10\%)$-values for 30.36 GHz are systematically higher than those for the other two frequencies; *(ii)* the (CLS) local maximum for $n_0^3\omega$ near 1 is broader for 30.36 GHz than for the other two frequencies.

Despite these detailed differences, the non-CLS local maxima for $n_0^3\omega$ near 1.3 and just below 1.6 in the 30.36 GHz data set are even more striking than those in the data sets for the other two frequencies.

These experimental results show that the scar-induced stability mechanism for non-CLS in these cases was not affected by changes in the experimental pulse shape, although it should be emphasized that all three of those shown in Fig. 2 had comparable initial rise- and final fall-times. The theoretical work of Breuer *et al.* [30] showed that in 1d quantal Floquet calculations on a basis of bound states that the scar-induced stability (explained by these authors via the relative stability of 'continuously connected' Floquet states) for n_0=62 at 36.02 GHz would be established for pulse rise times longer than about five field oscillations. We note that all of the pulse shapes in Fig. 2 fulfill this condition. These authors also found that the comparable time scale for the stability at $n_0^3\omega$ near 1/1, *i.e.*, for n_0=57 at 36.02 GHz, was about a factor of ten longer. This may help to explain the remark made earlier about the local maximum in the 30.36 GHz scaled 10%-thresholds for $n_0^3\omega$ near and, particularly, above 1 being wider than the corresponding ones near $n_0^3\omega$=1 in the 26.43 GHz and 36.02 GHz data sets.

One should emphasize that Refs. [86,123] present a convincing argument that the cases of non-classical local stability observed for $n_0^3\omega$ near 1.3 and just below 1.6 are important for the driven 3d atom, not just for the model 1d atom. These results suggest that enhanced stability due to scarred wavefunctions is a general feature of the periodically driven 3d hydrogen atom, whose classical counterpart moves for linearly polarization on surfaces in a 5-dimensional extended phase space. It is entirely an open theoretical/computational question about how one can find, visualize, and understand the invariant structures in such a high-dimensional divided phase space. Then, they need to be quantized. The recent experimental results [86,123] point to the desirability of confronting this theoretical challenge.

10 Regime–V: The High-Frequency Regime

When n_0^3 rises above 2, measured "ionization" and quench threshold amplitudes were shown in [52] to rise systematically above the results of 3d classical simulations. These measurements provided the first experimental confirmation of an earlier theoretical prediction of this quantal effect for high frequencies that was presented in a series of papers whose author lists included various combinations of the Milano-Pavia/Novosibirsk collaborators Casati, Chirikov, Guarneri, Shepelyansky and their students. See [125,38,39] for reviews and many references to their work. The prediction arose from their theory for the effect of *dynamical localization* in the quantum kicked-rotator (QKR) model system, as supported by numerical calculations for the rotator, for the quantum Kepler map model of the driven atom, and for sinusoidally driven 1d and 2d hydrogen atoms.

There is a large and still growing literature on the classical standard (or Chirkov-Taylor) map and its quantum counterpart, the QKR. Much attention has been focussed on the phenomenon of "dynamical localization" in the QKR and other models. For many details and references, the interested reader should consult elsewhere in this volume the article by Chirikov and the article by Prange.

Our laboratory observation [52] of this quantally enhanced stability in Regime–V was confirmed in a subsequent hydrogen experiment [10] with waveguide frequencies in the range 12–18 GHz, but see [83,84,85,106] for the care that should be exercised when different sets of experimental data and theoretical calculations are compared in order to decide whether or not experimental data confirm a theory.

Contrary to conclusions presented in [10], Refs. [83,84,85,106] found that photonic localization theory does not give good quantitative agreement with all experimental data obtained thus far for microwave "ionization" of hydrogen atoms at high scaled frequencies. Some of this may be seen in Fig. 11. The n-cutoff-corrected quantum delocalization borders (ncqdb) shown in Fig. 11 were calculated using the procedures given in [33,40], see also [43], for how to correct formulae for the "quantum delocalization border" for infinite n-cutoff in order to account for the finite n-cutoff in the "ionization" experiments. To show the sensitivity of the ncqdb to the actual value of the n-cutoff, frame (a) shows three different calculations for the ncqdb that span the range of n_c^q=86–92 in the 36.02 GHz data set obtained with the AD method, see Sec. 2.2. Similarly, frame (b) shows five different calculations for the ncqdb that span the range of n_c^i=160–190 in the 36.02 GHz data set obtained with the ED method, see Sec. 2.2.

Particularly for $n_0^3\omega \geq 2$, the calculated ncqdb's in Fig. 11a pass reasonably well through the middle of the experimental data whose up-and-down undulations signal the various classical and nonclassical local-stability mechanisms discussed elsewhere in this article. The ncqdb's calculated for values of n_c^i that lay within the experimental range 160–190, however, do not pass through the data in Fig. 11b for $n_0^3\omega \geq 2$: these ncqdb's are systematically above the experimental 10%-thresholds.

Refs. [83,84,85,106] noted this behavior and used it as evidence for the photonic localization theory not giving the correct distribution of excited bound atomic states (particularly far from the initial state) for excited hydrogen atoms driven by a strong, high-frequency microwave field. The time-averaged, stationary distribution of localized bound states used in the photonic localization theory was "borrowed" [39] from that derived for the quantal kicked rotator (QKR) model system.

Ref. [96] examined how well this distribution function borrowed from the QKR actually described bound-state distributions obtained in their 1d quantal numerical calculations that were designed to model the experimental results reported in [10]. The authors of [96] concluded that their numerical results were not well fitted quantitatively by the distribution function borrowed from the QKR. Therefore, the detailed, quantitative behavior of the distribution function remains an open question.

Another aspect of this distribution function in the photonic localization theory needing further work is its time dependence. A central feature of the dynamical localization theory is that only up to a "quantal break time" t_b after the interaction with the strong field begins does the quantal wavepacket spread, *e.g.*, in angular-momentum space for the QKR, or in the (positive or negative) number of absorbed photons for the the driven hydrogen atom treated by photonic localization theory [39]. For $t > t_b$, it is said that the time-averaged distribution function of states relevant to the particular system localizes, *i.e.*, freezes and no longer evolves in time. (An exception to this must be made for the driven hydrogen atom, which unlike the QKR, has an ionization continuum. Flux can leak out of the localized time-averaged distribution into the continuum. If the "localization length" in number of photons, say, is small compared to the number of photons needed to reach the continuum, or to reach the n-cutoff when one corrects for this effect, then the small amount of leakage is not supposed to alter the picture greatly. However, such ideas have not yet been made at all precise.)

Different microwave "ionization" experiments with excited hydrogen atoms, however, use pulses with different shapes and durations, not to mention different values of the n-cutoff or the presence in some experiments, *e.g.*, [10], of a static electric field superimposed with the microwave field. (As was pointed out in [85,96], the added static electric field can have important consequences that should not be ignored.) Photonic localization theory has not yet been adapted to deal with the problems of different pulse shapes and durations or the presence of a static electric field.

New microwave "ionization" data for $n_0^3\omega > 2$, *i.e.*, in Regime–V, have been obtained in our Stony Brook laboratory as part of B.E. Sauer's recently completely Ph.D. dissertation work. Though these results are still being written up for publication, we may make some remarks based on them. The new experiments were carried out for 3d hydrogen atoms exposed to the frequencies and pulse shapes shown in Fig. 2. Separate data sets for n_0-values as high as 84 were obtained with the AD and ED methods, see Sec. 2.2, for the respective n-cutoffs $n_c^g \simeq 92$ and $n_c^i \simeq 114$. For 36.02 GHz, the data extended up to $n_0^3\omega$=3.25, a bit higher than the 2.8 reached in previous experimental investigations [52,10]. The new 26.43 GHz and 36.02 GHz data obtained with the flat-topped pulse shapes in Fig. 2 generally confirm the behavior shown in Fig. 11. Specifically, when one compares these data to the ncqdb's calculated for the respective values of the n-cutoffs, the same behavior that was discussed above emerges: the ncqdb calculated for $n_c^g \simeq 92$ passes reasonably well through the center of corresponding experimental data, whereas the ncqdb calculated for $n_c^i \simeq 114$ is systematically higher than all corresonding experimental data.

For the 30.36 GHz data obtained with the three-lobed pulse shape shown in Fig. 2, however, whose 20% higher center lobe should control the "ionization" behavior, the behavior is opposite to that discussed in the previous paragraph. For this pulse shape and interaction time (which was similar to that used in the 12–18 GHz waveguide experiments reported in [10]), the ncqdb calculated for $n_c^q \simeq 92$ lies systematically *below* the corresponding experimental data, whereas the ncqdb calculated for $n_c^i \simeq 114$ passes more or less through the corresponding experimental data.

The only major difference between the three new data sets from Sauer's work is the pulse shape and significantly shorter near-peak-amplitude interaction time for the 30.36 GHz case. Based on these observations we conclude that if dynamical/photonic localization theory is to be of quantitative use to experiments, one must extend this theory to take the interaction time and pulse shape into account.

We may remark that similar high-frequency behavior, *i.e.*, a near-linear rise of scaled "ionization" thresholds with scaled frequency, has also been observed by Arndt *et al.* [4] with rubidium Rydberg atoms prepared in $(np)^2P_{3/2}$ states and driven with linearly polarized fields in a waveguide with frequencies between 10–15 GHz. However, because of unresolved questions concerning proper definitions of scaled frequency and scaled amplitude for Rydberg states of multi-electron atoms, it is unclear how to use theory developed for hydrogen atoms to interpret experimental results for such atoms.

The interested reader may wish to explore how the theoretical work by Casati and coworkers, reviewed or presented in, for example, [38,39,33] and references therein, has stimulated lively discussions at conferences and in the literature. Other authors have commented on and offered clarifications of this work, and some have advanced their own explanations for the enhanced quantal stability of sinusoidally driven, excited hydrogen atoms in what we call Regime–V; see, for example, [99,102,94,45,72,96] and references therein. It should be emphasized that Casati *et al.* first sharply focused our attention on the 'high-frequency' regime of the dynamics. Though their dynamical localization theory, thus far, has been incapable of reproducing the effects of resonances that are such a prominent feature of our experimental data, *e.g.*, in Fig. 11, other of their calculations [40] using the quantum Kepler map (see also [106]) reproduced fairly well some of the resonant structure displayed in the 36.02 GHz data from our group [52], see also Fig. 11, for both values of the experimental n-cutoff in those data, n_c^q=86–92 and n_c^i=160–190, respectively. This was so not only for the data in Regime–V, but also for data discussed earlier in this paper for Regime–III and Regime–IV. In particular, the quantum Kepler map reproduces the the non-classical local stability of n_0=62 at 36.02 GHz.

As was shown in [95], however, the quantum Kepler map is only a fair *local* approximation to the 1d quantal dynamics, *i.e.*, for transitions to states not too far from the initial state; because each iteration of the quantal map corresponds to a different time-scale for each quantal level, it is a poor approximation for transitions to very highly excited states, those near the ionization continuum. Therefore, it is reasonable that the quantum Kepler map may be a fair approximation for resonances that involve only a group of levels near the initial one. For nonclassical local stability caused by scarred wavefunctions whose Husimi distributions project onto only a relatively narrow band of classical actions, *e.g.*, as in Fig. 13, this limitation of the quantum Kepler map is less serious.

11 Conclusions

Beginning in Sec. 1, conclusions have appeared throughout this paper. In the author's opinion, insights gained from classical and semiclassical dynamical treatments of strongly perturbed Kepler problems involving many strongly-coupled quantal states are having an enormous impact on the recent progress being made in these problems. This review focussed on one of them, the ionization of excited hydrogen atoms by a linearly polarized microwave field. Other systems are given in Sec. 1, from which we emphasize the diamagnetic Kepler problem and the semiclassical quantization of doubly-excited helium atoms.

Central has been recognizing the importance of classical scaling relations for the Kepler problems. For microwave ionization this involves the concepts of scaled frequency and scaled amplitude, which has played such a key role in organizing our understanding of the different regimes of the dynamical behavior in this system. Time-dependent quantum mechanical systems subject to non-perturbative forces are generally *complicated*. Anything that simplifies such a problem, breaking it down into smaller pieces in which the important physical effects can be isolated and understood, is of great use. In the present case, we now generally know how the dynamics changes as the scaled frequency grows from very small values to large values. What we do not yet know is how to understand many of the important details of behavior in each of these regions.

One example is the importance of scarred wavefunctions in explaining nonclassical local stability observed in our experiments. The phenomenon appears to be a general feature in at least a few of the dynamical regimes identified for this problem, suggesting that it will be a general feature of related problems. This is already known theoretically for the diamagnetic Kepler problem. This naturally leads one to expect that scarred states should supply a new kind of basis. If one can see the stabilizing effect of such states, why not study them more directly. Do spectroscopy on them, drive transitions between them, perturb them: do all the things that atomic and molecular physicists do to more "normal" states of atoms and molecules.

These notions are still a bit speculative, because we still do not have a general theory of scars. For example, what is the set of classical phase space structures which can support scarred wavefunctions. This goes to the knotty problems of the topology of a divided classical phase space. The nonlinear classical dynamics is hard enough; then one has to quantize it. We know how to quantize classically regular regions, even those in systems with many degrees of freedom. The recent flurry of theoretical activity seems to have taught us how to do semiclassical quantization of fully chaotic regions, at least in model systems. The remaining challenge, therefore, is the semiclassical quantization of problems with a divided phase space.

General progress on this problem may come slowly, but it will come as a combined result of analytical and numerical treatment of model problems and of real physical systems. Only in the latter case can real experiments be done. The present author has experienced the joy of seeing experimental discoveries of unsuspected phenomena in a simple system stimulate intense theoretical activity that helps to lead to making sense of the discovery. It also goes the other way, with theoretical predictions or explanations stimulating new experiments. And so on, back and forth. There is still plenty of room left for these strong interactions between experiments and theory in the careful study of simple quantal systems whose classical counterparts are chaotic. Atomic and molecular physics will continue to play a cental role in these interactions.

Acknowledgements

This paper is dedicated to the memory of Jean-Claude Gay, who died far too young in Paris in the Spring of 1992. I was priveleged to have known him personally and professionally for over ten years, ever since we became acquainted during a sabbatical leave I spent in Paris in 1980-1. Jean-Claude became one of the trailblazers in quantum chaology among atomic physicists, though neither of us knew when we met that this is where we would end up. How nice it was to trade papers and compare notes occasionally as we both began to realize where we were heading. This was already becoming clear by 1982, at the Aussois conference in the French Alps that he co-organized. His early career as an experimenter evolved later toward a theoretical one, in his ongoing and powerful collaboration with Dominique Delande. Together they successfully tackled both analytic work and large scale calculations, never an easy combination, but Jean-Claude's background with experiments always made him aware of what was important to calculate and what needed to be understood deeply. He did not work on the problem of microwave ionization of hydrogen atoms, but he was always keenly interested in our latest progress and was always eager to tell me about his latest progress in his work, which concentrated on the diamagnetic Kepler problem. He and I both understood the relationship between these two problems that we were each enjoying so much. His good editorial taste and hard work paid off in the the articles on *Irregular Atomic Systems and Quantum Chaos* that he collected from many workers in this field. They recently appeared in journal form and as a collection organized into a book [53], that will endure as his gift to all of us. Death came too early, but as Jean-Claude often said, "Well, it is like that." I know that Dominique Delande and his colleagues will continue the fine work accomplished with Jean-Claude. A bientôt, Jean-Claude.

I appreciate many discussions with my experimental colleagues at Stony Brook and with others working around the world on the specific problem of microwave ionization and the general problem of quantum chaology. Their names are to be found among the lists of authors in the list of references and particularly among the author lists of papers of which I have been coauthor. The United States National Science Foundation has supplied continuous primary funding for this work, most recently under grant number PHY-8922506. Additional financial support is being supplied by Schlumberger-Doll Research and New York State.

References

[1] For the classical, non-relativistic hydrogen atom, the principal action is $I_n = I_r + I_\theta + I_\phi$; the total angular momentum is $I_\ell = I_\theta + I_\phi$; and the component of the angular momentum onto the polar axis is $I_m = I_\phi$, with $2\pi I_i = \oint p_i dr_i$ for $i = r$, θ, and ϕ (and no sum convention); see Refs. [22,111]. The quantization conditions become $I_n = n\hbar$, $I_\ell = k\hbar$, and $I_m = m\hbar$. In the old quantum theory $n = 1, 2, \cdots$ is the principal quantum number, and k is the subsidiary (or azimuthal) quantum number [22,124]. This is not $k = 1, 2, \cdots, n$, however, because modern quantization methods (Einstein-Brillouin-Keller) require $k = \ell + 1/2$ with $\ell = 0, 1, 2, \cdots, n$, in which the 1/2 refers not to the spin but to the Maslov (or Morse) index in semiclassical quantization. This index counts the number of classical turning points α encountered by a closed trajectory in the classical phase space; for a more general description in terms of caustics, see [12]. At each turning point a phase loss of $\pi/2$, which is equivalent to one-quarter of a wave, has to be taken into account. Continuity of the phase of the wavefunction then leads to $I = (n + \alpha/4)\hbar$. As one example, which is a case is usually associated with Wentzel-Kramers-Brillouin, it is well known that the one-dimensional harmonic oscillator, which has two turning points per period, has a 'zero-point' energy, *i.e.*, a non-zero energy when v=0 in the quantized energy expression $E_v = (v + 1/2)\hbar$, where $v = 0, 1, \cdots$. For further reading and to see how this may be understood from conditions to be satisfied under a coordinate transformation of a path integral from Cartesian into spherical coordinates, requiring a new term $\sim \hbar^2/4$ in the classical Hamiltonian to be added to the angular momentum part $|L|^2 = \ell(\ell + 1)\hbar^2$, giving $(\ell + 1/2)^2\hbar^2$, see p. 203 and p. 212 *etc.*, of Ref. [59].

[2] Secondary resonances are defined in a better way as the result of the perturbation Hamiltonian, and they give rise to secondary island chains in Poincaré sections of the phase space. An important difference between these resonances is that the strength of primary resonances depends only weakly on ϵ, namely as $\epsilon^{1/2}$, whereas for secondary resonances, it decreases much faster than this for small ϵ, where ϵ is the (linear) interaction coupling parameter. See Chap. 2.4b of Ref. [97].

[3] Andrei, E.Y. Yücel, S., and Menna, L. Phys. Rev. Lett. (1991) 67, 3704-7.

[4] Arndt, M., Buchleitner, A., Mantegna, R.N., and Walther, H. (1991) Phys. Rev. Lett. 67, 2435-8.

[5] Banks, D. and Leopold, J.G. (1978) J. Phys. B 11, 37-46 and 2833-43.

[6] Bardsley, J.N., Szöke, A., and Comella, M.J. (1988) J.Phys B 21, pp. 3899ff.

[7] Bayfield, J.E. and Koch, P.M. (1974) Phys. Rev. Lett. 33, 258.

[8] Bayfield, J.E., Gardner, L.D., and Koch, P.M. (1977) Phys. Rev. Lett. 39, 76.

[9] Bayfield, J.E. and Pinnaduwage, L.A. (1985) Phys. Rev. Lett. 54, 313-6.

[10] Bayfield J.E., Casati G., Guarneri I., and Sokol D.W. (1989) Phys. Rev. Lett. 63, 364-7.

[11] Bayfield, J.E. (1991) CHAOS 1, 110-113.

[12] Berry, M. (1981) 'Semiclassical Mechanics of Regular and Irregular Motion in Chaotic Behavior of Deterministic Systems', in *Les Houches Lectures XXXVI*, Eds.: G. Iooss, R.H.G. Helleman, and R. Stora, (North Holland, Amsterdam), pp. 171-271.

[13] Berry, M. (1989) Physica Scripta 40, 335-6.

[14] Berry M.V. (1989) Proc. Roy. Soc. London A423, 219-31.

[15] Berry, M.V. (1991) in *Chaos and Quantum Physics*, Eds.: M.-J. Giannoni, A. Voros, and J. Zinn-Justin (Elsevier, Amsterdam).

[16] Blümel, R. and Smilansky, U. (1987) Z. Phys. D 6, 83-105.

[17] Blümel, R. and Smilansky, U. (1988) Phys. Rev. Lett. 60 (1988) 477-80.

[18] Blümel, R. and Smilansky, U. (1988) in *The Structure of Small Molecules and Ions*, Eds.: R. Naaman and Z. Vager, (Plenum Press, New York), pp. 319-31.

[19] Blümel, R., Graham, R., Sirko, L., Smilansky, U., Walther, H., and Yamada, K. (1989) Phys. Rev. Lett. 62, 341-4; Blümel, R., Buchleitner, A., Graham, R., Sirko, L., Smilansky, U., and Walther, H. (1991) Phys. Rev. A 44, 4521-40.

[20] Blümel, R. and Smilansky, U. (1989) Physica Scripta 40, 386-93.

[21] Blümel, R. and Smilansky, U. (1990) J. Opt. Soc. Am. B 7, 664-79.

[22] Born, M. (1924) *The Mechanics of the Atom*, (republished by Frederick Ungar Publishing Co., New York, 1960).

[23] Breuer, H.P., Dietz, K., and Holthaus, M. (1988) Z. Phys. D 8, 349-57.

[24] Breuer, H.P., Dietz, K., and Holthaus, M. (1988) Z. Phys. D 10, pp. 13ff.

[25] Breuer, H.P., Dietz, K., and Holthaus, M. (1989) J. Phys. B 22, 3187-96.

[26] Breuer, H.P. and Holthaus, M. (1989) Z. Phys. D 11, 1-14.

[27] Breuer, H.P. and Holthaus, M. (1989) Phys. Lett. A 140, 507-12.

[28] Breuer, H.P., Dietz, K., and Holthaus, M. (1990) Nuovo Cimento 105B, pp. 53ff.

[29] Breuer, H.P., Dietz, K., and Holthaus, M. (1990) J. Phys. France 51, pp. 709ff.

[30] Breuer, H.P., Dietz, K., and Holthaus, M. (1991) Z. Phys. D 18, 239-48.

[31] Breuer, H.P. and Holthaus, M. (1991) J. Phys. II 1, 437-49.

[32] Breuer, H.P. and Holthaus, M. (1991) Annals of Physics 211, 249-91.

[33] Brivio, G.P., Casati, G., Perotti, L., Guarneri, I. (1988) Physica D 33, 51-7.

[34] Broyer, M., Delacrétaz, G., Ni, G.-Q., Whetten, R.L., Wolf, J.-P., and Wöste, L. (1989) Phys. Rev. Lett. 62, 2100-3.

[35] Buchleitner, A., Sirko, L., and Walther, H. (1991) Europhys. Lett. 16, 35-40.

[36] Burgdörfer, J. and Bottcher, C. (1988) Phys. Rev. Lett. 61, 2917-20; Kemmler, J., Burgdörfer, J., and Reinhold, C.O. (1991) Phys. Rev. A 44, pp. 2933ff.

[37] Casati, G., Chirikov, B.V., Guarneri, I., and Shepelyansky, D.L. (1987) Phys. Rev. Lett. 59, 2927-30.

[38] Casati, G., Chirikov, B.V., Guarneri, I., and Shepelyansky, D.L. (1987) Phys. Rep 154, 77-123.

[39] Casati, G., Guarneri, I., and Shepelyansky, D.L. (1988) I.E.E.E. J. Quantum Electron. 24 1420-45.

[40] Casati, G., Guarneri, I., and Shepelyansky, D.L. (1990) Physica A 163, pp. 205ff.

[41] Chen, Y., Halle, S., Jonas, D.M., Kinsey, J.L., and Field, R.W. (1990) J. Opt. Soc. Am. B7, 1805-1815.

[42] Chirikov, B.V. (1979) Phys. Rep. 52, 263-379.

[43] Chirikov, B.V. (1991) in *Chaos and Quantum Physics*, Eds.: M.-J. Giannoni, A. Voros, and J. Zinn-Justin (Elsevier, Amsterdam), pp. 443-545.

[44] Chu, S.-I. (1988) Adv. Chem. Phys. 73, pp. 2799ff; see also (1985) Adv. At. Mol. Phys. 21, pp. 197ff.

[45] Dando, P.A., and Richards, D. (1990) J. Phys. B. 23, 3179.

[46] Delande, D. and Gay, J.C. (1987) Phys. Rev. Lett. 59, 1809-12.

[47] Feingold, M., Littlejohn, R.G., Solina, S.B., Pehling, J.S., and Piro, O. (1990) Phys. Lett. A 146, 199-203.

[48] Ford, J., Mantica, G., and Ristow, G.H. (1990) in *Chaos / XAOC*, Proceedings of a Soviet-American Conference, Ed.: D. K. Campbell (American Institute of Physics, New York), pp. 477-493.

[49] Friedrich, H. and Wintgen, D. (1989) Phys. Rep. 183, 37-79.

[50] Fu, P., Scholz, T.J., Hettema, J.M., and Gallagher, T.F. (1990) Phys. Rev. Lett. 64, 571-4.

[51] Gajda, M., Grochmalicki, J., Lewenstein, M., and Rzążewski, K. (1992) Phys. Rev. A 46, 1638-53.

[52] Galvez, E.G., Sauer, B.E., Moorman, L., Koch, P.M., and Richards, D. (1988) Phys. Rev. Lett. 61 2011-4.

[53] Gay, J.-C., Editor, *Irregular Atomic Systems and Quantum Chaos* (1992) (Gordon and Breach, Montreux). This is a reprinting of articles that appeared in the two combined numbers of Vol. 25 of the journal Comments in Atomic and Molecular Physics (Nos. 1–3, 1991 and Nos. 4–6,1992).

[54] Goldstein, H. (1950) *Classical Mechanics*, (Addison-Wesley, 12th printing, 1977).

[55] Gomez Llorente, J.M., Taylor, H.S., and Pollak, E. (1989) Phys. Rev. Lett. 62, 2096-9.

[56] Gontis, V. and Kaulakys, B. (1987) J. Phys. B 20, 5051-64.

[57] Graham, R. (1988) Europhys. Lett. 7, 671-75.

[58] Grimes, C.C., Brown, T.R., Burns, M.L., and Zipfel, C.L. (1976) Phys. Rev. B 13, 140-47.

[59] Gutzwiller, M.C. (1990) *Chaos in Classical and Quantum Mechanics*, (Springer Verlag, New York).

[60] Halbach, K. and Holsinger, R.F. (1976) Part. Accel. 7, 213-22.

[61] Hasegawa, H., Robnik, M., and Wunner, G. (1989) Prog. Theor. Phys. Suppl. 98, pp. 198ff. The volume in which this review article appears was devoted to *New Trends in Chaotic Dynamics of Hamiltonian Systems* and contains other articles on related problems.

[62] Heller, E.J. (1984) Phys. Rev. Lett. 53, 1515-8.

[63] Howard, J. (1991) Phys. Lett. A 156, 286-92.

[64] Howard, J. (1992) Phys. Rev. A 46, 364-72.

[65] Iu, C.-h., Welch, G.R., Kash, M.M., Kleppner, D., Delande, D., and Gay, J.C. (1991) Phys. Rev. Lett. 66, 145-8.

[66] Jensen, R.V. (1982) Phys. Rev. Lett. 49, 1365-8.

[67] Jensen, R.V. (1984) Phys. Rev. A 30, 386-97.

[68] Jensen, R.V. (1987) Physica Scripta 35, pp. 668ff.

[69] Jensen, R.V., Susskind, S.M., and Sanders, M.M. (1989) Phys. Rev. Lett. 62, 1476-9.

[70] Jensen, R.V., Sanders, M.M., Saraceno, M., and Sundaram, B. (1989) Phys. Rev. Lett. 63, 2771-4.

[71] Jensen, R.V. and Sundaram, B. (1990) Phys. Rev. Lett. 65, 1964-7.

[72] Jensen, R.V., Susskind, S.M., and Sanders, M.M. (1991) Phys. Rep. 201, 1-56.

[73] Kleppner, D., Littman, M.G., and Zimmerman, M.L. (1983) in *Rydberg States of Atoms and Molecules*, Eds.: R.F. Stebbings and F.B. Dunning (Cambridge University Press, New York), pp. 73-116.

[74] Koch, P.M. and Mariani, D.R. (1980) J. Phys. B 13, L645-50.

[75] Koch, P.M. and Mariani, D.R. (1981) Phys. Rev. Lett. 46, 1275-8.

[76] Koch, P.M. (1982) J. Physique Colloq. 43, C2-187—C2-210.

[77] Koch, P.M. (1983) in *Rydberg States of Atoms and Molecules*, Eds.: R.F. Stebbings and F.B. Dunning (Cambridge University Press, New York), pp. 473-512.

[78] Koch, P.M., in *Fundamental Aspects of Quantum Theory*, Eds.: V. Gorini and A. Frigerio, (Plenum, New York, 1986).

[79] Koch, P.M., Leeuwen, K.A.H., Rath, O., Richards, D., and Jensen, R.V. (1987) in *The Physics of Phase Space*, Eds.: Y.S. Kim and W.W. Zachary, Lecture Notes in Physics vol. 278 (Springer-Verlag, Berlin), pp. 106-13.

[80] Koch, P.M. (1988) in *Electronic and Atomic Collisions*, Eds.: H.B. Gilbody, W.R. Newell, F.H. Read, and A.C.H. Smith, (North-Holland, Amsterdam), pp. 501-16.

[81] Koch, P.M., Moorman, L., Sauer, B.E., and Galvez, E.J. (1989) *Classical Dynamics in Atomic and Molecular Collisions*, Eds.: T. Grozdanov, P. Grujić, and P. Kristić, (World Scientific, Singapore), pp. 348-67.

[82] Koch P.M., Moorman L., Sauer B.E., Galvez E.J., and Leeuwen K.A.H. van (1989) Physica Scripta T26, 51-7.

[83] Koch, P.M. (1990) in *Chaos / XAOC*, Proceedings of a Soviet-American Conference, Ed.: D. K. Campbell (American Institute of Physics, New York), pp. 441-475.

[84] Koch, P.M. (1990) in *The Ubiquity of Chaos*, Ed.: S. Krasner (American Association for the Advancement of Science Press, Washington), pp. 75-97.

[85] Koch, P.M., Moorman, L., Sauer, B.E. (1990) Comm. At. Mol. Phys. 25, 165-83.

[86] Koch, P.M. (1992) CHAOS 2, 131-144.

[87] Koch, P.M. and Leeuwen, K.A.H. van (1992), submitted for publication.

[88] Kolmogorov, A.N. (1954) Dokl. Akad. Nauk. SSSR 98, pp. 527ff.; Moser, J. (1962) Nachr. Akad. Wiss. Göttingen Kl. II, n°1, 1; Arnold, V.I. (1963) Usp. Mat. Nauk. SSSR 18, pp. 13ff, and English translation in (1963) Russian Math. Surv. 18, pp. 85ff.

[89] Landau, L.D. and Lifshitz, E.M. (1960) *Classical Mechanics*, (Pergamon Press, Oxford, 1977 printing).

[90] Landau, L.D. and Lifshitz, E.M. (1977) *Quantum Mechanics*, 3rd edition, (Pergamon Press, Oxford), pp. 120ff.

[91] Leeuwen, K.A.H. van, Oppen, G.v., Renwick, S., Bowlin, J.B., Koch, P.M., Jensen, R.V., Rath, O., Richards, D., and Leopold J.G. (1985) Phys. Rev. Lett. 55, 2231-4.

[92] Leopold, J.G. and Percival, I.C. (1978) Phys. Rev. Lett. 41, 944-7.

[93] Leopold, J.G. and Percival, I.C. (1979) J. Phys. B 21, 2179-204.

[94] Leopold, J.G. and Richards, D. (1989) J. Phys. B 22, 1931-61.

[95] Leopold, J.G. and Richards, D. (1990) J. Phys. B 23, 2911-27.

[96] Leopold, J.G. and Richards, D. (1991) J. Phys. B 24, 1209-40.

[97] Lichtenberg, A.J. and Lieberman, M.A. (1983) *Regular and Stochastic Motion*, (Springer-Verlag, New York).

[98] Lombardi, M., Pique, J.P., Labastie, P., Broyer, M., and Seligman, T. (1992) in Ref. [53].

[99] MacKay, R.S. and Meiss, J.D. (1988) Phys. Rev. A 37, 4702-7.

[100] Main, J., Wiebusch, G., and Welge, K.H. (1991) Comm. At. Mol. Phys. 25, 233-51.

[101] Meerson, B.I., Oks, E.A., and Sasorov, P.V. (1979) Pis'ma Zh. Eksp. Teor. Fiz. 29, 79-82 [Sov. Phys.–JETP Lett. 29, 72-5].

[102] Meiss, J.D. (1989) Phys. Rev. Lett. 62, 1576.

[103] Moorman, L., Galvez, E.J., Sauer, B.E., Mortazawi-M, A., Leeuwen, K.A.H. van, Oppen, G.v., and Koch, P.M. (1988) Phys. Rev. Lett. 61, 771-4.

[104] Moorman, L., Galvez, E.J., Sauer, B.E., Mortazawi-M., A., Leeuwen, K.A.H. van, Oppen, G.v., and Koch, P.M. (1989) in *Atomic Spectra and Collisions in External Fields*, Vol. 2, Eds.: K.T. Taylor, M.H. Nayfeh, and C.W. Clark, (Plenum Press, New York), pp. 343-57.

[105] Moorman, L. (1991) in *The Electron*, Ed.: A. Weingartshofer, (Kluwer, Dordrecht).

[106] Moorman, L. and Koch, P.M. (1992) in *Quantum Nonintegrability*, Eds.: D.H. Feng and J.-M. Yuan, Vol. 4 in *Directions in Chaos*, Series Editor: Bai-Lin Hao (World Scientific, Singapore, in press).

[107] Nauenberg, M. (1990) Phys. Rev. Lett. 64, 2731; and (1990) Europhys. Lett. 13, 611-6.

[108] O'Connor, P.W., Gehlen, J.N., Heller, E.J. (1987) Phys. Rev. Lett. 58, 1296-9.

[109] Oppenheimer, J.R. (1928) Phys. Rev. 31, pp. 66ff.

[110] Ozorio de Almeida, A.M. (1988) *Hamiltonian Systems: Chaos and Quantization*, (Cambridge University Press, Cambridge).

[111] Percival I.C. and Richards, D. (1975) Adv. At. Mol. Phys. 11, 1-82.

[112] Radons, G. and Prange, R.E. (1988) Phys. Rev. Lett. 61, 1691-4.

[113] Rath, O. and Richards, D., in preparation.

[114] Reinhold, C.O., Burgdörfer, J., and Kemmler, J. (1992) Phys. Rev. A 45, R2655-8.

[115] Richards, D. (1987) J. Phys. B 20, 2171-92.

[116] Richards, D., Leopold, J.G., Koch, P.M., Galvez, E.J., Leeuwen, K.A.H. van, Moorman, L., Sauer, B.E., and Jensen, R.V. (1989) J. Phys. B 22, 1307-33.

[117] Richards, D. (1990) in *Aspects of Electron-Molecule Scattering and Photoionization*, Ed.: A. Herzenberg, AIP Conference Proceedings 204, 45-64.

[118] Richards, D. and Leopold, J.G. (1990) in *The Physics of Electronic and Atomic Collisions*, Eds.: A. Dalgarno, R.S. Freund, P.M. Koch, M.S. Lubell, and T.B. Lucatorto (AIP Conf. Proc.205), pp. 492-8.

[119] Ruff, G.A., Dietrick, K.M., and Gallagher, T.F. (1990) Phys. Rev. A 42, 5648-51.

[120] Sanders, M.M., Jensen, R.V., Koch, P.M., and Leeuwen, K.A.H. van (1987) Nucl. Phys. B (Proc. Suppl.) 2, 578-9.

[121] Sauer, B.E., Leeuwen, K.A.H. van, Mortazawi-M., A., and Koch, P.M. (1991) Rev. Sci. Instr. 62, 189-97.

[122] Sauer, B.E., Yoakum, S., Moorman, L., Koch, P.M., Richards, D., and Dando, P.A. (1992) Phys. Rev. Lett. 68, 468-71.

[123] Sauer, B.E., Bellermann, M.R.W., and Koch, P.M. (1992) Phys. Rev. Lett. 68, 1633-6.

[124] Series, G.W. (1988) *The Spectrum of the Hydrogen Atom: Advances*, (World Scientific, Singapore), pp. 20-4.

[125] Shepelyansky, D.L. (1985) in *Chaotic Behavior in Quantum Systems*, Ed.: G. Casati, (Plenum, New York), NATO ASI Series B, Vol.120, pp. 187-204.

[126] Stevens, M.J. and Sundaram, B. (1989) Phys. Rev. A 39, 2862-77.

[127] 'Irregular' here means chaotic, in the sense of exponential instability, for a finite time duration, as opposed to an infinite time duration; see *e.g.*, T. Tél, 'Transient Chaos', in (1990) *Directions in Chaos*, Vol. 3, Ed.: Hao Bai-lin (World Scientific, Singapore), pp. 149ff.

[128] Wang, K. and Chu, S-I. (1989) Phys. Rev. A 39, 1800-8.

[129] Waterland, R.L., Yuan, Jian-Min, Martens, C.C., Gillilan, E., and Reinhardt, W.P. (1989) Phys. Rev. Lett. 61, 2733-6.

[130] Wintgen, D. and Hönig, A. (1989) Phys. Rev. Lett. 63, 1467-70.

[131] Wintgen, D., Richter, K., and Tanner, G. (1992) CHAOS 2, 19-33.

[132] Wolf, A., Swift, J.B., Swinney, H.L., and Vastano, J.A., (1985) Physica D 16, 285-317.

[133] Yamazaki, Y., Stolterfoht, N., Miller, P.D., Krause, H.F., Pepmiller, P.L., Datz, S., Sellin, I.A., Scheurer, J.N., Andriamonje, S., Bertault, D., and Chemin, J.F. (1988) Phys. Rev. Lett. 61, 2913-6.

Topics in Quantum Chaos

R. E. Prange
University of Maryland
College Park, MD 20742

I. Introduction

A. Philosophy

A philosophical stance known as *reductionalism* is favored by certain physicists. The holy grail of the reductionalist is the *theory of everything*. An extreme reductionalist wants to find the *equations* for the theory of everything and is only interested in *solving* the equations to the extent required to establish their validity. Solving the equations of an already established theory is considered to be an inferior activity.

We quantum chaologists do not, of course, accept this view of things. While agreeing that finding equations is nice, we also think that actually reducing the theory of everything to theories of somethings is intellectually very challenging. It is not as if there in a nice little index to the theory of everything so that one can look up to see how the surf breaks on the shore or how a living cell divides.

Indeed, if the concepts of surf and cells were manifestly part of the theory of everything, the reductionalists would not be very happy. What they want is a couple of equations, a few simple rules only. Concepts like brains, moons, crystals, nucleons, must *emerge* somehow in the process of solving the equations.

The solution of the equations of a theory can be accomplished in several ways, e.g. numerically or by some version of mean field theory. Complicated things like photosynthesis probably will not be understood by using mean field theory directly on the theory of everything, however. Often most insight is gained by the identification of a small parameter. In that case perturbation theory might work but perturbation theory is also quite limited.

There are, however, other important possibilities if a small parameter is available. Sharp new *concepts* can be introduced in the limit of a small parameter. Perhaps the most famous case of this type is the thermodynamic limit. In this case the thermodynamic functions acquire singularities at critical values of parameters in the limit $L \to \infty$, where L is the system size. Often we do this almost subconsciously. E.g. the earth is a sharply defined astronomical concept provided the scale thickness of its surface is small compared with its radius and the radius is small compared with the distance to other comparably massed astronomical bodies.

More cogent to the subject of this school is the example of the Lyapunov exponent. In this case a double limit is involved. The Lyapunov exponent for a dynamical system is defined as

$$\gamma = \lim_{t \to \infty} \lim_{|z'(0)-z(0)| \to 0} \frac{1}{t} \log[|z'(t) - z(t)|/|z'(0)-z(0)|] \quad (I.1)$$

where z(t), z′(t) are initial neighboring phasespace points evolving with time t. Classical *chaos* is usually defined as $\gamma > 0$ for a sufficient number of points z(0) with z(t) sufficiently bounded.

A second more extreme possible use of a small parameter is that in some limit, the parent theory gives rise to a *daughter* theory based on *entirely new* "daughter" *concepts*. If, some day, the theory of everything is found, it may be that some chain of limits will show it to reduce to superstring theory, which in turn will, as the energy/string mass $\rightarrow 0$ become [in some sectors] quantum chromodynamics, which in turn will be reduced in different low energy limits to one or more versions of nuclear theory. These three theories are based in turn on strings, quarks and gluons, and for example, nucleons and mesons, or nucleons and nucleon-nucleon potentials.

The example of prime concern in this school is of course the classical limit of quantum mechanics. This limit is mathematically, for many purposes, equivalent to the short wavelength limit of any number of wave theories, e.g. the geometrical optics limit of Maxwell's equations. However, we stick to quantum terminology for brevity. We think of $\hbar$ as the dimensionless small parameter defined as $\hbar = \hbar/PA \cong \lambda/A$, where P,A are a characteristic classical momentum and length respectively, $\hbar$ is Planck's constant, and λ is the wavelength. In addition to the fact that there are many applications, it is philosophically advantageous to study this system because both the parent theory and the daughter theory are comparatively well understood.

We now define *emergent, submergent,* and *persistent* concepts. These are defined in the limit $N \rightarrow \infty$, say, while the daughter theory is defined for $\varepsilon \rightarrow 0$. An emergent concept exists in the daughter theory but not in the parent, a submergent concept exists in the parent but not the daughter, a persistent concept exists in both. A daughter concept occurs for $\varepsilon \rightarrow 0$ and finite N. An example is phase space, a concept valid in classical mechanics but not in quantum mechanics.

It may seem impossible that an emergent concept could exist, since the daughter theory is completely defined by the parent. However, if the two limits of N and ε are distinct, then if the limit $N \rightarrow \infty$ *does not commute* with the limit $\varepsilon \rightarrow 0$, then either an emergent or a submergent concept exists, or possibly both. In the persistent case, the two limits *commute.*

An example of a persistent concept is that of *phase transition* which appears in the thermodynamic limit in both the quantum and classical cases, and probably exists even in string theory. Tunnelling is a *submergent* concept. [Consider a particle initially confined by a potential barrier. If $p(t,\hbar)$ is the probability of finding the particle behind the barrier then lim $t \rightarrow \infty$ $\hbar \rightarrow 0$ of $p(t,\hbar) = 1$, while $p = 0$ in the commuted limit.]

The prime example of an emergent concept for us is, of course, chaos. In the definition I.1, one must *first* let two phase space points approach each other arbitrarily close, a meaningful operation only if $\hbar$ has already gone to zero. Thus, chaos exists, by the definition I.1, only if $\hbar$ vanishes before t becomes large.

Naturally, if one is miffed at the lack of chaos in quantum mechanics, one is free to invent other definitions of chaos. Getting people to pay much attention to your definitions is another matter. It is fair to say that no definition of chaos not based on I.1 has been widely accepted to date, therefore there is no chaos in quantum mechanics by current definitions.

B. Time scales

Let $\mathcal{O}_q(t,K,\hbar)$ be a quantum observable, where K describes the remaining parameters. We can always arrange things so that the parameters K also have classical meaning. Then there is a corresponding classical observable $\mathcal{O}_c(t,K) = \mathcal{O}_q(t,K,0)$. If there is a transition to chaos, for example, it could happen that $\mathcal{O}_c(\infty,K)$ becomes singular at a critical value of the parameter $K = K_c$. However,

$$\mathcal{O}_c(\infty,K) = \lim_{t\to\infty}\lim_{\hbar\to 0} \mathcal{O}_q(t,K,\hbar) \neq \lim_{\hbar\to 0}\lim_{t\to\infty} \mathcal{O}_q(t,K,\hbar).$$

For large t and small $\hbar$, $\mathcal{O}_q(t,K,\hbar)$ will approximate $\mathcal{O}_c(\infty,K)$ only if $t \ll \tau(\hbar)$, where $\tau(\hbar) \to \infty$ as $\hbar \to 0$. We shall encounter different formulas for τ. For example, the so-called *log-ħ time*[1] is

$$\tau(\hbar) = (1/\gamma)\ln(1/\hbar) \qquad \text{(I.2)}$$

We shall also encounter a case $\tau \propto 1/\hbar^2$, as well as fractional powers of $\hbar$, both rational and irrational. These different cases correspond to different degrees of chaos, as well as to different measurements. It is obvious that some measurements are less sensitive to interference effects than others. For example, measuring the unit operator is the same classically and quantally for all time.

The reason for I.2 is intuitively quite clear. Consider a wavepacket starting off at $t = 0$. Corresponding to the wavepacket there will be classically a bunch of phase space points. For short enough times the wavepacket spreads just as the phase space points do. Namely, in the chaotic case, the packet spreads exponentially fast. However, before long the size of the system or other scales determined by nonlinear effects is reached and interference effects set in. Since the size of the initial packet is limited by $\hbar$, the uncertainty relation, the estimate of Eq.I.2 follows.

As an instructive example consider the question: *How long, according to quantum theory, can a pendulum be balanced vertically in the unstable position?* Clearly, *too* great an effort to make the angle $\theta = \pi$ will be bad because, by the uncertainty principle, a large angular momentum will result. The best compromise, $\delta\theta \propto \delta p \propto \hbar^{1/2}$ gives Eq.I.2, with γ replaced by ω_0, the frequency of small oscillations, the only time scale in this simple problem. Chaotic systems have unstable fixed orbits, like the top of the pendulum. In fact, the simple generalization of this to unstable *periodic* orbits is central to our understanding of chaotic systems. Thus an instability gives rise to a remarkably short time before quantum effects take over.

The *quantum* version of the pendulum has discrete eigenstates, some of which spend most of their time nearly balanced in the unstable equilibrium state. There are corresponding states in many chaotic systems. These are called *scars of unstable periodic orbits.*[2]

The theory of these scars has not made much progress. It is not hard to see why from the pendulum analogy. The scar wavefunction is largest near the unstable equilibrium, since it spends most of its time there. However, the *quantization condition* is a global condition that is determined from how the wavefunction matches at the *bottom*. Replacing the pendulum by a particle on a frictionless loop of more or less arbitrary shape, we see that there will be quantum states spending lots of time near the unstable point at the top of the loop, whose exact quantization depends on the details of the shape of the loop far away from the top. The *large part* of the states, in other words, doesn't depend much on the rest of the loop, but the exact quantization *does*.

The exact quantization is a long time observable. We shall have more to say about the "theory" of scars later on.

C. The quasiclassical approximation

In addition to finding what $\tau(\hbar)$ is, there is the problem of finding the behavior of the system for $t > \tau(\hbar)$. By definition, the classical approximation fails. However, one may hope that other approximations work, perhaps even approximations based on the smallness of $\hbar$. Enter the quasiclassical approximation, which we shall refer to as the WKB approximation, for short.

There are many part-quantum-part-classical approximations in physics. For example, the Sommerfeld model of metals treats the electrons as classical charged points, *except* that the exclusion principle is put in, by hand, *and* there is a scattering time that is only sensibly calculated in quantum theory, *and* there is a quantum mass renormalization, etc. This kind of theory is basically classical with parameters calculated quantally.

The quasiclassical or generalized WKB approximation, on the other hand, uses the full quantum apparatus of wavefunctions and/or propagators, but approximates these wavefunctions by purely classical constructs, *together* with a single and very nontrivial appearance of $\hbar$. To remind you, a WKB wavefunction, in its simplest incarnation, has the form

$$\psi(q) = A(q)e^{iS(q)/\hbar} \tag{I.3}$$

Here S is a classical action, and $|A(q)|^2$ has a classical probabilistic interpretation. This action can generally be related to various classical orbits in phase space. Eq.I.3 must be generalized a bit, as follows:

$$\psi(q) = \Sigma_a \; A_a(q)e^{iS_a(q)/\hbar} \tag{I.4}$$

where a is some integer label. Without getting into the complications at this stage, we may say that classical chaos is characterized by the *number of terms in the sum over* a *increasing exponentially* as larger and larger actions are encountered. If I.4 represents a wavefunction evolving in time, then there is an exponential increase in the number of terms with time.

The same questions may be asked about the quasiclassical approximation as about the classical. Does the WKB admit of a definition of chaos? In one sense it obviously does, since the S_a's are purely classical constructs, and so they know about chaos. On the other hand, since the theory is completely quantum in structure, one can't define a phase space, and so chaos cannot directly be defined.

We can also ask for the range of validity of the WKB in the presence of chaos. This is a much deeper question, rarely discussed in the literature. We shall argue that the WKB is valid up to much longer times than the classical theory, which usually fails at the log-$\hbar$ time. On the other hand, there is more than one implementation of the WKB and the more popular ones fail at a short time in generic situations. In particular, we find that the WKB theory of scars fails generically in this way. Special situations are known however where the WKB is exact. However, we shall argue that generically the WKB fails at a time $\Theta(\hbar)$ which goes as $1/\hbar^2$ in the best studied situation.

Aside from questions of principle, because of the proliferation of terms in the WKB approximation, it generically becomes useless for calculations already at the log-$\hbar$ time. In special cases there are reorganizations of the sum I.4 which effectively allow it to be used for much longer times. Much of the recent progress in quantum chaology[3] is associated with the reorganization of sums like I.4. Smilansky has described some of this exciting work at this meeting. This work effectively solves the problem of quantum chaos for certain special "hard chaos" cases which have been historically important.

D. Pseudorandom matrix theory

The second main line of approach to this class of problems is not based on the smallness of $\hbar$, but rather on the underlying chaos itself. The idea here is that if one formulates the quantum problem appropriately, the chaos will show up already in the Hamiltonian itself. Thus, if the Hamiltonian is regarded as a matrix, its elements will be chaotic, or better put, pseudorandom. [We don't find sensitivity to initial conditions, just something that seems random.] Forswearing interest in further details, one can replace these elements by random entries and study statistical properties of random matrices.

As you shall be hearing much on this subject these weeks, I shall be brief. The original point of view of Wigner was that, in a many particle nucleus, we knew very little indeed about the Hamiltonian. Taking a minimalist approach, it was almost obvious that one should choose the representation of the Hamiltonian at random as well, since one has no idea which representation to prefer. Only major symmetries should be respected, such as time reversal invariance. Thus the Gaussian orthogonal ensemble, the GUE, etc. were introduced.

More recently, it was recognized that a specific Hamiltonian might have properties very similar to, say, the GOE. The Hamiltonian then, in some specific representation, might be regarded as a rather typical large matrix member of the ensemble. Just as in solid state physics, where a specific sample of material with many many impurities at definite irregular sites has most of the properties of an ensemble of samples, the specific instance shares the properties of the ensemble.

However, there is reason to be more ambitious than the nuclear physicists were in this case. We may know much more about the Hamiltonian than just its symmetry under time reversal. In fact, the usual case is that there is a natural representation in which the Hamiltonian looks very simple indeed, and we know all about that. For example, we could be considering the stadium model. This means that there can indeed be special classes of representations. In one of these the Hamiltonian can perhaps be regarded as pseudorandom. As far as I know, little or no work has distinguished these special representations mathematically. They are usually fairly obvious physically however.

An example of a system with a special representation is provided by an electron in an alloy or random lattice. Here, in the position representation, it is physically reasonable to suppose that the potential is disordered. P. W. Anderson in his Nobel prize winning work of 1958 introduced the simplest model along this line, known as the Anderson model.[4] This Schrödinger equation is

$$T_n u_n + \sum_m W_{n-m} u_m = E u_n \qquad (I.5)$$

where u_n is the amplitude for finding the electron on site n. Anderson assumed that the site diagonal energy T_n is randomly chosen from some distribution. For example, if there are two kinds of atoms in the alloy, T_n can take on two values. Somewhat unphysically, but mathematically conveniently, the hopping term W is chosen to depend only on the distance between sites n and m, rather than also being random. From the conceptual point of view, this approximation of *site diagonal disorder* turns out to be insignificant.

Note that Eq.I.5 encompasses a wide range of very interesting models. For example, T_n independent of n is the simplest tight binding model, T_n periodic in n with integer period is a tight binding model with a basis, T_n periodic with noninteger period is a model of motion in an incommensurate potential, etc.[5] We shall introduce a class of models in which T_n is a *pseudorandom* function of n. The question then arises: How random does T_n need to be that the results are statistically the same as if T_n is random?

The hopping W is usually assumed to be *short ranged,* indeed it is usually taken to be nearest neighbor. This means that the infinite matrix representation of I.5 has some special properties. In particular, in one dimension, the matrix is band diagonal, with random elements along the diagonal. Such an ensemble of matrices has much different properties than a GOE. In particular, its spectrum is dense pure point, and its eigenstates are *localized,* i.e. if u^0 is an eigenstate, then $|u^0_m| \cong \exp(-\gamma|m - n_0|)$, i.e. the wavefunction drops off exponentially asymptotically far away from its nominal center n_0. The exponent γ is the inverse localization length, and turns out to be very much like a Lyapunov exponent.

One way, therefore, to improve the original random matrix theory if the Hamiltonian is known, is to find a representation in which the system appears to be pseudorandom. We shall give an example later.

A second method is as follows: Try to express the desired statistical quantities within the WKB approximation. If the quantities studied are smeared over a significant width ΔE of energy, it means that in effect, only short times are considered. Then the WKB works and good results are obtained, which however differ from random matrix theory and depend on specific classical orbits. Longtime quantities, like the nearest level repulsion, have not been amenable to the WKB, but there random matrix theory works. Thus, one can try to interpolate.

E. Types of chaotic systems

The (classical) systems under study can be categorized in several ways. One very fundamental distinction is between conservative, dissipative and noisy systems. Quantum mechanics cannot be easily generalized to dissipative and noisy systems. More precisely, adding dissipation and noise requires adding a "heat bath" with an infinite number of degrees of freedom, while the chaotic systems of interest have just a few freedoms. The question is whether these degrees of freedom can be "integrated out", to leave tractable equations with just a few variables. This important subject is treated by Robert Graham in these lectures. However, it should be remembered that it is really more than one subject. By this, I mean to point out that adding dissipation to Maxwell's equations is quite different than adding dissipation to Schrödinger's equation.

I emphasize that dissipation and/or noise critically affect the longtime behavior of a classical system, just as quantum effects do. There are characteristic times τ_d, τ_n after which a nearly conservative system crosses over to a system where dissipation and/or noise is prominent. These times depend on the strength of the coupling of the system to the heat bath. A realistic system will have finite $\tau_{d,n}$ and dissipation and noise can only be neglected if $\tau_{d,n} >> \tau(\hbar)$.

In experimental systems, however, to study quantum chaos without getting involved in noise and dissipation, is not generally easy. One also must get $\tau(\hbar)$ in a range where interesting things are going on classically which will be modified by quantum effects. We have proposed some experiments which if carried out could complement the beautiful experiments on hydrogen. Fortunately, you will hear several lectures on actual and proposed experiments.

Chaotic systems can also be categorized as autonomous (time independent) and driven systems. Usually periodically driven systems are discussed, but quasiperiodic driving is also quite interesting. Classically or for that matter quasiclassically there is little to distinguish between autonomous and driven categories from the point of view of chaos. The reason is that the long time behavior is best discussed in terms of *maps*. That is, rather than focussing on continuous time, one looks at a sequence of particular discrete times. For periodically driven systems, it is natural to look stroboscopically once every period. One may find a map which takes the orbits forward one period and concentrate on studying it. The subject then becomes mathematically the study of the iteration of these maps. For autonomous systems, it is possible to define Poincaré surfaces of section in phase space, and construct Poincaré maps carrying each orbit forward to its next passage through the surface of section. The Poincaré maps can be iterated just as can the maps in the stroboscopic case: Indeed these can be regarded as Poincaré maps as well.

In the quantum case this doesn't work, strictly speaking. The reason is that a surface of section in phase space cannot be precisely defined. However, for small $\hbar$, a WKB approximation surface of section map can be defined. Furthermore, since a single map carries the system forward a short time, (except perhaps for some pathological cases), the WKB approximation should be very good for $\hbar$ small enough. So, one can study iterates of this approximate map. Whether the small corrections accumulate in some bad way after many iterations is an open question, however.

We may also distinguish between closed, bounded systems, and open or unbounded systems. In the former case, e.g. the billiard, the spectrum is trivially a point spectrum. Driven systems are usually unbounded, at least in momentum and energy since energy is not conserved.

Unbounded systems may be further distinguished as scattering systems or infinite media systems. A scattering system has a finite scattering region connecting asymptotic regions of incoming and outgoing particles. A scattering system automatically has a continuous spectrum. An example of an infinite medium is a very large piece of metal with random scattering centers. In this case, the nature of the spectrum may well turn out to be point as we mentioned. In that case, we speak of Anderson localization.

[The nature of the spectrum[6] of a Hamiltonian depends on the nature of the function $\mu(E; \psi) = <\psi|\theta(E - H)|\psi>$, where ψ is an arbitrary normalizable (and normalized) state, (not necessarily an eigenstate) and θ is the step function. μ is called the spectral measure in mathematics. In physics, N(E) = $d\mu/dE$ is the local density of states, (i.e. the density of states weighted by the probability that the state is contained in ψ). μ monotonically increases from 0 to 1 as E goes from $-\infty$ to $+\infty$. If μ has jumps (corresponding to δ functions in N(E), the spectrum has a pure point component. If μ has a nontrivial continuous component, so does the spectrum. To each point in the point spectrum corresponds one (or more) normalizable eigenstates. Points of the continuous spectrum have no corresponding normalizable eigenstate. One may also prove that the continuous spectrum may be divided into absolutely continuous and singular continuous. The absolutely continuous spectrum corresponds to an N(E) which is an ordinary function. I will not discuss the singular case, although it does come up in certain instances in quantum chaos.]

A second way of categorizing chaotic systems is by how chaotic they are. Here, we have *hard* and *generic* chaos. In hard chaos, there are no stable phase space points. To achieve Hamiltonian hard chaos, infinities or discontinuities or impossible curvatures are required. Billiards with sharp boundaries are an example. The Baker's map is another. Cases are known in which the WKB is exact for hard chaos. All the cases where resummation of an expression like I.4 works are hard chaos.

Hard chaos can be further subdivided. I shall mention only the hyperbolic or Axiom A systems, in which *all* phase space points are unstable in the sense of the inverted pendulum. This is fairly extreme, and relatively well understood.

Generic chaos arises in sufficiently high dimensional nonlinear systems without special properties. It is characterised by the separation of phase space into three regions: A region resembling hard chaos in most respects, a region nearly integrable, and a boundary region between the two. The major problem is that the boundary region is fractal. We shall spend some time discussing this situation.

F. Summary and outline

The remainder of my lectures will be devoted to the following subjects. The first lectures will be devoted to various models of localization and the quantum suppression of diffusion in dynamical systems. We shall be quite brief, since there are recent reviews already in existence, and also Professor Chirikov will be discussing it.

We shall next discuss the situation in generic chaos. In particular, we will recount the scaling theory of the transition to global chaos, as well as the somewhat simpler period doubling transition to chaos. After that, we shall see how this phase transition is extended to the quantum case, within the framework of the WKB approximation. Finally, we study the validity of various versions of the WKB approximation as it is applied to the theory of scars, for example.

II. Quantum longtime behavior and localization

We have seen that for long enough times, quantum behavior can differ from the corresponding classical behavior. Sometimes this could just be a numerical difference, but the most interesting cases arise when there is a qualitative distinction in the two behaviors. There are basically two simple and striking cases, either the quantum system goes where it classically shouldn't, or it doesn't go which it classically should. The former case is tunnelling, the latter is known as *localization,* or more specifically, *Anderson localization.*

In what follows, we restrict consideration to low dimensional systems, with phasespace either 3 or 4 dimensional. [Anderson localization is strongly dependent on dimension, and not much is known about higher dimensions in the context of QC.]

A. The kicked rotor

Now is a good place to introduce the specific system most often studied, the kicked rotor. It is identical to that discussed at length in other lectures. Slightly more generally, let us consider a Hamiltonian,

$$\mathcal{H} = p^2/2m + kV(q/a)\Delta(t) \qquad \text{(II.1)}$$

where $\Delta(t) = \Sigma\delta(t\text{-}nT)$ is the periodic δ function. Now let $q = a\hat{q}$, $t = T\hat{t}$, $p = (ma/T)\hat{p}$.

Then $\mathcal{H} = (ma^2/T^2)[\hat{p}^2/2 + KV(\hat{q})\Sigma\delta(\hat{t}\text{-}n)]$, where $K = kT/ma^2$ is the only classical parameter.

In the quantum case there is an extra parameter, because one cannot rescale q and p independently. Namely, we must replace $p = \frac{\hbar}{i}\frac{\partial}{\partial q}$ and the Schrödinger equation becomes

$$\bar{h} i\frac{\partial}{\partial t}\psi = [-\tfrac{1}{2}\bar{h}^2\frac{\partial^2}{\partial q^2} + KV(\hat{q})\Sigma\delta(\hat{t}\text{-}n)]\psi \qquad \text{(II.2)}$$

with $\bar{h} = \hbar T/ma^2$, the dimensionless Planck's constant. Sometimes it is convenient to use $k = K/\bar{h}$ and $\tau = \bar{h}$ as the parameters. From now on we drop the ^'s, [or we can use units such that a,T,m are unity.] This model corresponds to a one-dimensional random system. The class of models is generic, and aside from a few idiosyncracies, is believed to be typical of most chaotic systems. The kicked rotor more narrowly defined takes $V(q) = \cos q$, where q is regarded as an angle. The kicked particle has the same equation, but q is not identified with $q + 2\pi$.

The classical Hamilton's equations for this system may be integrated for one period to give the standard, or Chirikov-Taylor map:

$$q_{n+1} = q_n + p_{n+1}\ ;\quad p_{n+1} = p_n - KV'(q_n) \qquad \text{(II.3)}$$

where the first equation is understood to be mod 2π. Here p_n is the momentum just before the n'th kick.

Similarly, the quantum Schrödinger equation may be integrated for one period to yield the time evolution operator for one period,

$$\psi_{n+1}(q) = \int dq' U(q,q')\psi_n(q') \qquad \text{(II.4)}$$

where $U = \exp(-ip^2/2\hbar)\exp(-iV(q')/\hbar)$ and $p = (\hbar/i)\partial/\partial q$. Expressed in the q,q′ notation,

$$U(q,q') = [-iS_{12}(q,q')/2\pi\hbar]^{1/2}\exp\{iS(q,q')/\hbar\} \qquad \text{(II.5)}$$

where in this case

$$S(q,q') = [(q-q')^2/2 - V(q')] \qquad \text{(II.6)}$$

[We have used a shorthand notation strictly valid for the kicked particle and not the rotor. If q,q′ are restricted to the interval $[0,2\pi]$, then the function U must be periodically continued, i.e. $\exp[i(q-q')^2/2\hbar]$ is replaced by $\Sigma_r \exp\{i(q-q'-2\pi r)^2/2\hbar\}$. For the rotor, the integral in II.4 is over the interval $[0,2\pi]$. However, one may dispense with the sum over r used to extend U periodically, and instead do the integral over $[-\infty,\infty]$.]

We have written II.5 in a more general form, by introducing in the prefactor S_{12}. The subscript *i* means the partial derivative with respect to the i'th argument. Thus for the S of II.6, $S_{12} = -1$. However, the form II.5 is the general form of the WKB approximation to the time evolution operator.[7] To see this, note that, according to Feynman's path integral formulation, the time evolution operator $U(q,t,q't')$ is given by

$$U(q,t,q't') = \int d(\text{Paths}) \exp[(i/\hbar)\int \mathscr{L}(v(\tau),x(\tau),\tau)d\tau] \qquad \text{(II.7)}$$

where the functional integral is over paths $x(\tau)$, such that $q = x(t)$, $q' = x(t')$. In our case, t' is just before the *n*'th kick, $t = t'+1$, and there is no *n* dependence left. The WKB approximation is the stationary phase approximation to II.7. The stationary phase approximation assumes the dominance of the classical path [which makes the Lagrangrian $\mathscr{L}$ stationary]. The exponential in II.7 is expanded to second order about the classical path, and the resulting integral is a [complex] Gaussian which gives the prefactor in II.5. The function S is just the action integral in II.7 evaluated along the classical path. If there is more than one path, then there will be a func- tion S for each path, and a sum over distinct classical paths, similar to I.4.

It should be noted that, for the kicked rotor, the WKB form II.5,6 is exact, for one step. In general, the WKB approximation to the unitary operator U is not unitary, although it satisfies the unitarity condition

$$\int U^{\dagger}(q,q'')U(q'',q')dq'' = \delta(q-q') \qquad \text{(II.8)}$$

if the integral is evaluated in stationary phase approximation. Also remark that for a single time increment in the kicked particle problem there is just one path [with velocity $q - q'$].

B. Tunnelling and KAM torii

Tunnelling is of course an essential quantum characteristic. Usually, there is an energy barrier forbidding the classical motion, but nevertheless the quantum particle gets out, i.e. it tunnels. To describe this in the WKB approximation is painful: It requires continuing the classical orbits into imaginary time. We shall assume there is no tunnelling of this type, or shall deal with it in other ways. As far as I know, there is no ordinary tunnelling in the kicked rotor.

Something similar occurs in the rotor, however, because it is a KAM system. That is, for small enough K, there are invariant torii. These torii are just distortions of the torii which appear in the integrable system at K = 0, namely, they are the lines p = constant. [For the rotor, these lines

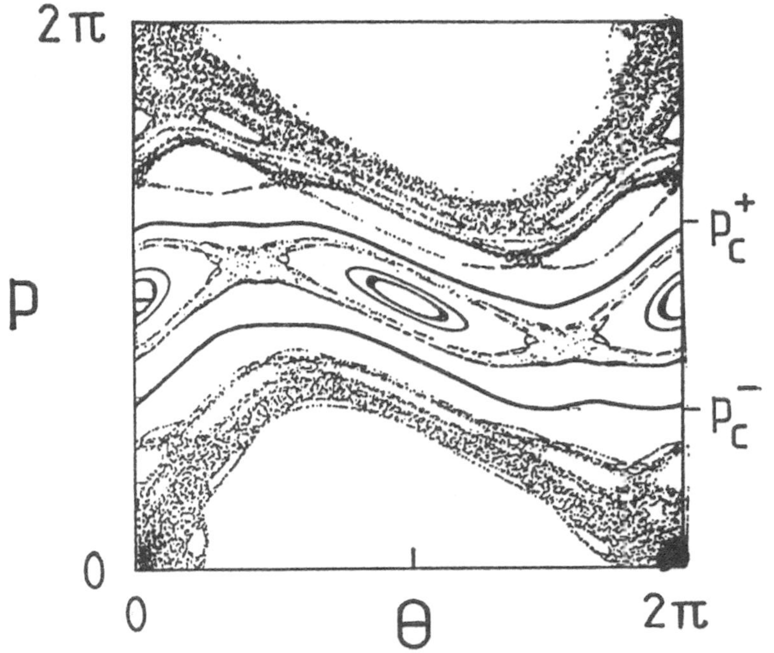

Fig. II.1. Some classical orbits for the kicked rotor at $K = K_c$. The last KAM torii are shown near the labels p^+_c, p^-_c. Other prominent orbits are a chaotic orbit extending out from the unstable fixed point at 0,0 [all four corners are equivalent], a stable period two orbit at π,0, surrounded by elliptic islands, and the chaotic orbit related to the period two unstable points which alternate with the stable ones. Little regions where the chaotic orbits do not enter usually have stable period island inside, but these are not shown.

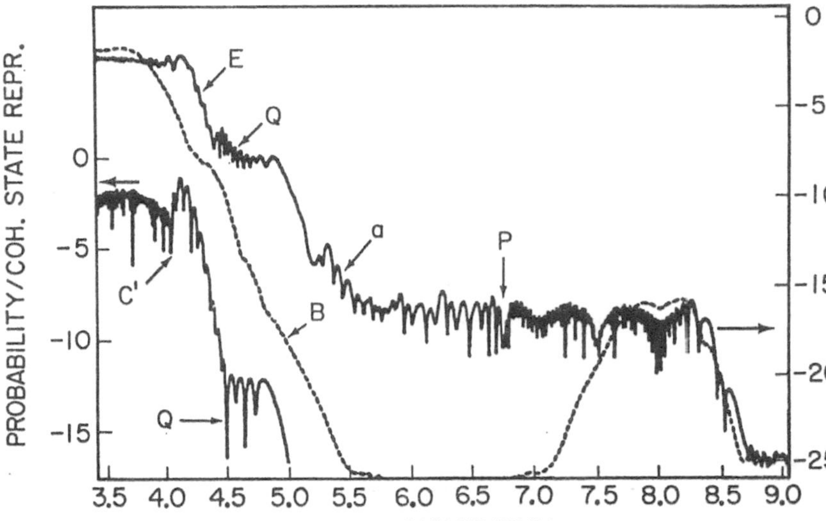

Fig. II.2. The longtime probability of finding momentum p (abscissa), where p is above p_c^+ of Fig. II.1, if the initial state was a state of definite momentum $p = \pi$, also in Fig. II.1. This figure is discussed in more detail in section III.

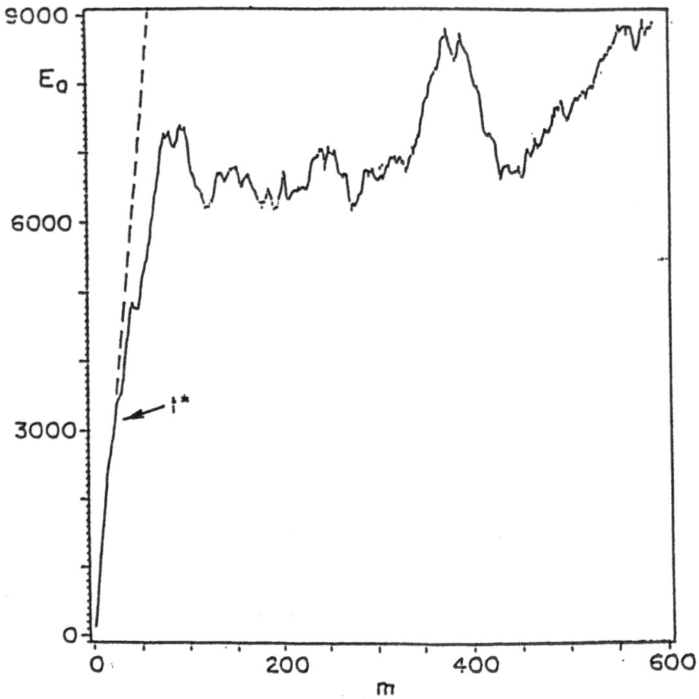

Fig. II.3. Kinetic energy of the kicked rotor vs integer time for K > K_c. The classical rotor shows diffusive behavior, (dotted line), while the quantum rotor shows the suppression of diffusion, i.e. localization in momentum, given by the solid line. The quantum and the classical agree until a time t^* (called $\tau(\hbar)$ in the text).

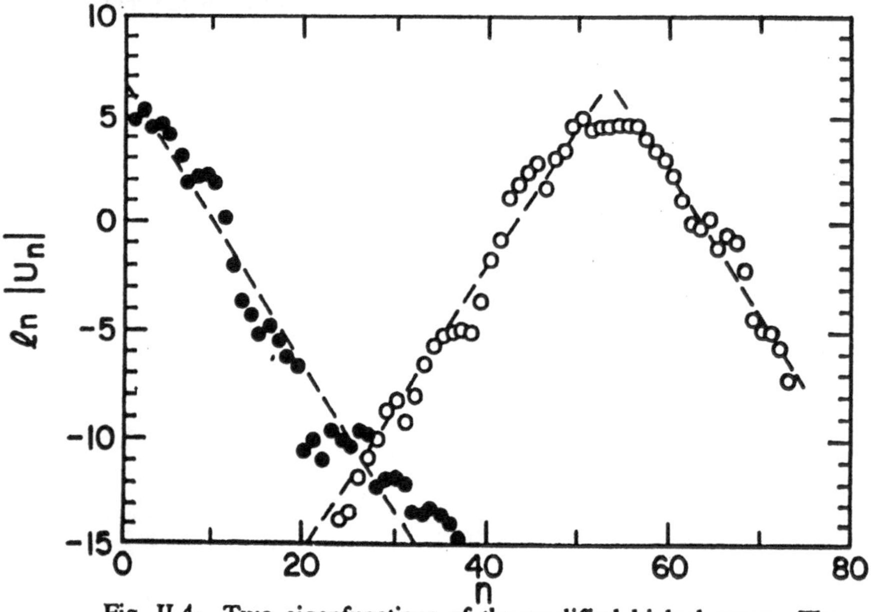

Fig. II.4: Two eigenfunctions of the modified kicked rotor. The kicking is such that the corresponding solid-state problem has nearest neighbor hopping. The logarithm of the absolute value of the functions are shown. The straightline slopes are given by the Ishii-Thouless-Lloyd formula, Eq. II.32.

are circles, since $q = 0$ is identified with $q = 2\pi$.] For $K \neq 0$ the lines p rational are destroyed (resonant torii), but for K sufficiently small, the torii corresponding to the most irrational p are preserved. These act as a dynamical barrier, which confines the motion. Motion started on one side of the KAM invariant torus never gets to the other side, classically, in this low dimensional phase space. This is a remnant of the conservation law. At $K = 0$, momentum is conserved, so a particle of a given momentum cannot change to another momentum. It is in that sense similar to energy barrier controlled localization. For finite K, momentum is not conserved, but the KAM theorem controls the degree to which the conservation law fails.

A particularly well-studied and striking case is that of the last KAM torus. We shall have more to say about this is the third chapter. The last KAM torus corresponds at $K = 0$ to a momentum equal the golden mean. This torus [actually two of them, by symmetry] itself disappears at a critical value $K_c \cong 1$. In Fig. II.1 we show several orbits at $K = K_c$, including the last KAM torii. If a classical trajectory is started between the two torii, it will never get out.

However, a quantum wavepacket does leak out. What happens is that a wavepacket started in, say, a chaotic region controlled by a bounding KAM surface oscillates around in that region with tails over the KAM surface. Unlike the standard case of tunnelling through a barrier, the packet does not increase its probability of getting out as time goes on. That is, the probability of being outside saturates to some small average value, rather than increasing to unity. So the packet remains *localized,* with exponential tails in the forbidden region.

In Fig. II.2, is shown the longtime probability of finding a particle in a wave packet, initially of momentum p_0 lying between the two KAM torii, which show exponentially decaying tails in the forbidden region. This is discussed further in III.

As the parameter K is increased, and the last KAM torii disintegrate, there is a dramatic and singular change in the longtime classical behavior. Namely, for $K < K_c$ a phasespace point remains forever bounded, but for $K > K_c$ the typical phasespace point diffuses out to infinitely high energy at long time. There is a transition to *global chaos* at $K = K_c$. This transition is a true phase transition, and it is possible to define critical exponents, do scaling and all that.

The quantum kicked rotor does not undergo this transition, however. Even for $K > K_c$, a quantum wavepacket is confined forever. Nothing singular happens at K_c, although near K_c one can discuss scaling in the quantum context, as we shall see in the third chapter.

C. Dynamic localization[8]

Let us take the case K considerably larger than K_c. Fig.II.3, similar to those shown by Chirikov and others, compares the time evolution of the energy, $<p^2(t)>$ for a classical bunch of phase space points, and a corresponding quantum packet. Again, we see that there is a time $\tau(\hbar)$, which in this case goes as $\hbar^{-2}$. We shall study this case below, and show that it is related to Anderson localization.

However, the example shows that there is a continuum of types of localization, starting with the familiar localization to satisfy conservation laws, through perturbative localization, when K is small, through localization because of KAM torii or cantori which evolves for large K to something like Anderson localization.

There are other numerically discovered instances of localization, where the quantum results differ dramatically from the classical. In particular, the "bouncing ball" states in the stadium problem have been extensively studied. It is not known whether these can be fit into the continuum of localization types found in the rotor.

We now discuss Anderson localization. The usual context for this is motion in a disordered potential or lattice. The main result is that it is possible for the eigenstates of a particle in a disordered potential to be exponentially localized even though it is energetically possible for the particle to go arbitrarily far on the lattice. The result depends on dimension. In one dimension, there is always localization. In two dimensions, the same is true, but this is a marginal case, and the localization length is exponentially large for weak disorder. In three dimensions, there is always localization at the band edges, but away from the edges, there is localization only for sufficiently strong disorder. For weaker disorder, there is one or more mobility edges, a definite energy at which the localization length becomes infinite, and which separates the localized states from the extended states.

This is a big subject, so I shall just give some handwaving arguments. Take, for example, a model potential [in one dimension] consisting of barriers of random heights, with zero potential between the barriers. At fixed energy, the wave function in the *n*th interval is

$$A_n e^{ikx} + B_n e^{-ikx}.$$

There is a transmission and reflection coefficient t_n, r_n taking and incident wave $A_n = 1$ into a reflected wave $B_n = r_n$, $A_{n+1} = t_n$, $B_{n+1} = 0$. Clearly,

$$|t_n|^2 + |r_n|^2 = 1.$$

Let us now revert to classical thinking. Namely, we argue that when the particle is in the n'th interval, it has a probability $T_n = |t_n|^2$ of going to the n+1'st, and a probability T_{n-1} of going to the n-1'st interval, and so on. This would lead to a random walk, or diffusion. Note, as emphasized by Chirikov, that a completely classical treatment does not lead to diffusion in one dimension. But this idea, that in a complicated situation, with random potentials, we can treat the quantum effects one scattering center at a time, evidently had great appeal historically, even though there are many cases in which it is completely wrong.

We now go on to an approximation not assuming that the wave function "collapses" each time it passes a barrier. From time reversal invariance, [assume the barriers are real], we know that the complex conjugate wavefunction is also a solution, ie. $A_n = r_n^*$, $B_n = 1$, $A_{n+1} = 0$, $B_{n+1} = t_n^*$. Let $\mathcal{M}_n$ be the matrix connecting the coefficients:

$$\begin{bmatrix} t_n \\ 0 \end{bmatrix} = \mathcal{M}_n \begin{bmatrix} 1 \\ r_n \end{bmatrix}, \quad \begin{bmatrix} 0 \\ t_n^* \end{bmatrix} = \mathcal{M}_n \begin{bmatrix} r_n^* \\ 1 \end{bmatrix} \qquad \text{(II.9)}$$

These four equations may be readily solved for $\mathcal{M}_n$ to give

$$\mathcal{M}_n = \begin{bmatrix} 1/t^* & -r^*/t^* \\ -r/t & 1/t \end{bmatrix} \tag{II.10}$$

Next consider the transmission between two successive barriers. This will be given by the product of two matrices, $\mathcal{M}_n\mathcal{M}_{n+1}$. The transmission probability is given by the inverse of the absolute square of the (1 1) element. Namely, it is

$$\frac{1}{T_{12}} = \frac{1}{TT'}\left[1 + RR' + 2Re\ r^*r't'^*/t'\right] \tag{II.11}$$

Here we have put $T_n = T$, $T_{n+1} = T'$, $R = 1-T$, etc. One way of trying to deal with this is by a renormalization type of argument. Namely, reduce the lattice by a factor of two by replacing pairs of barriers by a single barrier, and integrating over the random parameter in one of the barriers. This suggests that the *Re* term above averages to zero, and thus

$$T_{12} < TT' \tag{II.12}$$

Thus, we expect that the transmission coefficient of a large number of barriers decreases exponentially in the number of barriers. Since this must be true in both directions, the state is localized.

To make this more rigorous, one may invoke Fürstenberg's theorem.[10] This states, if $\mathcal{M}_n$ are unimodular matrices depending on a random number in such a way that they do not systematically commute, then

$$\lim_{N\to\infty} \frac{1}{N} \ln|\mathcal{M}_N\mathcal{M}_{N-1}\ldots\mathcal{M}_1 u_o| = \gamma > 0. \tag{II.13}$$

Here u_o is some initial vector, and the theorem is true for almost all choices of u_0 and for almost all realisations of the random numbers. This means more intuitively that the largest eigenvalue of the matrix

$$\mathcal{M}_{(N)} = \mathcal{M}_N\mathcal{M}_{N-1}\ldots\mathcal{M}_1 \tag{II.14}$$

is $\propto \exp(\gamma N)$, the other eigenvalue being $\exp(-\gamma N)$. Substituting the transmission matrix in this formula we find that $T_{(N)} \propto \exp(-\gamma N)$.

We remark that this same Fürstenberg's theorem was used to justify the concept of Lyapunov exponent. In that case, *M* was the tangent matrix, and it wasn't random, but pseudorandom. For this reason, γ, the inverse localization length, is also called the Lyapunov exponent here too.

Now consider the original Anderson model of diagonal disorder but in one dimension. Let u_n be the amplitude of the 'electron' on a tight binding lattice, and let the Schrödinger's equation be

$$\mathcal{H}u_n = T_n u_n + \Sigma W_r u_{n+r} = Eu_n \tag{II.15}$$

Take the case $W_r = W$, $r = \pm 1$, $W_r = 0$, else; i.e. nearest neighbor hopping. Let $\mathcal{M}_n$ be the matrix

$$\mathcal{M}_n = \begin{bmatrix} (E-T_n)/W & -1 \\ 1 & 0 \end{bmatrix} \tag{II.16}$$

We may use the $\mathcal{M}$'s to solve for the u's according to

$$\begin{bmatrix} u_{n+1} \\ u_n \end{bmatrix} = \mathcal{M}_n \begin{bmatrix} u_n \\ u_{n-1} \end{bmatrix} \tag{II.17}$$

This time, for random T_n, Fürstenberg's theorem shows that the u's blow up for large n. Well, not quite. For almost all u_0, u_1, that will be the case, but there is one pair for which the amplitude decreases exponentially to the right. Of course, going to the left, it will still blow up ...except at special values of the energy E, in which case it can decay in both directions.

What this heuristic argument shows is that a) there are discrete values of E giving localized states, b) the localization length is $1/\gamma$, and for other values of E, the equation gives exponentially blowing up solutions which are unphysical.

Finally, let us give the following consistency argument which gives the nature of the wavefunctions. Let's take the Anderson Hamiltonian. Let's consider the eigenstates to be localized. Now consider one state, centered near the origin and an energy E, and another state, centered a long distance L away. [$L\gamma >> 1$]. Choose this state to have an energy E′ very close to E. Now change one of the parameters in the Hamiltonian, say T_0. Then E will be a relatively strong function of T_0 but E′ will be almost independent of it. We could thus try to make E cross E′ and become degenerate with it. However, there will also be an off diagonal coupling between state E and state E′, of order $\Delta T_0 \exp(-\gamma L)$. Thus the states will repel, the crossing will be avoided, and there will be a minimum splitting $\propto \exp(-\gamma L)$.

Right at the crossing, however, the eigenstates will mix, becoming approximately equal parts E and E′. They thus are more extended. One might think that surely if we go far enough, we can find such a resonant pair of states corresponding to any given state E. What this takes is a special parameter lying in a range of order $\exp(-\gamma L)$. The number of parameters which can be special grows with L, but only linearly with L, so in fact for a random system, one cannot find resonating pairs of states at large distances.

This type of argument shows that localization in one dimension is the rule rather than the exception. That is, it takes a *special rule*, so that distant sites will have exactly the same energy and environment and thus have a sequence of resonating states going infinitely far away. Periodicity, i.e. a crystalline structure, is the only simple rule doing this. Unfortunately, given a rule, i.e. given T_n in the Anderson model as a function of n, it is generally very difficult to prove that the states are localized, and even more difficult to prove that they are extended. Only in a few cases is the answer known with mathematical rigor.

Whether one wants to call these cases Anderson localization is a matter of taste. I like to use that name, because I learned first about what is going on in random crystals and then saw that there was an analogy. In more complicated cases, the analogy is somewhat strained, and some people prefer to use a new name, such as dynamical localization. This is alright, but it should be remembered that this type of localization occurs in other places than dynamical systems.

D. Connection of Anderson localization to quantum chaos

The kicked rotor is an example of a periodically driven system, $\mathcal{H}(t) = \mathcal{H}(t+1)$ and it follows that if $\psi(t)$ is a solution, so is $\psi(t+1)$. Therefore

$$\psi_a(t+1) = e^{-i\omega_a}\psi_a(t) \qquad \text{(II.18)}$$

or we may take

$$\psi_a(t) = e^{-i\omega_a t}u_a(t),\ u_a(t+1) = u_a(t) \qquad \text{(II.19)}$$

in direct analogy to the treatment of Bloch states in a periodic potential. Here ω_a is the *quasienergy*, and *a* labels the eigenstates. Another way to say this is that the states are eigenvalues of the time evolution operator U with eigenvalues $\exp(-i\omega_a)$.

Recalling the explicit definition of U after II.4 we have

$$U = \exp[-ip^2/2\hbar]\exp[-iKV(q)/\hbar] \qquad \text{(II.20)}$$

Note the two exponentials cannot be combined into one. In the momentum representation U is a matrix, U_{mn}. In this representation, $p = \hbar m$, where m is an integer, and

$$U_{mn} = \exp[-i\hbar m^2/2]J_{mn} \qquad \text{(II.21)}$$

where $J_{mn} = (2\pi)^{-1}\int dq\ e^{i(m-n)q}e^{-iKV(q)/\hbar}$. If $V(q) = \cos q$, then $J_{mn} = (-i)^{m-n}J_{m-n}(K/\hbar)$, i.e. J_{mn} is essentially Bessel's function of the first kind. This drops off rapidly for $|m-n| > K/\hbar$.

The structure of U thus given is of a unitary matrix whose nonvanishing elements lie along the diagonal. The eigenvalues of U are $\exp(-i\omega_a)$, or thinking of U as having the form $U = \exp(-i\mathcal{H}_q)$, the quasienergy operator $\mathcal{H}_q$ has eigenvalue ω_a.

However, if there are theorems regarding the quasienergy operator, I don't know them. Instead, it is convenient to introduce new operators W,T

$$\exp[-ikV(q)/\hbar] \equiv (1-iW)/(1+iW) \qquad \text{(II.22)}$$

$$\exp[-ip^2/2\hbar + i\omega_a] = (1-iT)/(1+iT) \qquad \text{(II.23)}$$

Then

$$u_a^+ = [(1-iW)/(1+iW)][(1-iT)/(1+iT)]u_a^+ \qquad \text{(II.24)}$$

where the + indicates just after the kick. Introduce $\hat{u} = (1+iT)u^+$ where one can show $\hat{u} = (u^+ + u^-)/2$. Now multiply through from the left by (1+iW). Then

$$(1+iW)(1+iT)\hat{u} = (1-iW)(1-iT)\hat{u} \qquad \text{(II.25)}$$

or

$$(T+W)\hat{u} = 0. \qquad \text{(II.26)}$$

In the momentum representation this is

$$T_m\hat{u}_m + \Sigma W_r\hat{u}_{m-r} = E\hat{u}_m \qquad \text{(II.27)}$$

where $E = -W_0$ and the sum is over nonzero r. This is exactly of the form of the Anderson Hamiltonian, I.5.

We may solve for T and W, and find,

$$T_m = \tan[\tfrac{1}{2}(\hbar m^2/2-\omega_a)], \qquad \text{(II.28)}$$

$$W(q) = \tan(\frac{K}{2\hbar}V(q)). \qquad (II.29)$$

One may think of E as the eigenvalue and ω_a as a parameter, or *vice versa*, there is a one-one relation between them as $dT_m/d\omega_a < 0$. For $K/\hbar$ small, $W \cong KV(q)/\hbar$, so this is shortrange hopping. If the argument of W ever reaches $\pi/2$ the transformation doesn't work or at least loses its simplicity. An alternative is to choose a V so that $W(q) = K'\cos q$, which is the form of V corresponding exactly to nearest neighbor hopping.

The T_m of II.28 is not random. We shall show that it is a *pseudorandom* function of m however, however, given certain restrictions on $\hbar$. Thus there is a correspondence of the quantum chaos problem to a lattice problem as follows:

Angular momentum $\hbar m \Leftrightarrow$ Lattice site m

Angle q $\Leftrightarrow$ Lattice momentum

Kicking function V $\Leftrightarrow$ Hopping function W

Kinetic energy $\frac{1}{2}p^2 \Leftrightarrow$ Site energy T_m

Angular momentum function $u^+ \Leftrightarrow$ Lattice wavefunction $\hat{u}$

E. Pseudorandomness of T_m.

We now discuss how T_m can be random. If the pure number $\hbar/4\pi$ is a rational, p/q, then $T_{m+q} = T_m$, so this case corresponds to a regular lattice with a basis of q sites so this case is not pseudorandom at all! This is known as *quantum resonance*. In the original units, $\hbar = 4\pi p/q$ is $\hbar T/2m = 2\pi p/q$, i.e. $bq(\hbar^2/2m) = bp\hbar(2\pi/T)$. Here b is an integer. The unit of energy level spacing in this case is $\hbar^2/2m$, and for any q, there is a spacing of bq units, (in particular for b = q). Thus if the quantum resonance condition is satisfied, the driving frequency or one of its harmonics will always be exactly in resonance with a level spacing, and the system will be driven to high energies. Correspondingly, the eigenstates of the lattice problem will be *extended Bloch states*.

If $\hbar/4\pi$ is irrational however, all is well. The tangent reduces its argument mod π. [This is the same thing that is used in random number generators on computers.] One can discuss the correlations, etc, such as $\Sigma T_n T_{n+r}$ and it is found that this function is very similar to one in which the T's are replaced by random numbers from a computer. Since the rationals are a set of measure zero, one can conclude that generically, the function T_n can be treated as random.

F. An aside on Liouville numbers.

Actually, there is a class of irrational numbers related to the so-called Liouville numbers which can be shown to give extended states. Liouville numbers are irraRtionals which are extremely close to rationals in a precise sense. They are much more numerous than the rationals, being uncountable while the rationals are countable. However, they are still a set of measure zero. Intuitively, consider a number represented by an infinite series

$$L = \Sigma(\pm)2^{-2^n}.$$

Truncation of the series at any large n will give an extremely accurate approximation. The sum is over a 'random' choice of $\pm$, so there are a continuum of numbers L so represented. If $\hbar = 4\pi L$, then the lattice has period 256 to good approximation. This starts to be spoiled on a larger scale, but 2^{16} is *very* accurately periodic. The eigenstates have a peculiar structure. Supposing they are localized on the scale of unity, they then resonate with states 256 sites away, forming a group of resonating levels. This group then resonates with a nearly identical group at sites 64K distant, this group resonates with another at a distance 3.6M, etc. This kind of structure is characteristic of what is called a singular continuous spectrum, which is intermediate between localized and extended.

G. Comparison of pseudorandom and truly random cases

The distribution of T_n of II.28 is given by $p(T) = \int d\omega\ \delta(T-\tan(\omega)) = 1/\pi(1+T^2)$. This is a Lorentzian, or more properly a Cauchy distribution. The long tails correspond to the infinities of the tangent. As it happens, if T_n is distributed *randomly*, with a Cauchy distribution, one can find various things analytically about the corresponding model, which is called the Lloyd model.[11] The simplification is that the average Green's function, $\langle G_{nn}\rangle = \langle 1/(E - T - W - i\varepsilon)_{nn}\rangle$ where $\langle\rangle = \Pi_i\int dT_i p(T_i)$, can be found exactly. Expand

$$G_{nn} = G^o{}_{nn} + G^oWG^o + G^oWG^oWG^o + \ldots \qquad \text{(II.30)}$$

The integral can be done term by term and just results in replacement of T_n by i everywhere. If T is constant, the eigenstates are plane waves, and the energies are T + W(q). The site diagonal average Green's function is thus $\int dq\ 1/(E - i - W(q))$, and the imaginary part of this is the density of states. So, $2\pi\rho(E) = \int dq\ 1/[(E - W(q))^2 + 1]$.

In the case of nearest neighbor hopping, there is the Ishii-Thouless[12] formula which gives the inverse localization length by

$$\gamma(E) = -\int dE'\rho(E')\ln|E-E'| \qquad \text{(II.31)}$$

This can be integrated to give, for the nearest neighbor Lloyd model

$$2W \cosh\gamma = [(E - W)^2 + 1]^{1/2} + [(E + W)^2 + 1]^{1/2} \qquad \text{(II.32)}$$

Thus the localization length for the pseudorandom model can be numerically compared with that of the truly random case. The two agree. In Fig. II.4 is shown a numerically determined wavefunction, together with the slope determined by the Ishii-Thouless-Lloyd formula II.32.

H. Numerical solutions

A word might be said on how to solve these quantum equations. One way is to treat the matrix form of the Anderson Hamiltonian, truncate the matrix, and diagonalize it. If nearest neighbor hopping is assumed, rather large matrices can be handled, as the matrix will be in tridiagonal form. A better way than simple truncation is to choose $\hbar = 4\pi\ p/q$ with q large. Then the true states are extended, and there are bands. However, the bands will be very narrow, of order $\exp(-\gamma q)$ in width. This shows how large denominator rationals are physically similar to irrationals.

It is not so bad to treat the original problem directly. Start with an initial state $\psi(0)$, say $\psi_n(0) = \delta_{n0}$. Keep operating with U to find $\psi(n)$. If the states are localized,

$$\psi(n) = \Sigma\ u_{a0}{}^* u_a e^{-i\omega_a n}.$$

Then Fourier analyse ψ. This gives the quasienergies and the eigenstates. An operation with U is apparently a matrix multiplication, taking N^2 steps for a matrix of size N. However, one factor of U is ordinary multiplication in momentum space, the other in angle space. Angle space transforms to momentum by Fourier transform, and the fast Fourier transform takes $N \log_2 N$ steps. There is no great problem therefore taking $N = 1024$ or even quite a bit bigger.

I. Relationship of the localization length to classical diffusion

We repeat the Chirikov argument. Suppose the localization length $\xi = 1/\gamma$ is rather long. Then if the system is started at a definite m value, say m = 0, it will follow the classical motion for a while. This classical motion is diffusive. So, $p^2 \cong Dt$. This will go on until time $\tau(\hbar)$, and at this time $p^2 \cong \hbar^2\xi^2$. [Note ξ is given in dimensionless lattice variables.] We may also estimate $\tau(\hbar)$ as follows. There will be of order ξ states in the expansion of the wavefunction, and therefore of order ξ quasienergies distributed over the interval $0 \to 2\pi$. The typical spacing between quasienergies is thus of order $1/\xi$. The time $\tau(\hbar)$ is just the time that is needed before one can tell that these spacings are not zero, i.e. $\tau(\hbar) \cong \xi$. Thus $\hbar^2\xi^2 \cong D\xi$ or

$$\xi \propto \tau(\hbar) \propto D/\hbar^2 \qquad (II.33)$$

This shows that in this case, for small $\hbar$, the $\hbar^{-2}$ dependence of $\tau(\hbar)$.

III. Transitions to chaos

A. Introduction

One of the factors that caused many physicists to become interested in chaos was the discovery by Feigenbaum[13] that the transition to chaos by means of period doubling was universal and could be described as a phase transition using ideas of scaling and renormalization. It was subsequently found that the transition to global chaos at the disappearance of the last KAM torus could be described in similar terms. We shall describe these ideas for the case of period doubling in the logistics map. The cases of period doubling in a Hamiltonian map, as well as the last KAM transition are quite similar, but a bit more complicated.

We shall then discuss the quantum case. Since quantum mechanics does not admit chaos, finite $\hbar$ destroys the transition to chaos. One may discuss how this happens. It transpires that finite $\hbar$ affects the dynamical transition in much the same way that finite H ($\equiv$ magnetic field) destroys a thermodynamic magnetic phase transition.

B. The logistics map

This map is a particular case of a one dimensional noninvertible map,[13]

$$x_{n+1} = f(\mu, x_n) \qquad (III.1)$$

where μ is a parameter. Noninvertible means there are two or more values of x_n giving the same value of x_{n+1}. One way of writing the logistics map is

$$f(\mu, x) = \mu(1-x)x \qquad (III.2)$$

[The term μx linear in x gives, by itself, an exponential "population growth" with rate μ. The quadratic term can be regarded as describing a decrease in the rate of growth proportional to the population itself. Rescaling the variables makes both coefficients the same.] However, any quadratic map may be put into this form. Alternative parameterizations are

$$f(x,C) = 2Cx + 2x^2, \; f(x,\lambda) = 1 - \lambda x^2. \qquad (III.3)$$

We shall see that the main feature of importance is that there is only one hump with a quadratic maximum -- otherwise the shape is unimportant.

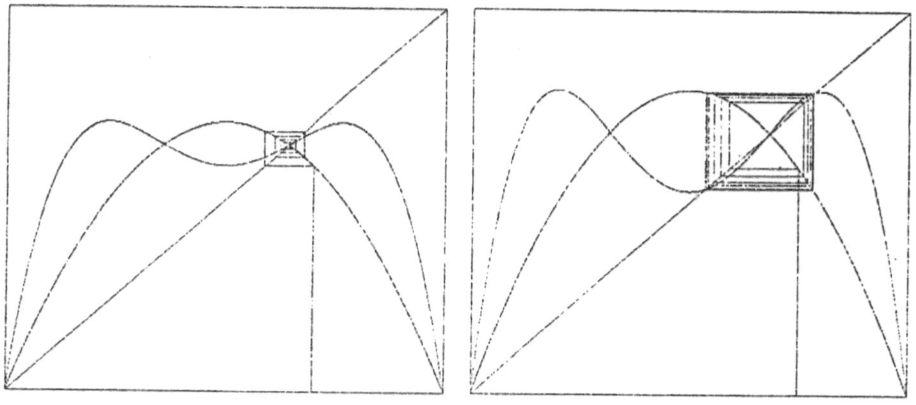

Fig. III.1. The logistics map $f(\mu,x)$ vs x for μ less than $\mu_1 = 3$ (a), and $\mu_1 < \mu < \mu_2$ (b). The first iterate $f^2(\mu,x)$ is also plotted. Iterations of f can be geometrically followed, by following the sequence of straight lines up from an initial value at x = 0.7. If the fixed point is stable, as in (a), the iterations converge to it. If the fixed point is stable, the iterations converge to a limit cycle, that is, to the stable fixed points of $f^2(\mu,x)$, as in (b).

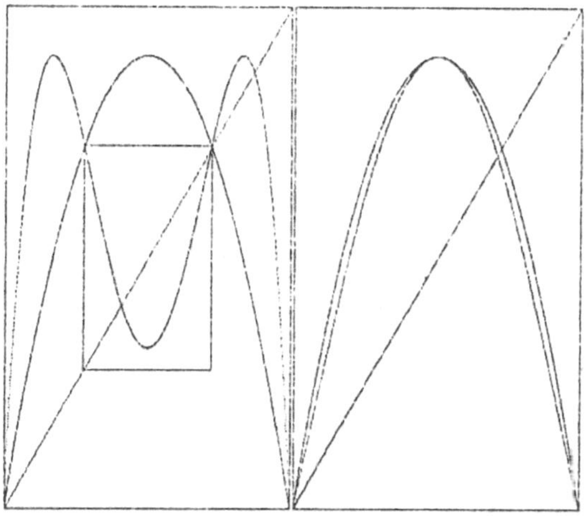

Fig. III.2. The logistics map and its first iterate at an approximation to the critical value $\mu = 3.602 \cong \mu_c$. The iterated map inside the inner square on side (a) is similar to the map itself. This is made clearer by scaling up the inner square, flipping it about the diagonal, and plotting that piece of f^2 and comparing it with f itself, in (b). The outer curve in (b) is f, the inner curve f^2.

The *fixed points* solving $x = f(x)$ are at $x = 0$ and $x = y = 1 - 1/\mu$. For $\mu < 4$, the interval $0 < x < 1$ is mapped into itself. For $\mu < 1$, $x = 0$ is a stable fixed point, and y is not in the interval [0,1] to which we confine attention.

We are interested in iterating this mapping many times, i.e. studying $f^n(x) \equiv f(f(f(...(f(x))...))))$ where the iteration is repeated n times. If a fixed point is stable, and x is close enough to it, $f^n(x)$ will approach the fixed point. Fig. III.1 shows a graphical method of performing this iteration. It is easy to see from the graphical analysis that the fixed point x will be stable if $|f'(x)| < 1$, unstable if $|f'(x)| > 1$.

What happens when a fixed point becomes unstable as μ is changed? What happens is generally a sort of pitchfork bifurcation, namely there comes into existence two new fixed points. However, these new fixed points are of higher order, namely they are mapped into each other. They *are* fixed points of $f^2(x) \equiv f(f(x))$, that is to say, they are periodic points of period two. We now note that $f^2(x)' = f'(f(x))f'(x)$ by the chain rule. At the fixed point we had $f' < -1$, so $f^{2\prime} > 1$. Thus we expect $f^2(x)$ to look as in Fig. III.2. The two new period two points also have the same slope by this argument. All the periodic points are chained together in this way, so it is only necessary to keep track of one of them. We shall concentrate on the point closest to $x = 1/2$.

C. Period doubling sequence

The first period doubling is at $\mu = 3 \equiv \mu_1$, where $f'(\mu_1, 1-1/\mu_1) = -1$. As μ increases to μ_2, the new order 2 fixed points become unstable and two order 4 fixed points appear.

As μ increases through a sequence μ_n successive order 2^n fixed points become unstable. However, the sequence μ_n approaches a finite limit called μ_c. For the logistics map $\mu_c \cong 3.5700$. A very important point is that this sequence is very much like a phase transition. The transition is to *chaos,* because for $\mu < \mu_c$ the Lyapunov exponent is negative, and it becomes positive for $\mu > \mu_c$. Near the transition, there are all the trappings of a critical phase transition, namely *scaling, critical exponents, and universality.* The universality is such that *all smooth single humped maps show the same transition, with the same scaling and critical exponents.*

Next remark that the region near $x = 1/2$ is invariant: Once a point gets into that region, it stays [under mappings iterated 2^n times] The invariant region extends from the unstable fixed point symmetrically about 1/2. Now note that f^2 restricted to this region looks very much like f itself. Make this more quantitative. Taking f to be the logistics map, consider the region near $x = 1/2$ defined by the square in Fig. III.2. The upper right corner of the square is at $1-1/\mu, 1-1/\mu$. The side of the small square is thus $1 - 2/\mu$. To scale this up to the original size requires a factor $\alpha = 1/(1 - 2/\mu)$. The distance of the minimum at $x = 1/2$ to the top of the square is $1-1/\mu - \mu^2(1-\mu/4)/4$. If this is rescaled by α and inverted one obtains the renormalized height $(\mu^4/16 - \mu^3/8 - \mu/4 + 1/2)$ which can be compared with the original height $\mu/4$.

The rescaled curve is not, of course, a parabola - it is a quartic. We can ask the question, however: Is there a parabola-like curve $g^*(x)$ which *exactly* goes into itself under the procedure

• restrict region, • iterate, • rescale.

We call this the *renormalization procedure*. If g^* exists, it will satisfy the equation

$$g^*(x) = -\alpha g^*(g^*(x/(-\alpha))) \qquad \text{(III.4)}$$

The factor $-\alpha$ in the argument is shorthand for the rescaling of the x-axis, the minus sign indicating we also reflect from right to left about the point $x = 1/2$. It also is a shorthand for the restriction of the region from one of unit length to one of length $1/\alpha$. The overall factor rescales and inverts the vertical axis. [One can shift the origin of coordinates to make these factors simple multiplication, and we pretend we have done that, i.e. $g(x) + 1/2 = f(x+1/2)$.]

In fact there is [essentially] a unique solution to III.4, and α is determined by the requirement that there be a solution. We can get a simple estimate of $g(x)$ by supposing it to be approximated by $f(\mu,x)$ for some value of μ. One estimate determines which μ by the condition that the height of the new curve obtained by the renormalization procedure above, namely $\mu^4/16 - \mu^3/8 - \mu/4 + 1/2\ \mu/4 = \mu/4$, the height of the original curve.[14] The solution of this, $\mu_c = 3.604$ is not a bad approximation to the exact value 3.5600. The value of $\alpha \cong 1/(1-2/\mu)$ found in this way is 2.25 [exact = 2.502]. Let us also imagine that we start with a value of μ slightly too small, i.e. $\mu = \mu_c - \nu$. Upon iteration, the new height will be $\mu_c - \nu'$ and in linear approximation we will have $\nu' = (\delta)\nu$ where $\delta = 5.14$ [the exact value being 4.669].

One might hope that a way to solve for g^* is as follows: Start with a reasonably good approximation for it, and iterate the renormalization procedure. If all is well, the result will converge to $g^*(x)$. The numbers of the preceding paragraph suggest that a parabola with $\mu = 3.6$ is not a bad starting point. However, it also strongly suggests that $\delta > 1$. In that case it is clear that iteration of the renormalization procedure is unstable -- if we start with $f(x)$ for $\mu \neq \mu_c$ and iterate the renormalization procedure in the hopes of finding $g^*(x)$, we get farther and farther from the solution $g^*(x)$ for which we are searching.

The following procedure does work. Let us suppose that instead of starting with f we have started with a renormalized f^m where $m = 2^n$. We then fix μ so that the 2^n iterated map has such and such a property. For example, we choose it as the point at which a new period is born. Let μ_n be the value of μ in the original map which has the chosen property in the m'th iterate. This procedure of readusting μ does converge and $\mu_c = \lim \mu_n$. In fact, we find that $\mu_n \cong \mu_c - A/\delta^n$ for large n.

To formalize this a bit, let $\mu = \mu_c$. Then the reduced rescaled map approaches a limit, i.e.

$$\lim_{n\to\infty} (-\alpha)^n g^{2^n}(\mu_c, x/(-\alpha)^n) = g^*(x) \qquad \text{(III.5)}$$

It is easy to prove from this that g^* satisfies the functional relation III.4. III.4 does not determine the overall scale of g^*, i.e., if g^* satisfies III.4 then so does $h(x) = bg^*(x/b)$. We may use this to normalize the solution of III.4 so that $g^*(0) = 1$. Then it follows that $g^*(1) = -1/\alpha$. If we also insist that g^* is symmetric about $x = 0$ the solution is unique. [Remember we have shifted the origin of x.]

Let us parametrise the general one-humped function $g(x)$ by an (infinite) set of parameters. For example, we could expand (a symmetric) $g(x) = 1 - \Sigma c_k x^{2k}$. The logistics map has $c_1 = 4\mu/(\mu-2)$ and all other c_i's zero. The renormalization procedure of III.4 can be regarded as a mapping of a point in the infinite dimensional space whose coordinates are c_k to a new point c_k', i.e. $\mathbf{c}' = \mathcal{R}\mathbf{c}$. The function $g^*(x)$ corresponds to a fixed point $\mathbf{c}^* = \mathcal{R}\mathbf{c}^*$.

Without loss of generality, we may choose the parametrization so that $\mathbf{c}^* = 0$. The next step is to linearize $\mathcal{R}$ for small $\mathbf{c}$. Thus, for small $\mathbf{c}$ we have $\mathbf{c}' = T\mathbf{c}$ where T is a linear operator, which may be diagonalized. In the eigendirections, we label the coordinates d_b and the renormalization gives

$$d'_b = \lambda_b d_b \qquad \text{(III.6)}$$

There will be exactly one direction u for which λ_u is greater than one in magnitude. This gives the tangent to the unstable manifold of the fixed point $\mathbf{c}^*$. All other directions give the stable manifold of codimension one.

Why do we say the dimension of the unstable manifold is one? It is because we have found that the limit III.5 exists if we adjust carefully enough just one parameter. That is, we have found that by changing only c_1 to a value $4\mu_c/(\mu_c-2)$ and keeping all other $c_i = 0$, $i \neq 1$ fixed, we put c onto the stable manifold. One could, of course, have fixed c_1, say at its critical value, and then varied c_2, keeping all other c_i's zero. One would then find the critical value of c_2 would be zero. Values of c_{2n} analogous to μ_n exist, and these would be $c_{2n} = B/\delta^n$, with the same δ but a new B.

The variable $\mu_c-\mu$, if small, is proportional to d_u, as it measures how far we are from the stable manifold. Such variables are called *relevant* variables in renormalization group lingo. Under repeated action of the renormalization operation, they grow. Variables in the stable manifold are called irrelevant variables. On iteration of the renormalization they approach the fixed point. It's a good idea to keep track of relevant variables explicitly. Therefore we introduce

$$\lim_{n\to\infty} (-\alpha)^n g^{2^n}(\mu_c-B/\delta^n, x/(-\alpha)^n) = g(B,x) \qquad \text{(III.7)}$$

and

$$g(B,x) = -\alpha g^2(B/\delta, x/(-\alpha))) \qquad \text{(III.8)}$$

Knowing g(B,x) tells one everything about the function f^m [$m = 2^n$]for large n in the neighborhood of the fixed point. For example, let μ_n be the value of μ for which $g^m(0) = 0$ The point $g^{m/2}(\mu_n,0)$ is mapped into 0 by $g^{m/2}$, so we let $z_n = g^{m/2}(\mu_n,0)$. The distance z_n decreases rapidly with n. For large n, we may write, $g^m(\mu_n,0) = 0$ as $g((\mu_c-\mu_n)\delta^n,0) = 0$. If $g(B_s,0)=0$, we have $\mu_c-\mu_n = B_s/\delta^n$. Then we find that $z_n = g(B_s/\delta,0)/(-\alpha)^{n-1}$.

An alternative notation is as follows. Replace the iteration number by a "time" variable, i.e. $t = 2^n$. Let $g^t = g(B,x,t)$. Then for large t there is a scaling relation

$$g(B,x,t) = (-b)^z g(b^{-u}B, (-b)^{-z}x, bt) \qquad \text{(III.9)}$$

which is valid for all b. This is the same as III.7 provided $z = \ln\alpha/\ln 2$ and $u = \ln\delta/\ln 2$. Then $g(B,x) = g(B,x,1)$. To see this, put $b = 1/t = 2^{-n}$. Then $b^{-z} = 2^{nz} = (2^z)^n = \alpha^n$, etc. The critical exponents z,u are equivalent to the scaling factors α and δ.

All single humped maps with a quadratic maximum satisfying certain

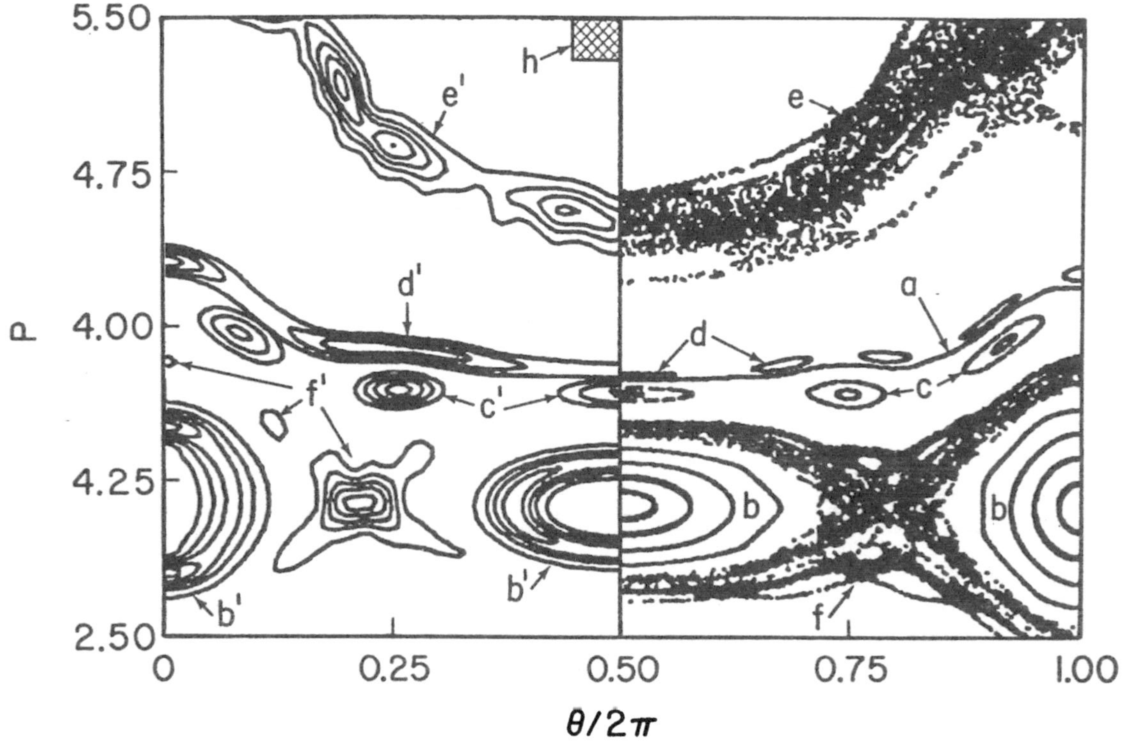

Fig. III.3. Classical orbits, (right), and quantum eigenstates in the coherent state representation (left), of the kicked rotor at $K = K_c$. The size of *h*, Planck's constant is indicated. The last KAM trajectory .*a*. is approximated by island chains in the Fibonacci sequence 2,*b*, 3,*not shown*, 5,*c*, 8,*d*,....The scaling region blown up in III.4 is at the intersection of the symmetry line $\theta = \pi$ and the KAM curve *a*.

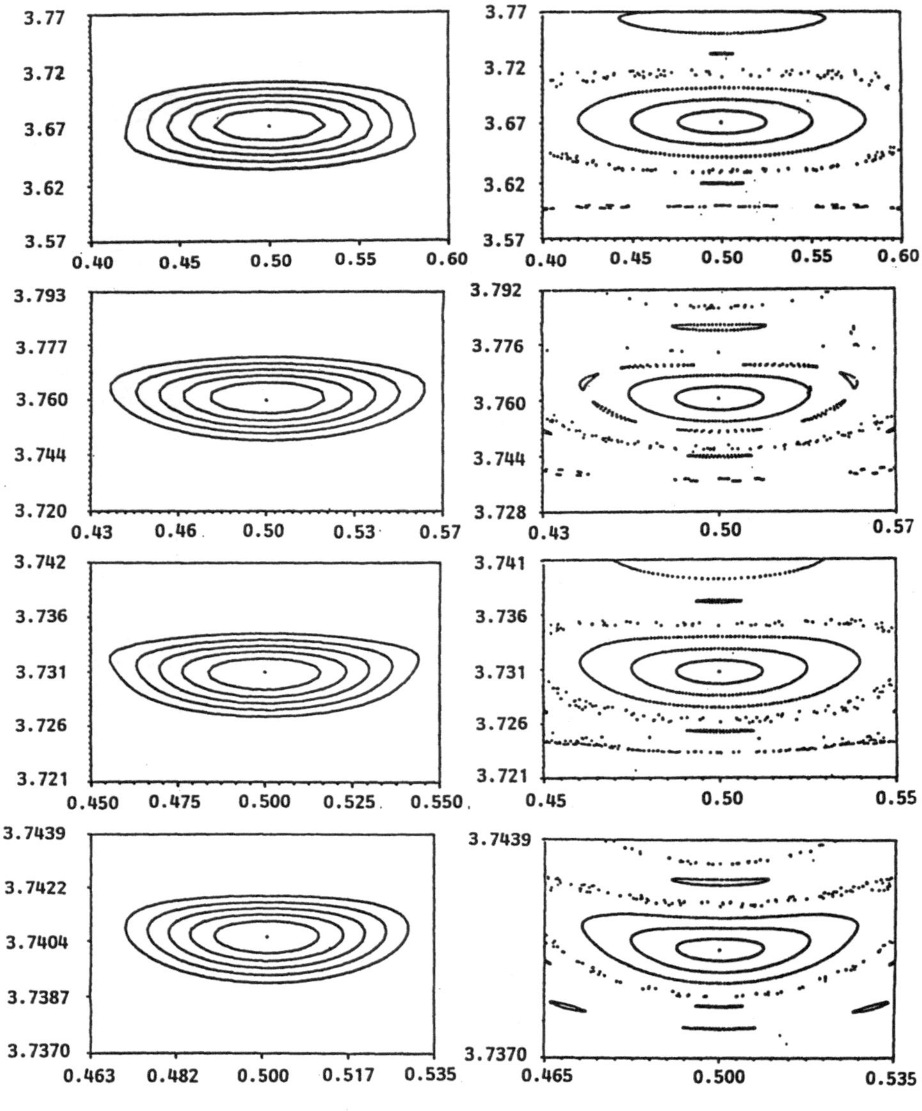

Fig. III.4. Classical orbits (right) and quantum eigenstates (left) in the scaling region (see Fig. III.3). Each successive picture, top to bottom, is a blowup of a piece of the preceding picture. The blowup is by a factor α in the q direction, β in the p direction, and the picture is inverted, top to bottom. For the quantum case, $\hbar$ is also reduced by a factor $\alpha\beta$ at each step. This demonstrates the classical scaling as well as the quantum scaling. The warping which develops could be avoided by a more sophisticated scaling procedure.

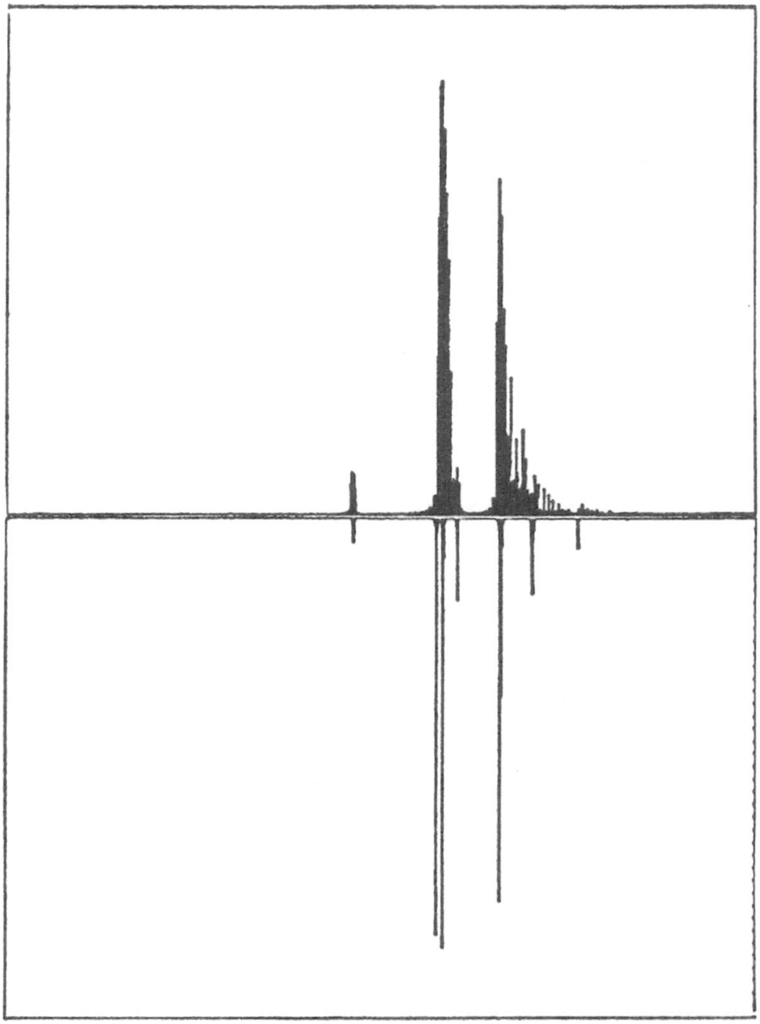

Fig. III.5. Spectrum of a wavepacket, (lower) started on an unstable period 5 point in a scaling region at the intersection of $q = 0$ and the KAM trajectory. This is compared with the spectrum of a scaled wavepacket started on a period 21 unstable orbit in the same scaling region, with $\hbar$ reduced by a factor $(\alpha\beta)^3$. The period 5 spectrum has been shifted by a constant amount, as only spectral differences matter. The period 5 spectrum smeared over a width of order $1/F_5$ is the same as the period 21 spectrum smeared over a width of order $1/F_{21}$. Thus the two wavepackets stay in the scaling region for the same time, after scaling, and behave in the same way.

technical conditions will lead to the same function g^* and the same exponents α and δ. That is, we change some parameter in the map so the map with that parameter lies on the stable manifold. Iteration of the renormalization then brings us to the fixed point. This is the idea of *universality*. Once we see that with many iterations we are essentially examining the properties of the small region near the maximum at $x = 1/2$, the fact of universality may be less surprising. That is to say, only the quadratic character of the maximum is of importance to the universal properties. The position, height and curvature of the maximum are absorbed into the nonuniversal parameters, position of scaling point, μ_c, and the factor of proportionality between the argument B and the normal coordinate d_u. Thus, $g(B,x)$ is a universal function up to a scale factor on B, i.e. $g(yB_s,x)$ is a universal function of x,y, the same for all one humped maps with a quadratic maximum.

D. Hamiltonian maps

Interesting Hamiltonian maps are on two dimensional phase space and conserve area. Period doubling also occurs in this case. For example, at $K = 4$, the stable periodic point at $p = 0$, $q = \pi$ in the standard map becomes unstable and two new stable points appear for K slightly greater than 4. Again, there is a point of accumulation, critical exponents as in III.9, etc.

The main formal difference is that the map is a vector and the argument $z = (p,q)$ is also a vector. The fundamental operation is again iteration. The scaling operation, represented by multiplication or division by $-\alpha$ above must now be generalized. Let Λ be a rescaling operator on z. Then III.7,8 is generalized to

$$\lim \Lambda^n T^{2^n}(\mu_c-B/\delta^n,\Lambda^{-n}z) = T^*(B,z) = \Lambda T^*(T^*(B/\delta,\Lambda^{-1}z))\qquad \text{(III.10)}$$

Just as α and δ must be chosen properly in order that III.8 have a solution, now Λ must be found. In this case there is a coordinate system $z = (x,y)$ such that $\Lambda z = (\alpha x,\beta y)$, i.e. different directions in phase space scale differently. There will be special points, here chosen as $z = 0$, so that this holds close enough to $z = 0$. Again, α, β and δ are universal, but their values have nothing to do with the dissipative case. In fact, it is found that $\alpha = -4.02$, $\beta = 16.4$, and $\delta = 8.72$ for Hamiltonian period doubling.[15] [These rather large values make numerical studies of this case rather inconvenient.] It is possible to go a little bit away from $z = 0$ by making Λ nonlinear, i.e. introducing distortions as well as a change of scale. This is necessary in practice if one wishes to show numerically that III.10 is nearly independent of n for not too large n.

E. Last KAM torus

The last KAM torus is the one associated with winding number w = the golden mean.[16,17] The golden mean is the least rational number, but it is approximated in some best sense by a sequence of rationals, the ratio of the Fibonacci numbers F_n/F_{n+1} where $F_{n+1} = F_n + F_{n-1}$. For rational winding numbers p/q the tori break up into island chains of alternating stable and unstable period q points.

The idea is that the last KAM torus is approximated by the sequence of island chains of period F_n, and this is confirmed numerically. There are two symmetry lines, $q = 0$ and $q = \pi$ on which the period F_n points of the island chains lie. We shall concentrate on the line $q = \pi$ which contains stable points of the chains and no unstable points. The chains alternate above and below the last KAM. Fig. III.3 shows some of the low order stable islands approximating the last KAM.

Taking the origin of coordinates at the intersection of the last KAM and the symmetry line $q = \pi$, [and using a symmetric version of the standard map corresponding to evaluating times halfway thru the kick instead of before or after the kick], a renormalization operation is found corresponding to III.10

$$\lim \Lambda^n \, T^{F_n}(K_c - \delta K/\delta^n, \Lambda^{-n} z) = T^*(\delta K, z) \qquad \text{(III.11)}$$

The parameters α, β and δ are of course different from the values encountered for period doubling, but are universal to all last KAM situations. They are in this case $\alpha = -1.41$, $\beta = -2.56$, $\delta = 1.628$ and Det $\Lambda = \alpha\beta = 4.34$.

Fig. III.4 shows[18], on the right side, some orbits calculated by the rescaled maps for finite n and $\delta K = 0$. The distortions which develop are due to our neglect of nonlinearity in the rescaling operator Λ.

F. Other relevant variables

A general renormalization transformation may well have unstable manifolds for its fixed points of dimension greater than one. For example, adding a magnetic field to a ferromagnetic phase transition destroys the transition because the magnetic field is relevant. To see the sharp transition, not only must the size of the system become large (corresponding to time n approaching infinity in our case), and the temperature be at criticality, (corresponding to $K = K_c$), but the field must vanish. Usually, adding a relevant variable enlarges the system in such a way that one can by symmetry know the value that puts the system on the stable manifold.

Adding a new variable such as a magnetic field means that one should add a new variable H/γ^n to the function corresponding to the map. Then there is a new scaling factor γ as well. Generally γ must be found numerically.

A natural extension of a deterministic map is noise.[19] For example, we could add to the logistics map a random term, i.e.

$$x' = f(x) + \sigma\xi \qquad \text{(III.12)}$$

where ξ is a random variable chosen from a distribution of unit width and σ is a small parameter. Adding noise takes us out of the space of deterministic maps, of course. One thing to do is to introduce a new function mapping a *distribution* of points $\rho(x)$ into a new distribution $\rho'(x)$. Let

$$\rho'(x) = \int dx_1 K(x,x_1,\mu,\sigma)\rho(x_1) \qquad \text{(III.13)}$$

At $\sigma = 0$, K will be just $\delta(x-f(\mu,x_1))$, but at finite σ it could be taken to be $(2\pi\sigma^2)^{-1/2} \exp[-(x - f(\mu,x_1))^2/\sigma^2]$ or some other distribution of width σ.

Iteration of the map K is possible. It will not show period doubling in the strict sense of course, since for finite σ there is no such thing as a periodic orbit. However, for small σ, one can be near a period 2^n orbit, and then diverge exponentially away from it. This introduces a new scaling factor γ, which has to be found numerically, but which is also universal, i.e. it is independent of exactly how we put in noise. The universality in this case depends on a sort of central limit theorem.

G. Planck's constant as a relevant variable

Going from classical mechanics to quantum also takes us out of the space of deterministic maps.[20] However, it is natural to generalize along the lines of III.13. Namely, introduce the quantum "distribution" of phase space points W(q,p) known as the Wigner function, discussed elsewhere in these lectures. We use the standard definition

$$W(q,p) = \int \frac{d\xi}{2\pi\hbar} \exp\left(\frac{ip\xi}{\hbar}\right) \psi^*\left(q+\frac{\xi}{2}\right) \psi\left(q-\frac{\xi}{2}\right) \qquad \text{(III.14)}$$

However, we find it more convenient[21] to use $w(q,\zeta)$, the Fourier transform of W with respect to p, which is

$$w(q,\zeta) = \psi^*\left(q+\hbar\frac{\zeta}{2}\right) \psi\left(q-\hbar\frac{\zeta}{2}\right) \qquad \text{(III.15)}$$

If w_n is the Wigner function just before the n'th kick, then w_{n+1} is given by a time evolution operator κ for the Wigner function, similar to K in III.13, which is simply expressed in terms of the evolution operator U for the wavefunction as

$$\kappa(q,\zeta,q'\zeta') = \hbar U^*(q+\hbar\tfrac{\zeta}{2},q'+\hbar\tfrac{\zeta'}{2})\,U(q-\hbar\tfrac{\zeta}{2},q'-\hbar\tfrac{\zeta'}{2}) \qquad \text{(III.16)}$$

The classical limit of III.16 is easy by use of II.5. It is

$$\kappa_{cl}(q,\zeta,q'\zeta') = (2\pi)^{-1}|S_{12}(q,q')|\exp[-i\zeta S_1(q,q')-i\zeta' S_2(q,q')] \qquad \text{(III.17)}$$

The function S introduced earlier is Hamilton's principal function, and it is known and may easily be shown that the classical map is given by its derivatives, i.e. S is the generating function for the map:

$$p = S_1(q,q'), \quad p' = -\, S_2(q,q') \qquad \text{(III.18)}$$

where we continue to use the subscript notation for derivatives with respect to the i'th argument. Then it easily follows that III.17 correctly gives the classical evolution of a Fourier transformed phase space distribution.

The map κ of the Wigner function may be iterated. This is obviously equivalent to the iteration of the evolution operator U. For small $\hbar$ we do this in the WKB approximation, in other words, we do the integral in

$$U^{(2)}(q,q') = \int dq''\, U(q,q'')\, U(q'',q') \qquad \text{(III.19)}$$

by stationary phase. The result is that $U^{(2)}$ is expressed exactly in the form II.6 with the classical $S^{(2)}$ corresponding to the two step map, namely,

$$S^{(2)}(q,q') = S(q,\underline{q}) + S(\underline{q},q') \qquad \text{(III.20)}$$

where $\underline{q}$ solves the stationarity condition

$$S_2(q,\underline{q}) + S_1(\underline{q},q') = 0 \qquad \text{(III.21)}$$

which in view of III.18 is just the condition that $\underline{q}$ be the intermediate point on the classical orbit from q' to q. In general, there is more than one such orbit, but we ignore that possibility for the moment.

It's clear then that κ_{cl} is just another notation for the mappings such as that in III.11. Under appropriate rescaling then, for $m = 2^n$, $S^{(m)}$ will approach a limit, which is the fixed point of the renormalization operation

$$\mathcal{R}[S](q,q') = \alpha\beta[S(q/\alpha,\underline{q})+S(\underline{q},q'/\alpha)] \qquad \text{(III.22)}$$

where $\underline{q}(q,q')$ extremizes the right hand side. [We have taken $K = K_c$ to avoid encumbering the notation.]

We are now in a position to discuss finite but small $\hbar$. We can do this by expanding the exponent in κ to the next higher order, which is $\hbar^2$. [We ignore the expansion of the prefactor which does not give anything interesting.] Thus, we approximate

$$\kappa(q,\zeta,q',\zeta',\hbar) \equiv \rho\exp(-i\phi) \cong$$

$$(2\pi)^{-1}|S_{12}(q,q')|\exp[-i\zeta S_1(q,q')-i\zeta' S_2(q,q')-i\hbar^2\Phi(q,\zeta,q',\zeta'))] \equiv$$

$$\rho\exp(-i\phi_0-i\hbar^2\Phi) \qquad \text{(III.23)}$$

The renormalization operation for ϕ which corresponds to III.22 is

$$\mathcal{R}[\phi\,](q,\zeta,q',\zeta') = \phi^{(2)}(q/\alpha,\beta\zeta,q'/\alpha,\beta\zeta') \qquad \text{(III.24)}$$

where $\phi^{(2)}$ is found from doing the integrals for $\kappa^{(2)}$, and keeping the leading terms in $\hbar$. If $\hbar = 0$, we may find $\phi^{(2)}$ by substituting $S^{(2)}$ in the expression III.23.

Let us now assume that ϕ_0 is at the fixed point of the classical period doubling renormalization, $\mathcal{R}[\phi^*] = \phi^*$. Then, for small $\hbar$, Φ will evolve according to the linearized renormalization, $T\Phi' = \Phi$. If $\hbar$ is a relevant perturbation, then the largest eigenvalue $\lambda_\hbar^2$ of T is greater than one and after repeated renormalizations, only the corresponding eigenvector $\Phi_\hbar$ is of importance. Therefore, near the fixed point, we have

$$\mathcal{R}[\phi\ ](q,\zeta,q',\zeta',\hbar) = \phi^{(2)}(q/\alpha,\beta\zeta,q'/\alpha,\beta\zeta',\hbar) = \phi(q,\zeta,q',\zeta',\lambda_\hbar\hbar) \qquad \text{(III.25)}$$

This says that the effective Planck's constant grows with iteration. The operator T in this case is moderately complicated, so we won't write it down.[21] Usually, one must find the eigenvalues of T numerically, and that can be done here.[22] However, one can here show that the eigenvalue is just

$$\lambda_\hbar = \alpha\beta \qquad \text{(III.26)}$$

Perhaps the easiest way to this result is to note that in the quantum case, the evolution operator κ factors into two operators which independently evolve, according to III.16. To leading nontrivial order in $\hbar$, this evolution is given in terms of the functions S according to the rules III.20, III.21, and the simplest expression of the renormalization is III.22.

Now, to start with Φ is given by a cubic polynomial in ζ,ζ' whose coefficients are 3rd derivatives of S. In fact, we can show that $T\Phi$ is given by the same construction with S replaced by $S^{(2)}$ and the rescaling. Once this is done, since S and its derivatives scale under renormalization, III.26 is easily found.

At first sight, this is somewhat surprising. Doing the integral III.19 by stationary phase involves expanding the phases to second order in $q'' - \underline{q}$, where $\underline{q}$ is defined in III.20, thereby involving only second derivatives of S. The function Φ on the other hand, involves third order derivatives of S. However, to leading order, the integral over the intermediate q'',ζ'' variables in the calculation of $\kappa^{(2)}$ can be done in the stationary phase approximation, and it is a simple change of variables to do this double integral as two one dimensional integrals like III.19. The stationary phase approximation is independent of a change of variables, so $\kappa^{(m)}$ can be constructed from the corresponding $S^{(m)}$, and therefore, we find the eigenvector Φ of the linearized renormalization operator T is given by the third derivatives of the function S satisfying III.22.

Thus, the eigenvalue of the linearized renormalization operator is found as always. In this particular case, because of the factorization property of the propagator for the Wigner function, the eigenvector of that operator can be constructed from the fixed point of the renormalization operation, and the eigenvalue is trivially related to the scaling factors.

H. Consequences of scaling

It is convenient to revert to the propagator for the Wigner function, rather than its Fourier transform. We introduce a time variable k by

$$W(z,t+k,\hbar) = \int d^2z' K(z,z',k,\hbar) W(z',t,\hbar) \qquad \text{(III.27)}$$

The asymptotic scaling for K is

$$\alpha\beta K(z,z',F_{n+1},\hbar) = K(\Lambda z,\Lambda z',F_n,\alpha\beta\hbar) \qquad \text{(III.28)}$$

Let P_n be a period F_n elliptic point in the scaling region S. If z is near P_n, then $z' = T(z,F_n)$ is also near P_n. For sufficiently small $\hbar$ there are one or more eigenstates of U^{Fn} localised in the elliptic region near P_n, and there are corresponding time independent Wigner functions $W_{nm}(z,\hbar)$ with support in S . Eigenstates of this kind are shown in the coherent state representation [CSR] on the left side of Fig. III.3. In that figure, c′ labels states centered on a period 5 orbit, b′ labels a state on a period 2 orbit, and d′ is on a period 8 orbit. Planck's constant is such that there is one period 5 state per island, 5 period 2 states, of which the fourth is shown, (and labelled by the subscript m), and the period 8 states barely exist.

There is a criterion for the existence of a state of this type: the state will exist only if $h < S_{nm}$, approximately 1/m of the island area. The scaling implies that $W_{n+1,m}(z,\hbar) = \alpha\beta W_{nm}(\Lambda z,\alpha\beta\hbar)$, and if S_{nm} is the criterion for the m'th state, then $S_{nm} = \alpha\beta S_{n+1,m}$. On the left side of Fig. III.4 we show CSR eigenstates centered on the period 5,8,13, and 21 elliptic orbits. The pictures for the successive n values have been scaled, as has $\hbar$. The quantum states follow the classical scaling, if $\hbar$ is also scaled. If $\hbar$ is not scaled, there will soon be not enough phasespace area to support a state, i.e. the value of $\hbar$ is so large that there is no classical-quantum correspondence.

There are actually F_n states which are nearly degenerate under U^{Fn}. This can be seen by noting that a wave packet could be formed on any one of the F_n islands. These states are not degenerate under U however. Their quasienergies therefore differ by $2\pi/F_n$. The quasienergies may then be written as $E_{nmj} \cong E^0{}_n + \hbar\omega_n(m+1/2)+jh/F_n$ for $h < S_{nm}$ where we have approximated the m states by harmonic oscillators.

Consider $\hbar$ so small that there are a fair number of states m in an island, n. Then a wavepacket can be formed which under U^{Fn} revolves about the central point P_n with frequency ω_n. There is a corresponding wave packet in the r'th island, at a rescaled $\hbar$, which revolves with frequency ω_r. Rescaling the time, we see that $F_n\omega_n \equiv F_r\omega_r \mod 2\pi$. The ω's are quasienergy differences which enter into the expression for a Wigner function of a wavepacket state and which is time dependent.

One may also start a wavepacket at hyperbolic points and again these must scale. Although there are "scarred" eigenstates at the hyperbolic points, these are in general not degenerate, and cannot be localised to remain in the vicinity of the hyperbolic point indefinitely. However, as long as the packet remains in the scaling region, it must scale: i.e. there must be another packet for a scaled $\hbar$ at a scaled hyperbolic point which behaves in the same way. In particular, this implies that the local quasienergy spectrum of two scaled wavepackets must be approximately the same. In Fig. III.5 we show the spectra of two scaled wave packets and it is seen that when smoothed they are similar.

I. Tunnelling thru KAM barriers

We now discuss Fig. II.2 which shows tunnelling through the last KAM torus.[18] This probably has nothing to do with the scaling considerations, but it is quite instructive. What is plotted is the long-time average of the absolute square of the transport amplitude $\mathcal{M}$ defined as

$$\mathcal{M} = \langle f | U^M | i \rangle = \Sigma \langle f | b \rangle \langle b | i \rangle e^{-iM\omega_b} \qquad \text{(III.29)}$$

where b labels the eigenstate of U. We take i to be a momentum eigenstate, on one side of the KAM, and f a momentum eigenstate on the other side of the KAM. The plot is for fixed i as a function of f.

The eigenstates are large only on one or the other side of the KAM with exponential tails on the other side. Thus, one of the matrix elements in III.29 will be very small, and there is a very wide distribution of values of these matrix elements. It generally happens that only one or a few states b contribute appreciably for a given i,f although which state contributes can be quite a sensitive function of i and f. This is similar to conduction through a random network of large resistances, where one path tends to dominate, but as some parameter changes there can be large changes in the resistance as the dominating path switches. Which states b contribute can be found by time Fourier transform of $\mathcal{M}$.

For Fig. II.2, the state i is a momentum (=3.14) passing through the period 5 island chain just below the KAM in Fig.II.1. [This description is for just after a kick, not half-way through a kick. The numbers cannot be read off directly from Fig. III.3 therefore.] The first exponential drop, labelled E in the figure, is just a sequence of tails of wavefunctions associated with the period 5 chain. One of these states, is shown in the inset, namely absolute square of state c′ of Fig. III.3. This falloff scales, numerically, as $\exp(-fA/\hbar^{1/3})$, but it is not understood why.

The plateau Q comes from a single intermediate state b which is a period 3 state, while the plateau P comes from the state labelled e′ in Fig. III.3. The curve B corresponds to i a coherent state of momentum 3.14 as before, but concentrated in the region $q \cong \pi$, where the state e′ is very small. Thus the plateau P disappears in that case. The plateau Q′ also comes from the period 3 state, but it is likely an artifact. Namely, our numerically technique does not allow to find the very small tails of wavefunctions accurately. Small amounts of the wrong wavefunction are sometimes mixed in. However, the calculation of $\mathcal{M}$ does not have this deficiency.

IV. Validity of the semiclassical approximation in quantum chaos

A. Introduction

In this section we make some remarks on the question of the validity of the semiclassical or WKB approximation, in quantum chaos systems. Most of this work has been done in collaboration with Yin Guo. Eq. I.4 expresses the general nature of the expressions encountered when this approximation is made. The most famous formula of this type in the present context is the Gutzwiller trace formula,[23] which expresses the trace of the Green's function at fixed energy as a sum of type I.4.

$$\text{Tr } G(E) = \Sigma_a \; A_a(E) e^{iS_a(E)/\hbar} \qquad \text{(IV.1)}$$

The imaginary part of this trace is the density of states, i.e. it is

$\Sigma_n\delta(E - E_n)$ for the commonly considered case of discrete energy levels. [We leave aside a term in the Gutzwiller formula that gives the average density of states.] In this case a labels periodic orbits, and $S_a(E)$ is the action integral on this orbit at energy E. The prefactor A most importantly is proportional (for long very unstable orbits) to $\exp(-\gamma/2)$, where γ is the Lyapunov exponent for that orbit, which is proportional to the length of the orbit. There are also certain phase factors (Maslov indices) which must be included, but which we do not write.

Sometimes the Gutzwiller formula is exact. For spaces of negative curvature this is so and the formula then is known as the Selberg trace formula. However, it should not be thought that the Gutzwiller type of formula is absolutely convergent for real E. In a chaotic situation, the number of terms in the sum increases exponentially fast with increasing period. In fact the number of terms with Lyapunov exponent approximately equal to γ goes as $\exp(+\gamma)$. If one gives E a sufficiently large positive imaginary part, the sum will converge, as that effectively restricts the length of the periodic orbits. An analytic continuation can then in principle be made to put E back on the real axis.

Much of the most exciting recent work in quantum chaos[3] has been on ways to do this analytic continuation, i.e. to rearrange formulas of the Gutzwiller type, so as to reduce them to finite, convergent or at least asymptotic formulas which are numerically useful with relatively few terms to calculate. The several great successes in this effort have been for cases of hard chaos. It is not too much to say that the essential key to quantum effects in hard chaos has been found, at least in two dimensions.

This lecture is not on this subject, however. Rather, it is on a version of the question in which issues of convergence do not enter, i.e. the number of terms in the series I.4 is finite from the beginning, albeit very large. Then it could happen that even these terms are given incorrectly, i.e. to an untenably bad approximation, by WKB. There is one important work[24] calling into question the validity of the WKB in this case, suggesting that it is barely more satisfactory than the classical approximation itself.

Because of the successes we have just mentioned, however, the intuition is that the WKB approximation has a far greater range of validity than suggested. There is in fact numerical evidence[25] that indicates that some form of the WKB is better in a couple of special cases than naive interpretation of references [24] suggests.

This preamble indicates that it is a complicated question, not easily posed, to ask for the "validity" of "the" "semiclassical approximation", just as it is not so straightforward to discuss the validity of the classical approximation. There, as we discussed in section I, one must introduce the time scale $\tau(\hbar)$, and one can not interchange the limits time $\rightarrow$ infinity with $\hbar \rightarrow$ zero. Also, τ depended on which experiment was done, as well as, of course, which system is studied.

Our hypothesis is that much the same picture holds for the WKB. For a given observable, there is a WKB expression $\mathcal{O}_{WKB}(t,K,\hbar)$, similar to the expressions introduced for the classical and full quantum case in section I.B. This expression will, for small $\hbar$, adequately approximate $\mathcal{O}_q(t,K,\hbar)$ for $t < \theta(\hbar)$, but it fails at later times. The characteristic time θ depends not only on $\hbar$, but on what the measurement is, how chaotic the system is, and which version of the WKB is used. So, in cases where the Selberg trace formula applies, $\theta = \infty$, but probably in some other cases θ is as small as the log-$\hbar$ time.

B. Quantum maps

Expressing an observable $\mathcal{O}$ exactly in quantum mechanics by a Feynman path integral, the most straightforward version of the WKB is obtained by evaluation of the integral by stationary phase. [SΦ] The simplest problem which encounters the essential difficulties is expressed by a *quantum map*[24], i.e. a unitary time evolution operator U(q,q′) given by II.5, and repeated here,

$$U(q,q') = [-iS_{12}(q,q')/2\pi\hbar]^{1/2}\exp\{iS(q,q')/\hbar\} \qquad \text{(IV.2)}$$

but where S(q,q′) may be more general than given in II.6. For a quantum map, IV.2 is regarded as the exact propagator for unit time step in an integer time system. The *classical* map $z = T(z')$, $[z = (p,q)]$ is generated by S according to $S_2(q,q')=-p'$, $p=S_1(q,q')$ and we assume that S is such that these equations determine uniquely $T(z')$.

We study the M time-step propagator, or return amplitude

$$\langle 0|U^M|0\rangle = \int..\int dq_0 dq_1..dq_M \langle 0|q_M\rangle U(q_M,q_{M-1})..U(q_1,q_0)\langle q_0|0\rangle \qquad \text{(IV.3)}$$

which is a "path" integral of *finite* dimension. We take $|0\rangle$ to be a localized wavepacket, e.g. a coherent state. [One could also consider off diagonal matrix elements of U^M or the trace of U^M.] A matrix element like IV.3 is more sensitive to interference effects than is the trace and corresponding measurements will have the shortest times $\tau(\hbar)$. The "energy" version of this involves an infinite Fourier sum over M, and therefore gets into convergence questions. The use of a map rather than a continuous time avoids a true path integral for which mathematical proofs are difficult or nonexistent.

The SΦ contribution of all integrals except q_0 and q_M, e.g. that over q_i is determined by expanding q_i in the exponent to second order about $\underline{q}^a_i$ which solves the stationarity condition

$$S_2(q_{i+1},\underline{q}^a_i) + S_1(\underline{q}_i,q^a_{i-1}) = 0 \qquad \text{(IV.4)}$$

The label *a* is used to label possible multiple solutions of IV.4. The Gaussian integral resulting brings a factor $R^{-1/2}$ where $R = S_{22}(q_{i+1},\underline{q}^a_i) + S_{11}(\underline{q}^a_i,q_{i-1})$. The net result is quite neat: it gives the factor

$$U^{(2)}(q_{i+1},q_{i-1}) = \int dq_i U(q_{i+1},q_i)U(q_i,q_{i-1})$$

as explained in III.20. Namely $U^{(2)}$ is a sum over terms of the form IV.2 with S replaced in a given term by

$$S^{(2)a}(q_{i+1},q_{i-1}) = S(q_{i+1},\underline{q}_i^a) + S_1(\underline{q}_i^a,q_{i-1}) \qquad \text{(IV.5)}$$

and $\underline{q}_i^a(q_{i+1},q_{i-1})$ is a solution of (IV.4). [There is also a phase factor like a Maslov index which keeps track of the sign of R. We shall not write this factor to keep the notation simple.]

In the chaotic case, we expect multiple solutions corresponding to distinct classical paths from q_{i-1} to q_{i+1}. Suppose, for example, IV.3 always had two solutions. Then, considering the whole integral, we would expect about 2^M paths from q_0 to q_M, so the number of paths grows exponentially with M. The essential problem is to deal with this multiplicity of paths.

We next consider the q_0 and q_M integrals. The simplest case is to take the trace: this amounts to forgetting about $|0\rangle$, setting $q_0 = q_M$ and integrating over this variable too, in IV.3. Then the stationary points are the classical periodic orbits of period M. The number of these orbits grows exponentially with M. Alternatively, we may assume that $|0\rangle$ is such that the q_0 and q_M integrals may also be done by SΦ. This corresponds to classical orbits which are not necessarily periodic.

Thus in the WKB approximation, $\langle 0|U^M|0\rangle$ is a sum over terms indexed by points determined by solutions to III.5 together with two equations determining the first and last point. We may label this classical trajectory $\zeta^a=[q_0^a..q_M^a]$ as the point in M+1 dimensional space (i.e. the path in 1D) at which the phase of the integrand in (IV.3) is stationary.

There are several ways in which we can imagine doing the multiple integral in IV.3. One is decimation: Do every other integral, replacing in each case $U \circ U$ by $U^{(2)}$. Continue, replacing $U^{(2)} \circ U^{(2)}$ by $U^{(4)}$, and so on. That is a good way to study the period doubling transition, discussed in III.

Another technique is to start at one side, say q_0, and do successive integrals. This has the advantage that in some sense we will follow only one orbit. Of course we don't avoid the multiple solutions, they appear in a different form. We shall return to that technique later.

Now let us consider the integral as a multiple integral in M+1 dimensional space, rather than as a sequence of simple integrals. The SΦ condition then has solutions labelled by ζ^a. Around each such point we make a quadratic expansion of the phase of the integrand of IV.3 [call $\mathscr{W}$ the quadratic form in the variables $q_i - q^a_i$]. For SΦ to be accurate, this expansion must be adequate in a large enough region about ζ^a that the M+1 dimensional integral effectively converges. The volume of this region can be naively estimated as being of order $\hbar^{M/2}/|\det \mathscr{W}|^{1/2}$. Clearly, another point ζ^b cannot be in that region, since the SΦ condition on a quadratic function can only be satisfied at the origin, (if det $W \neq 0$). The number of points ζ^a grows exponentially with M, as b^M, say. The volume of M dimensional space grows also, as L^M say. This suggests that if $\hbar b^2 << L^2$ the SΦ points are separated, and SΦ is justified.

Even if many SΦ points are close together, all is not lost. The classic reason SΦ points are close is that there is a nearby caustic,[7] i.e. the quadratic form $\mathscr{W}$ is nearly singular and its determinant is very small. If only pairs [not n-tuples] of points are close, appropriate cubic terms can be kept in the expansion and uniform approximations can be made,[7] in principal if not in practice. However, for large M the contribution from ζ^a [*or* of a uniformly approximated pair of ζ's] goes inversely as the stability, i.e. essentially as $b^{-M/2}$. Thus if the number of close pairs of points increases more slowly than b^M, the contribution of the close pairs can be neglected, or crudely approximated.

For the SΦ to fail, then, one of two things must happen. There could be a systematic bunching of SΦ points in M-dim configuration space, for M large. No argument exists that this doesn't sometimes happen, but neither are cases known where it does.

The SΦ could also fail because it is after all not exact. Even if the M dimensional integral is formally dominated by the stationary points, up to $O(\hbar)$ corrections, these corrections might be significant for large M because many many orbits contribute corrections. There are many cancellations between the leading terms, and it might be possible that the correction terms could suffer fewer cancellations. However, all experience with the WKB for integrable systems, [and even for short time calculations in chaotic systems,] indicates the opposite, that the WKB is qualitatively correct, and numerically even somewhat better than formally expected.

C. Periodic Point Expansions

The version of SΦ approximation just discussed is practically useless, even though mathematically well defined. Not only do the number of points ζ^a proliferate exponentially, they depend on the states $|0\rangle$. [Indeed, for coherent states $|0\rangle$, the "phase" is complex and the SΦ points ζ^a are complex.] Some sort of intuition as to what the SΦ points mean is also desirable. Therefore a further approximation, which we call the assumption of periodic point dominance, [PPD] is standardly used,[26,27] although it is has not been discussed as such.

This goes as follows for the case that $|0\rangle$ is a localized state with support in some small region $\mathcal{A}$, of approximate area $\approx \hbar$. Clearly, only orbits leaving $\mathcal{A}$ and returning to it after M mappings are important. Any period M points living in $\mathcal{A}$ will satisfy this. Assume that *all* such returning points are organized by the period M points. That is, expand $S(q_M,q_{M-1})+..+S(q_1,q_0)$ about a period M point ζ^a whose end points $q^a_0 = q^a_M$ lie in $\mathcal{A}$. One may then perform the intermediate q_i [i = 1,M-1] integrals to obtain for the phase of the result

$$S^a(q_M,q_0) \cong S^a + p_0^a(q_M-q_0) + \Sigma S^a_{ij}\delta q_i \delta q_j/2. \qquad \text{(IV.6)}$$

Here $\delta q_1 = q_M-q_0^a$, $\delta q_2 = q_0-q_0^a$. The matrix S^a_{ij} is the second derivative of the sum of the S's with respect to q_M,q_0, where subscript 1 indicates derivative with respect to the first argument, q_M, etc. S^a_{ij} is equivalent to, and most easily calculated from, the tangent matrix to the periodic orbit. The tangent matrix M satisfies

$$\begin{bmatrix}\delta p_1\\ \delta q_1\end{bmatrix} = M \begin{bmatrix}\delta p_2\\ \delta q_2\end{bmatrix} \qquad \text{(IV.7)}$$

while S^a_{ij} satisfies

$$\begin{bmatrix}\delta p_1\\ -\delta p_2\end{bmatrix} = S^a \begin{bmatrix}\delta q_1\\ \delta q_2\end{bmatrix} \qquad \text{(IV.8)}$$

For an unstable periodic point, a small deviation δz_2 grows exponentially fast, which means that the matrix elements of M are large, proportional to $\exp(\gamma)$. The matrix M may be found as a product of matrices $M_M..M_1$ where M_i is the linearization of the classical map $\delta z^a_{i+1} = M_i \delta z^a_i$ for each of the M steps around the a'th periodic orbit. S^a is given in terms of M by $S^a_{12} = -1/M_{21}$, $S^a_{11} = -M_{22}/M_{21}$, $S^a_{22} = -M_{11}/M_{21}$. A very unstable orbit will therefore have the diagonal elements of S of order unity, while S^a_{12} is very small. It is not hard to check that if S^a_{12} can be neglected, the slope of the unstable manifold of the periodic point *a* is given by $\delta p_1/\delta q_1 = S^a_{11}$, while the slope of the stable manifold is given by S^a_{22}. Note also that the prefactors are proportional to the square root of S^a_{12}.

The introduction of S^a allows us to do the Gaussian integrals arising when $|0\rangle$ is a coherent state. Regarded as SΦ, the orbits are in this case complex, but there are many complications if all complex orbits are searched for. The net result for a coherent state centered at Q,P is that

$$\langle 0|U^M|0\rangle = \Sigma_a\ A_a\ e^{iS^a/\hbar}e^{-\mathcal{Q}^a/\hbar} \qquad \text{(IV.9)}$$

Here $\mathcal{Q}^a$ is a complex quadratic expression in the variables $Q - q^a_0$, $P - p^a_0$, and $\exp[-\mathcal{Q}^a/\hbar]$ decreases in Gaussian fashion when these variables are large. Thus we may include all periodic points in the sum, even if they do not fall near Q,P. The quadratic expansion about such points will not be accurate, but no significant error will arise. We have evaluated this expression in the case of the kicked rotor and compared it with the full quantum expression. We shall discuss the results later, but we do find cases where it fails badly.

We now take the opportunity to connect this formulation with a similar one of Berry[26], [which was carried out for autonomous systems of definite energy rather than a quantum map.] Instead of a diagonal element IV.2, the "Wigner" density is studied, namely

$$\mathcal{W}(Q,P) = \int d\xi\ e^{iP\xi/\hbar}\langle Q+\xi/2|U^M|Q-\xi/2\rangle \qquad \text{(IV.10)}$$

The assumption of PPD approximates the matrix element in IV.10 by

$$\langle Q+\xi/2|U^M|Q-\xi/2\rangle = \Sigma_a\ [-iS^a_{12}(q^M,q_0)/2\pi\hbar]^{1/2}\exp\{iS^a(q_M,q_0)/\hbar\} \qquad \text{(IV.11)}$$

$q_m = Q + \xi/2$, $q_0 = Q - \xi/2$ and S^a given in IV.6. As before, introduce $\delta q_1,\delta q_2,\delta p_1,\delta p_2$. The integral over ξ may be done (in SΦ) with the result, $\delta Q = (\delta q_1+\delta q_2)/2$, $\delta P = (\delta p_1+\delta p_2)/2$, where $\delta P = P - p^a_0$. In this case, since δp is determined by δq through the matrix S^a_{ij}, we may solve for the δq_i in terms of $\delta Q,\delta P$. The result for $\mathcal{W}$ is similar to IV.9:

$$\mathcal{W}(Q,P) = \Sigma_a\ A_a\ e^{iS^a/h}\ e^{i(\delta p_1\delta q_2\ -\ \delta p_2\delta q_1)/2\hbar} \qquad \text{(IV.12)}$$

In this case the prefactor has a simple dependence on parameters, namely $A_a \propto [1 + \mathrm{Tr}\ M_a]^{-1/2}$. Note that contributions to $\mathcal{W}$ from points far from Q,P are not very small: rather they oscillate violently for small changes of Q and P. IV.9 may be obtained from IV.12 by Gaussian smearing, of course.

Expressions like IV.9,12 are called scar expansions, since they give the contribution to a physical object sorted by periodic orbits. They don't tell why *eigenstates* often are numerically related to periodic orbits, however, and probably are inherently incapable of doing that in most cases. Returning to the example of the inverted pendulum of lecture I, exactly how states are quantized depend on parts of the wavefunction far from the periodic point. A quadratic expansion about that point cannot be expected to be accurate there. On the other hand, wave packets centered near a prominent periodic point can obviously be studied in this manner, at least if the sums don't become unwieldy or meaningless. The question then is, how long, to what value of M, will the expressions of the type IV.9 be meaningful.

D. Propagation of geometry

It is convenient to interpolate here another way of doing the integrals of IV.3, from right to left. We now consider states $|0\rangle$ of the form[24]

$$\langle q|0\rangle = \Sigma_a\ C|dt_a/dq|^{1/2}\exp[\frac{i}{\hbar}\int_0^{t_a(q)}dt\ \ p_0(t)\dot{q}_0(t)] \qquad \text{(IV.12)}$$

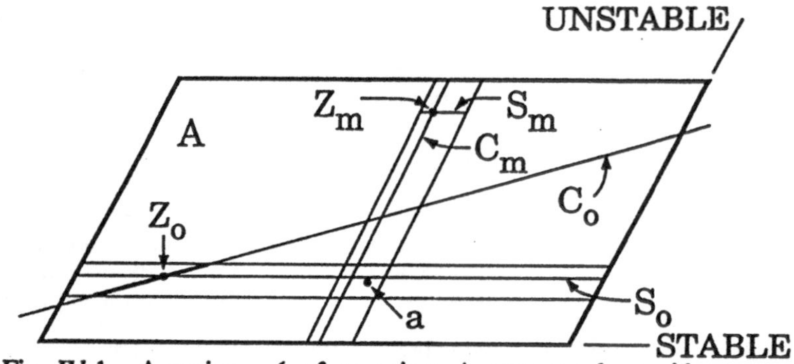

Fig. IV.1. A region $\mathcal{A}$ of an axiom A system whose sides are parallel to the stable and unstable manifolds. Point $\mathcal{Z}_0$ is mapped into $\mathcal{Z}_M$. The dark part of the line $\mathcal{C}_0$ is mapped into the line $\mathcal{C}_M$. Other parts of $\mathcal{C}_0$ will also be mapped back into $\mathcal{A}$ in general. The line S_0 is mapped into S_M. This defines narrow regions $\mathcal{B}_0$ and $\mathcal{B}_M$, where $\mathcal{B}_0$ is mapped into $\mathcal{B}_M$, whose intersection contains a periodic point a.

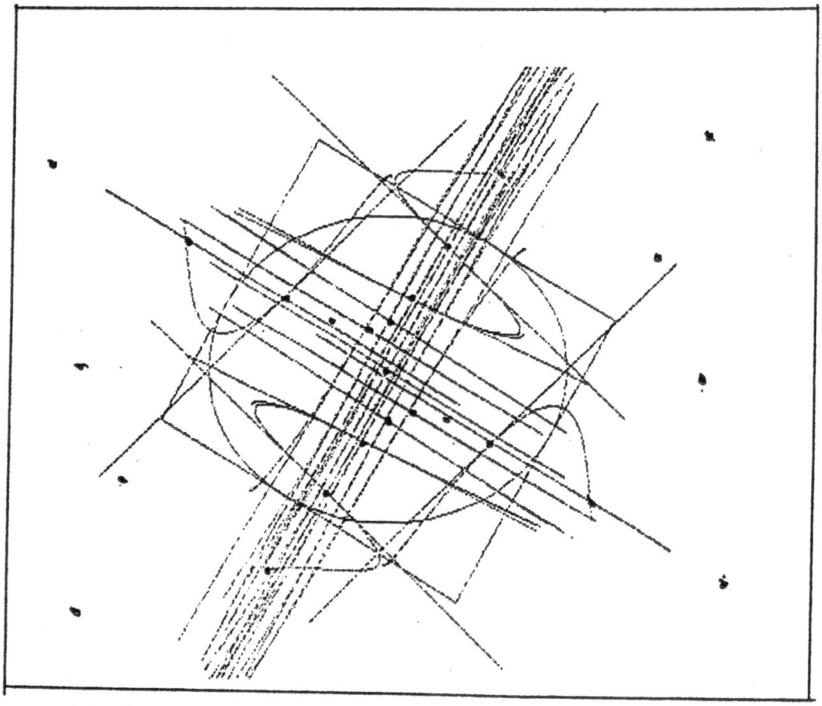

Fig. IV.2. A region $\mathcal{A}$ (parallelopid or circle) in the hard chaos region of the standard map (K = 1.5) at the origin. Twelve period six orbits in $\mathcal{A}$ are shown with the fixed point (also period 6 of course) at the center. Some additional period 6 points in the neighborhood are also plotted. With each point is also plotted its stable and unstable manifolds, (the latter from lower left to upper right). In most cases, these are nearly straight. For a number of points, these are loops. [These loops do not coincide, but in a couple of cases are very close together.] In the case of points with looping manifolds, the straightline approximation to the manifolds given by the PPD approximation is also shown which suggests the degree to which that approximation fails.

which are states of the WKB type. Here t is an arbitrary parametrization of the curve $\mathcal{C}_0$: $p_0(t),q_0(t)$, and a labels the solutions $t_a(q)$ of $q_0(t) = q$. [Again there are Maslov phases to be included and there is a quantization condition on the area enclosed by a closed curve, which we don't discuss. These factors are needed so that the different a's in IV.12 refer to pieces of a single continuous curve.] It is easy to show[24] formally that $\langle q|U^M|0\rangle$ takes exactly the same form, in SΦ, as (IV.12) with $\mathcal{C}_0$ replaced by $\mathcal{C}_M$, the image of $\mathcal{C}_0$ under T^M. [There also a phase coming from the end point at $t = 0$ which we suppress.] This technique reduces the quantum propagation of a wave function to the classical propagation of a curve $\mathcal{C}_M$ defining a wave function.

For a chaotic system (almost) any curve $\mathcal{C}_0$ is stretched and folded back and forth so that the number of points $t_a(q)$ where $q_M(t_a(q)) = q$ increases exponentially, and the areas of the folds decrease exponentially. This led to the suggestion[24] that the SΦ fails when the areas of the folds becomes smaller than $\hbar$, i.e. it fails at the log $\hbar$ time. We suggest that this is not necessarily the case. First we discuss:

E. Validity of the assumption of periodic point dominance

We now prove that for a hyperbolic (axiom A) system,[28] the PPD hypothesis is correct, in other words, that for a sufficiently small phasespace area $\mathcal{A}$ (i.e. small enough $\hbar$) the periodic orbits organize all the orbits returning to $\mathcal{A}$, no matter how large the period. The baker's map or a space of negative curvature are prime examples. All the points in $\mathcal{A}$ have nearly parallel stable manifolds and also nearly parallel unstable manifolds with a finite angle between them. [A stable manifold SM is defined for any point, not just a periodic point, as follows: A point y is in the stable manifold of x if and only if $|T^M(y) - T^M(x)| \to 0$ as $M \to \infty$. The unstable manifold UM is defined similarly for $M \to -\infty$. This implies that the stable manifolds of distinct points cannot intersect unless the two manifolds coincide.] The analysis is particularly simple if $\mathcal{A}$ is chosen to be a parallelpiped with sides parallel to the SUM's.

Consider a point $\mathcal{Z}_0$ in $\mathcal{A}$ whose image $\mathcal{Z}_M$ is also in $\mathcal{A}$. Let $\mathcal{C}_0$ be a line containing $\mathcal{Z}_0$ crossing $\mathcal{A}$ and not parallel to SM. Under the classical mapping T^M $\mathcal{C}_0$ will be carried into a new line $\mathcal{C}_M$ at least one part of crosses $\mathcal{A}$ and contains $\mathcal{Z}_M$. Call the instability b, so that the line is stretched by a factor of order b^M, which we assume to be large. The pieces of $\mathcal{C}_M$ in $\mathcal{A}$ will be very nearly parallel to the UM direction. Mapping the piece containing $\mathcal{Z}_M$ backwards, it is found to come from a small section of $\mathcal{C}_0$ of projected length $\mathcal{A}_U b^{-M}$ in the UM direction, where $\mathcal{A}_U$ is the side of $\mathcal{A}$ parallel to UM. A line through $\mathcal{Z}_0$ and parallel to SM will shrink under the mapping and will go to a little segment through $\mathcal{Z}_M$ of length $\mathcal{A}_S b^{-M}$ parallel to the SM. This is illustrated in Fig. IV.1 and is explained more fully in the literature.[30] These short line segments plus lines parallel to the SU and UM directions thus define two subregions $\mathcal{B}_0$, $\mathcal{B}_M$ of $\mathcal{A}$ associated with $\mathcal{Z}_0$, $\mathcal{Z}_M$ such that $\mathcal{B}_M = T^M\ \mathcal{B}_0$. There is a single periodic point, labelled a in $\mathcal{B}_0 \cap \mathcal{B}_M$.

The quadratic function $S^a(q_M,q_0)$ of IV.5 associated with the periodic point a generates a map carrying $\mathcal{B}_0$ into $\mathcal{B}_M$, [in the approximation that the SM and SU in the region are indeed straight lines] and it carries points in $\mathcal{A} - \mathcal{B}_0$ to points outside of $\mathcal{A}$. A wavefunction associated with $\mathcal{C}_0$ as in the previous section will then be correctly calculated in SΦ by the PPD. Furthermore, there is no bunching of period points giving rise to breakdowns

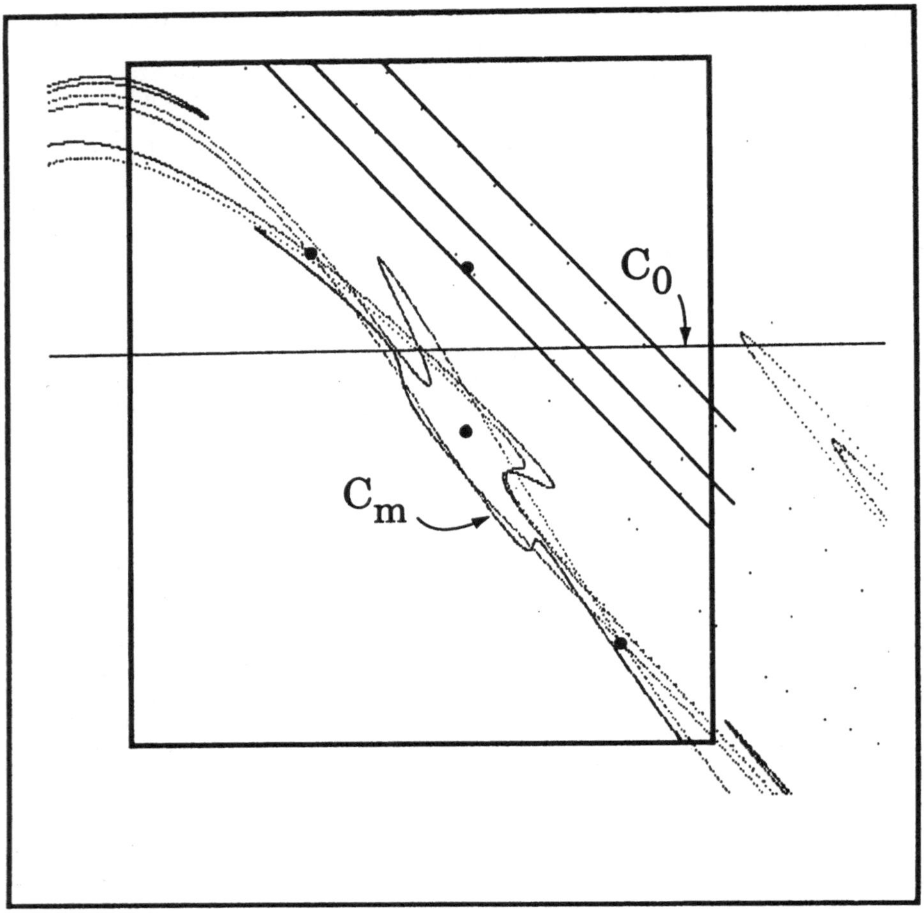

Fig. IV.3. A region $\mathscr{A}$ [inner square] in the transition region of the kicked rotor [parameters K = 1.5, period 7]. It contains three unstable points and a "central" stable point [dots]. Part of the image $\mathscr{C}_M$ of the horizontal line $\mathscr{C}_0$ is shown. [The density of dots of $\mathscr{C}_M$ is inversely proportional to the stretching of $\mathscr{C}_M$ in that region.] The central tangle clearly looks nothing like the axiom A case. However, a few returns are similar to the axiom A case, and are visible as widely spaced dots close to the diagonal solid lines in the upper right portion of $\mathscr{A}$. For $\hbar$ of order the area $\mathscr{A}$, [so that the three period seven points along the main diagonal must be included in the periodic orbit sum] formula IV.9 fails badly. There are indications that the full WKB approximation is quite accurate, however.

of SΦ. Note that even though for large M there are many period M points in $\mathcal{A}$ that the little regions $\mathcal{B}_0$ associated with them do not overlap. Errors coming from small curvatures of the manifolds will be smaller than h if the area between the curved and straightline approximations to the SUM's is << h.

Thus for time M, the classical action S is adequately represented by the the quadratic expansion about the period M points, summed over these points, insofar as one restricts consideration to the mapping of a small region to itself. We can first restrict the size of the region so as to neglect the curvature of the SUM's, and this determines how small h must be. Then M can be allowed to grow as large as one pleases.

We next consider what happens if $\mathcal{A}$ is in a generic hard chaos region. By this we mean that there are no stable points, but we cannot expect all the SUM's to be parallel. In fact, there will be points at which an SM is tangent to a UM. A typical situation, shown in Fig. IV.2., is that of a homoclinic tangle near a low order unstable periodic point. The SUM's of most of the periodic points in the region are parallel to the "infinitesimal" SUM's of the low order point, as must be the case since two SM's cannot intersect. The SUM's of the low order point eventually loop back, however. For visual clarity we show a case where the SUM loops have substantial area, by considering a relatively low order [M= 6], not too unstable [b = 3.5] region. There are 12 period M points in $\mathcal{A}$, plus the fixed point at the center, (q = 0, p = 0 of the standard map with K = 1.5) of which 4 have serious curvature in their stable manifolds. These curved manifolds would be approximated as the straight lines shown in the PPD approximation.

If a wavepacket represented by a line $\mathcal{C}_0$ were to be propagated , the image line would contain loops depending on how close it came to the stable manifolds of the "bad" points. (Only a little region $\mathcal{B}_0$ near the stable manifold of a periodic point is mapped forward back into $\mathcal{A}$.) These loops are incorrectly given in PPD. They would give rise to square root singularities at the first level of approximation in the full WKB, which would be smoothed out by still higher "uniform" approximations[7].

In spite of the fact that the PPD clearly fails for 6 of the 13 points in $\mathcal{A}$, it doesn't do too badly in the evaluation of IV.3 for states $|0\rangle$ centered in this region, when compared with the exact calculation. [We did not do the full WKB approximation.] This is probably because the area of $\mathcal{A}$ where it fails is actually fairly small, and the PPD gives some sort of approximation to the correct answer.

For large M, the loops get to be very tight. They visually look like a single line coming to a stop and retracing itself. In this case an Airy function approximation is not appropriate for the WKB calculation, but another simple approximation exists.

When such loops in an SM exist it is clear that the quadratic expansion of PPD then fails to give the correct classical mapping in the presence of loops, since such an expansion will have straight SUM's. It may be possible to generalise the PPD by somehow expanding about pairs of periodic orbits, but we haven't tried to do that. However, it still could happen that a negligible error is incurred in using the PPD just because the loops make a negligible contribution for large M.

The question is then: Given a $\mathcal{C}_0$, how many loops of $\mathcal{C}_M$ are there in $\mathcal{A}$ compared with the number of pieces of $\mathcal{C}_M$ which are straight segments parallel to the UM, as in the Axiom A case. The answer is not known, but I argue as follows. For large M, $\mathcal{C}_M$ will have of order b^M intersections with $\mathcal{A}$. Each intersection will be preceded and ended by a loop, most of which are not in $\mathcal{A}$. The mean length of a branch of $\mathcal{C}_M$ is fixed by the geometry of the phase space, and is independent of M. $\mathcal{C}_M$ may loop back at any point, more or less at random, so the number of loops of $\mathcal{C}_M$ occurring in $\mathcal{A}$ is proportional to $\mathcal{A}_U$, i.e. proportional to $h^{1/2}$. In other words, this suggests that the error in the hypothesis of PPD is fractionally of order $\hbar^{1/2}$ for generically hard chaos in a homoclinic tangle region. Our numerical calculations for $h \approx 0.01$, for period of order 6 or 7 in various hard chaotic regions of the kicked rotor, did show deviations of as much as several percent from the exact calculations. We cannot say how this varies with M or with h however. If this is correct, it suggests that PPD for generically hard chaos regions has an unusually large error, $\hbar^{1/2}$ rather than $\hbar$, but that it does not breakdown at finite time.

F. Generic chaos

Generically, chaotic Hamiltonian systems have both stable and unstable periodic points. Typically, their phase space can be divided into three parts. First, there are hard chaos regions similar to that described above. Second, there are pseudointegrable regions containing [a dense set of] tiny pockets of controlled chaos. The "integrable" regions are characterized by nonisolated KAM torii. Finally, there is a fractal boundary layer between the two types of behavior. There can also be fractal boundary regions separating two chaotic regions. The latter are associated with a cantorus or an isolated KAM torus. This last of course was the subject of lecture III.

We expect that in the "hard chaos" regions, PPD is ok as just discussed , while in the "integrable" regions, PPD also works, although it is not as convenient as simpler and more powerful methods which are available. For example WKB eigenstates can be found by searching for $\mathcal{C}_0$'s which are closed invariant curves of the quantized area about elliptic periodic points. It is in the boundary region that PPD is expected to fail. In this boundary region stable and unstable points are mixed together and for fixed area $\mathcal{A}$, i.e. fixed $\hbar$, one can always find a place and a period, where $\mathcal{A}$ contains several periodic points of both varieties. An example is shown in Fig. IV.3, for M = 7. Under such conditions, it is not possible that the SUM's be parallel, and PPD is likely to fail badly. We, for example, find that if $|0\rangle$ is a coherent state centered at the stable periodic point in Fig. IV.3, then $\langle 0|U^7|0\rangle_{PPD} \cong$ 1.4, exceeding substantially the limit, [unity] set by the unitarity of U, and even more the exact quantum result, 0.8. On the other hand, one may search the 8 dimensional phase space for *complex* SΦ points. Just one is found, [rather than the one per periodic point suggested by PPD], and the contribution of this single point well approximates the exact result. This is evidence that PPD is failing but not the WKB itself.

PPD therefore fails in part of phase space, namely the boundary region. The effective width of the fractal boundary region depends, as usual, on the measuring scale, which in the current study means it depends on $\hbar$. The period of the important orbits also scales with the area as one approaches the last KAM torus at the core of the boundary region. At level n the fractional area of the boundary region is β^{-n}, the effective area of stability regions is $(\alpha\beta)^{-n} \cong \hbar$, the period M is w^n. Thus at time $M \propto \hbar^{-\delta}$ we expect the PPD to fail in a region of relative area $\hbar^{1/\ln\alpha}$. This is, as mentioned, an oversimplification, since there are generally several boundary regions, and pseudo boundary regions. However, we can expect that the PPD will fail in a region whose area vanishes with $\hbar$, and at timescales increasing with an inverse power of $\hbar$.

G. Breakdown of the semiclassical approximation

We have thus come to the conclusion that a special variant of the WKB, the assumption of PPD, breaks down in the mixed chaos regions of a generic system, whereas there is no reason to think that the full WKB breaks down. However, the full WKB *must* breakdown in one interesting case: that of dynamic Anderson localization.

This phenomenon is most cleanly illustrated in the quantum kicked rotor, the archetypal generically chaotic quantum map. Consider the propagation of a wavepacket localized near the origin, say, in a hard chaos region. For sufficiently large kicking parameter K, this hard chaos region extends out to infinity in both positive and negative momentum directions with relatively small regular islands and boundary regions. A set of classical phasespace points corresponding to the quantum packet diffuses out to infinite $|p|$ at long time, but the quantum packet remains localized in a region Ξ of momentum size $\xi \cong D/\hbar$, which it fills after time $\tau(\hbar) \cong D/\hbar^2$. A natural set of classical points to consider is a loop $\mathcal{C}_0$ about the origin which specifies an initial wavepacket as in IV.12. Thus, after a time $M >> \tau(\hbar)$, the wavepacket, propagated semiclassically, is given by (6) with $\mathcal{C}_0$ replaced by $\mathcal{C}_M$. But this packet cannot be localized as the true quantum packet is, since the bulk of the curve $\mathcal{C}_M$ lies outside the localization region.

One might object: Perhaps the myriad contributions to the wavepacket at time M might interfere destructively outside Ξ, and constructively inside Ξ: This seems unlikely for a number of reasons. First, a small change in $\mathcal{C}_0$ has no effect on the exact localization, but it changes the details of $\mathcal{C}_M$ drastically. This indicates that only statistical properties of $\mathcal{C}_M$ can count. But the region outside Ξ is classically *identical* to the region inside. For the kicked rotor, there is a *classical* periodicity in the momentum p. However, the *quantum* basis states are not periodic in p, since they are given by $\hbar n$ with n integer. Assuming $\hbar$ is a generic irrational, the quantum states localized in different regions of p are slightly different. This slight difference is the essence of localization. If the states in different regions had exactly the same energies they would resonate and localization would be lost. Indeed, if $\hbar$ is rational, there is no localization. In fact, for this case the failure of the WKB is even more obvious, since the WKB cannot predict the linear in time spread of the wave packet for rational $\hbar$: $\mathcal{C}_M$ diffuses outward at best. The WKB essentially assumes that momentum is continuous. It is difficult to see how the WKB can show effects which are dependent on the number theoretic properties of $\hbar$. Put another way: the WKB cannot distinguish between two localized states some distance apart which have very slightly different energies, so it mixes them.

These arguments are not completely convincing. Unfortunately, none of the systems for which WKB is known to be correct, or for which modern resumation methods work exhibit localization. Therefore, numerical testing to see if the WKB gives localization is not possible at present. However, I would personally find it rather surprising if the WKB were capable of giving such delicate quantum phenomena as localization.

H. Conclusions and acknowledgements

We have presented arguments and evidence in favor of the following: One should distinguish between the semiclassical approximation and a popular further approximation to it: that of periodic point dominance. [PPD can be generalized to study off diagonal elements, $|0> \neq |1>$.] The WKB does not typically break down in a direct way, that is, the SΦ approximation gives the leading terms in all but a negligible fraction of the cases. PPD can and will break down in some small fraction of phase space, at long times, as $\hbar \to 0$. The WKB may not be able to account for phenomena which depends on the underlying quantum basis. The most striking of these are localization and ballistic propagation which replace classical defusion and depend in some cases on the number theoretic properties of $\hbar$. In any case, now that ideal quantum hard chaos problems have effectively been solved, the major effort must be directed toward generic situations, which are less tractable.

Many valuable discussions with S. Fishman, Jens Jensen, E. Ott, G. Radons and Y. Guo, among others, have helped me learn this material. This work was support by NSF Grant DMR-9114328.

V. References

1. G.P.Berman and G.M. Zaslavsky, Physica **A91**, 450 (1978).
2. E.G.Heller, Phys. Rev. Lett.**55**,1515 (1984).
3. P. Cvitanovic and B. Eckhardt, Phys. Rev. Lett **63**, 823 (1989); G. Tanner, P. Scherer,E. B. Bolgomony, B. Eckhardt, and D. Wintgen, Phys. Rev. Lett **67**,2410 (1991); G. S. Ezra, K. Richter, G. Tanner, and D. Wintgen, J. Phys. B (to be published).
4. P. W. Anderson, Phys. Rev. **109**,1492 (1958).
5. D.R.Grempel,S.Fishman, and R.E.Prange, Phys.Rev.Lett.**49**,833 (1982).
6. H.L.Cycon,R.G.Froese, W.Kirsch and B. Simon, *Schroedinger Operators,* (Springer,Berlin,1987).
7. L.S.Schulman,*Techniques and Applications of Path Integration* (Wiley, New York,1981).
8. S.Fishman,D.R.Grempel and R.E.Prange, Phys. Rev. Lett. **49**,509 (1982); D.R.Grempel,R.E.Prange and S.Fishman, Phys. Rev. **A29**,1639 (1984).
9. For a review of Anderson localization in the context of disordered solids see P. Lee and T. V. Ramakrishnan, Rev. Mod. Phys.**57**,287 (1985); D. Vollhardt, *Advances in Solid State Physics,***27**,63, P.Grosse,ed. Viking (Braunschweig,1987).
10. For a review of rigorous results see, B. Souillard, *Proceedings of the Conference on Random Media,* Minneapolis, 1986.
11. P.Lloyd, J.Phys.C2,1717 (1969).
12. D.J.Thouless, J. Phys. **C5**, 77 (1972).
13. M.J.Feigenbaum, J. Stat. Phys. **19**, 25 (1978); *ibid.* **21**, 669 (1979).
14. A similar argument may be found in A.J.Lichtenberg and M.A. Lieberman, *Regular and Stochastic Motion* (Springer, New York, 1983).
15. J.M.Greene,R.S.McKay,F.Vivaldi and M.J.Feigenbaum,Physica(Utrecht) **30**,468 (1981).
16. J.M.Greene,J.Math.Phys.**20**,1183 (1981).
17. R.S.Mackay,Ph.D.Thesis,Princeton University (1982);Physica**70**,283 (1983).
18. G.Radons and R.E.Prange, Phys. Rev. Lett. **61**,1691 (1988).
19. J.Crutchfield, M.Nauenberg, and J.Rudnick, Phys.Rev.Lett.**46**,933 (1981); B.Schraiman,C.E.Wayhne, and P.C.Martin,*ibid*,**46**,935 (1981); M.J.Feigenbaum and B. Hasslacher,*ibid*,**49**,609 (1981).
20. S.Fishman,D.R.Grempel and R.E.Prange,Phys. Rev. **A36**,289 (1987).

21. G.Gyorgyi,R.E.Prange and R.Graham,J.Stat.Phys. (to be published).
22. R.Graham, Europhys. Lett. **3**, 259 (1987).
23. M.C.Gutzwiller, J. Math. Phys. **8**,1979 (1967);**12**,343 (1971).
24. M.V.Berry,N.L.Balazs,M.Tabor, and A.Voros, Ann.Phys.(NY)**122**,26 (1979).
25. S.Tomsovic and E.Heller, Phys.Rev.Lett.**67**,664 (1991).
26. M.V.Berry, Proc. Roy. Soc. (London) **A423**,219 (1989).
27. E.Bolgomony, Physica **D31**, 169 (1988).
28. J.Guckenheimer and P.Holmes, *Nonlinear Oscillations, Dynamical Systems, and Bifurcations of Vector Fields,* (Springer-Verlag, New York, 1986).
29. C.Grebogi,E.Ott and J.Yorke,Phys.Rev. **A37**,1711 (1988).

Dynamical localization in open quantum systems *

Robert Graham

Fachbereich Physik, Universität Gesamthochschule Essen
D4300 Essen 1, Germany

1 Introduction

Dynamical localization is a form of Anderson localization [1] appearing in quantum systems whose classical limit is chaotic. It is a novel quantum coherence effect — perhaps the most important new physical effect appearing in the field of quantum chaos, which is concerned with quantized dynamical systems, which are chaotic in the classical limit. Like the familiar Anderson localization, dynamical localization is based on destructive interference of waves in random systems. What is new in dynamical localization is the fact that the randomness is not externally imposed e.g. by a random medium, but is instead produced dynamically by a simple and completely deterministic system. Here the parallel to chaos (i.e. stochasticity) in deterministic classical dynamical systems with few degrees of freedom is apparent.

The discussion in the present lectures, and, indeed in the overwhelming part of the available literature is restricted to Hamiltonian systems which are either autonomous with two degrees of freedom or externally driven periodically in time with one degree of freedom. In fact extending phase-space the latter case can be viewed just a special case of the former [2]. Classically,

*Lectures delivered at the Chris Engelbrecht Summer School in Theoretical Physics "Chaos and Quantum Chaos" January 1992 Odendaal, South Africa

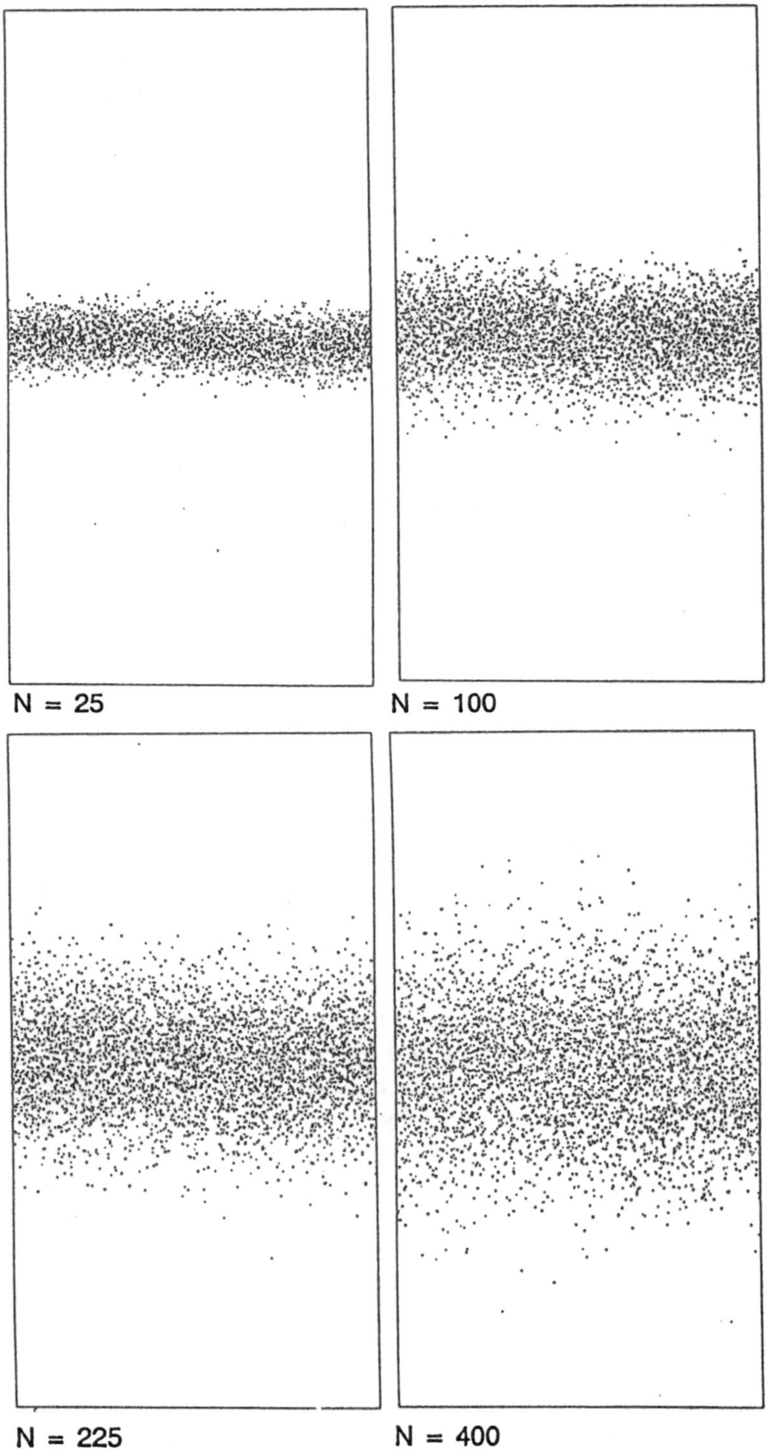

Fig. 1 Diffusion of an ensemble of phase points of the standard map, in the (q,p)-plane (arbitrary units), initially at $p=0$ and equidistributed over q (from [7]). N = number of kicking periods.

under conditions of chaos, the two action variables, describing the system together with the canonically conjugate angles, will undergo a diffusion process. We shall always assume that the chaotic part of phase space is sufficiently large to neglect boundary effects on the diffusion. Taking a Poincaré surface of section (ref. [2]) keeping the total energy and one of the angles fixed, and using as coordinates the remaining angle Θ and its canonically conjugate action I, one then finds the diffusion law

$$< (\mathrm{I}(n) - \mathrm{I}_\lambda)^2 >= Dn \tag{1.1}$$

where n is the discrete time, i.e. the number of iterations of the map and $< \ldots >$ denotes an average over the ensemble of initial conditions, for fixed initial value I_λ. Quantum mechanically the action variable I is quantized, $\mathrm{I} = \hbar l$, l integer. Dynamical localization is the phenomenon that the eigenstates φ_λ of the Hamiltonian (of the extended Hamiltonian in the periodically driven case) for large $|l - l_\lambda|$ decay exponentially

$$|\varphi_\lambda| \sim e^{-|l-l_\lambda|/\xi} \tag{1.2}$$

in a situation where the classical system is chaotic and obeys (1.1). Here ξ is the localization length of the state φ_λ. It is related (up to fluctuations depending on λ about which we have nothing to say here) to the classical diffusion constant D by

$$\xi = \alpha \frac{D}{\hbar^2}. \tag{1.3}$$

where α is a constant of the order of 1. We note that D in (1.1) may, in general, vary with I_λ and hence ξ may depend on l_λ. However, for simplicity we shall always assume, that the variation of ξ_λ over a range of l_λ of size ξ_λ is negligible. Otherwise (1.1)-(1.3) would have to be generalized [3]. Let us mention a few examples:

The kicked rotor [2, 4-6]: The Hamiltonian in convenient units, where $\hbar$ is dimensionless and denotes Plancks constant times the kicking period over the moment of inertia, is

$$H_S = \frac{p^2}{2} - K \cos q \cdot \delta^{(1)}(t) \tag{1.4}$$

where $\delta^{(1)}(t) = \sum_n \delta(t-n)$. For $K >> 1$, $\hbar < 1$ and irrational and generic initial values p_λ avoiding accelerator modes [2] (1.1)-(1.3) hold with $\mathrm{I} = p$, $\alpha = 1/2$ and

$$D = \frac{1}{2} K^2 (1 - J_2(K) + J_2^2(K) + O(K^{-2})) \tag{1.5}$$

where J_l is the ordinary Bessel function. The diffusion of the action variable p in a numerical experiment is shown in fig. 1 where an ensemble of initial states

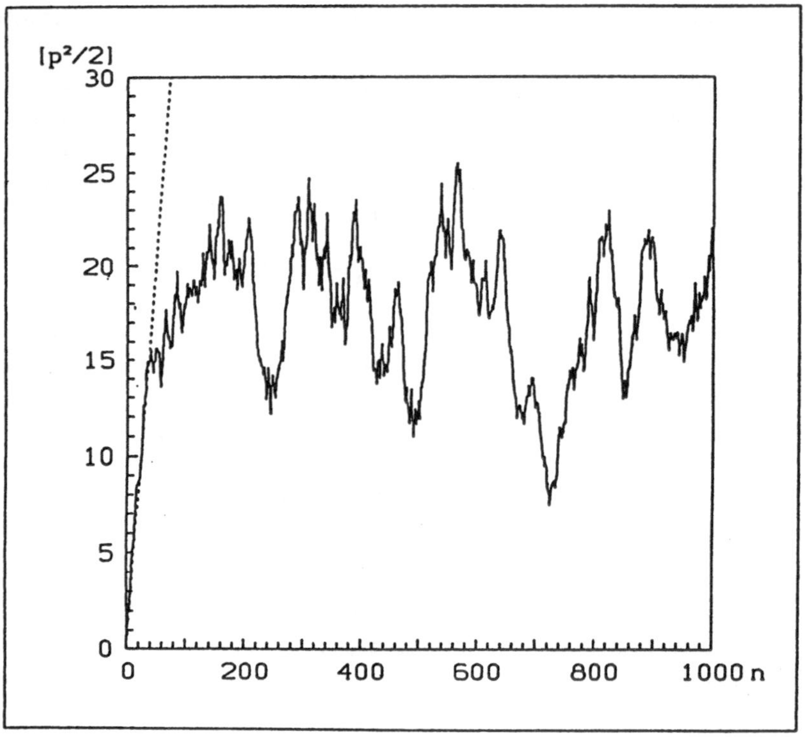

Fig. 2 Expectation value of the kinetic energy of the quantized kicked rotor as a function of kicking periods for $K = 10$, $\hbar/2\pi = 0.075(1+\sqrt{5})$, compared with the classical result (dashed line).

with $p = 0$ and equi-distributed q was started at time $t = 0$. The quantum version of this system has been studied in many papers. Principal sources are [8-12]. This is the prototype of all systems showing dynamical localization. The diffusive spreading of the ensemble is shown for the case $k = 5$ after $n = 25, 100, 225$, and 400 iterations of the map. In fig. 2 the evolution the quantized system is monitored via the expectation value of the kinetic energy of the rotor as a function of n. The initial state is given by the pure $p = 0$ state. The dashed line gives the classical diffusion. The full line gives the quantum evolution, which mimicks the classical diffusion up to a finite time and afterwards reveals the occurrence of dynamical localization with respect to p, when the coherent beating of the finite number of quasi-energy states contained in the $p = 0$ initial state replaces the classical diffusion. (A mathematical proof of this statement is, however, still lacking. In fact, since all available evidence is based on numerical work, it is quite possible, that dynamical localization, in this model, disappears on sufficiently long time scales, or if the Hamilton operator (1.4) is not truncated to some finite basis.)

Rydberg atoms in microwave fields (for a theoretical review see e.g. [13]): The Hamiltonian for a 1-dimensional version of this system, in atomic units where $\hbar = 1$, is

$$H_S = \frac{p^2}{2} - \frac{1}{x} + \varepsilon x \sin \omega t, \qquad x > 0 \tag{1.6}$$

and one obtains diffusion in the action variable $\mathrm{I} = \frac{1}{2\pi} \oint p dx$ according to (1.1) with $n = \omega t / 2\pi$ and the diffusion constant

$$D(\mathrm{I}_\lambda) = 2\xi_\lambda = 3.3 \frac{\varepsilon^2 \mathrm{I}_\lambda^3}{\omega^{7/3}} \tag{1.7}$$

depending explicitely on I_λ. The assumption that the localization length $\xi_\lambda \simeq D(\mathrm{I}_\lambda)$ does not vary over an interval of I_λ of size ξ_λ is satisfied if $3\xi_\lambda << \mathrm{I}_\lambda$, which is, indeed, satisfied in the cases of interest.

Other periodically driven nonlinear oscillators:
These can be described by Hamiltonians of the form

$$H_S = H_0(\mathrm{I}) + \varepsilon x \sin \omega t. \tag{1.8}$$

E.g. for multi-photon absorption by molecular vibrations, leading to excitation and dissociation, $H_0(\mathrm{I})$ is the Morse Hamiltonian [14,15]. In units where the period of small vibrations is 2π, the range of the potential is 1, and the size of the dissociation energy is 1/2, $H_0(\mathrm{I})$ takes the form

$$H_0(\mathrm{I}) = \begin{cases} \mathrm{I} - \frac{\mathrm{I}^2}{2} & H_0 < 0 \\ 1 - \mathrm{I} + \frac{1}{2}\mathrm{I}^2 & H_0 > 0 \end{cases} \tag{1.9}$$

and one obtains the I-dependent diffusion constant

$$D(\mathrm{I}_\lambda) = \frac{2\pi^2\varepsilon^2}{\omega} e^{-2\omega} \frac{1}{(1 - \frac{\mathrm{I}_\lambda}{N_0})^3} \tag{1.10}$$

where N_0 is the total number of bound vibrational states of the molecule. The associated localization length ξ_λ is given by (1.3) with $\alpha \simeq 0.5$ and $\hbar$ denoting Plancks constant, times the angular frequency of small vibrations, over twice the dissociation energy.

Another example is the driven pendulum realized e.g. in a quantum optical system [16] or in periodically driven Josephson junctions [17,18]. Here H_0 is the free pendulum Hamiltonian $H_0 = \frac{p^2}{2} - k\cos x$. The diffusion constant $D = < \Delta P^2 > /(2\pi/\omega)$, with $P = p - \frac{\varepsilon}{\omega}\cos\omega t$, in the chaotic domain $k > \sqrt{\pi\lambda}/40$, is obtained as

$$D = 2\hbar^2\xi = \frac{k^2\omega^2}{\varepsilon}. \tag{1.11}$$

The diffusion of P in the latter example arises from the crossing of the resonance $P = 0$ twice during each period of the driving field, and (1.11) holds if that crossing is fast $\varepsilon/k >> 1$.

Quantum optical examples:
The deflection of a laser-cooled beam of atoms crossing a standing wave laser-field in front of a vibrating mirror can be mapped on the periodically driven pendulum [16] and presents an example of dynamical localization in (chaotic) scattering. Another quantum optical example is a high-Q single-mode cavity containing a medium with $\chi^{(3)}$-nonlinear susceptibility [19], kicked by a periodic train of very short light-pulses of frequency ω. Its Hamiltonian is

$$H_S = \frac{\hbar}{2}\chi^{(3)}(a^+a)^2 + \frac{1}{2}\hbar\Omega_0(a^+e^{i(\omega_0-\omega)t} + ae^{-i(\omega_0-\omega)t})\delta^{(1)}(t) \tag{1.12}$$

where a, a^+ describe the cavity mode with frequency ω_0 in second quantization and the time-period of the kick has been normalized to 1.

The diffusion constant for the action $\mathrm{I} = \hbar a^+a$ obtained in this case is

$$D_{\mathrm{I}_\lambda} = 2\hbar^2\xi_\lambda = \frac{\hbar}{2}\Omega_0^2\mathrm{I}_\lambda. \tag{1.13}$$

Our assumption that the relative change of I_λ over a localization interval $\hbar\xi_\lambda$ remains small implies $\frac{\Delta I_\lambda}{I_\lambda} = \frac{\Omega_0^2}{4} << 1$.

In all the examples mentioned the dynamical localization, like any other quantum mechanical coherence effect, is limited by dissipation and (or) noise, i.e. by any inelastic interaction of the localizing system with its environment. Another physical reason for the occurrence of an inelastic interaction of the system with the "environment"is present if a quantum measurement is performed on the system, not just once, but continuously in time, e.g. by the measurement of the power emitted by the system. This inelastic interaction can be due to photons, like in the quantum optical examples or in Rydberg atoms, or due to phonons or electron-electron collisions etc. In the present lectures we wish to examine dynamical localization in the presence of such inelastic interactions with the environment. We shall proceed in several steps. First, for some typical working examples we review the basic formalisms for the treatment of the interaction with the environment: the Wigner-Weisskopf theory [20], the quantum Langevin equation in the Heisenberg picture [21,22], the master equation [23,24] and the influence functional technique [25,26,27]. Then we shall consider in detail dynamical localization in the presence of dissipation for the kicked rotor [28,29,30]. Very similar results apply for the kicked $\chi^{(3)}$-cavity oscillator which we shall briefly discuss. The influence of classical external noise on the kicked rotor has been analyzed in [31]. Then we consider Rydberg atoms in a noisy wave-guide, where theory has been compared with experiment [32,33]. Finally we discuss dynamical localization for the periodically driven pendulum and a proposed application to atomic beam deflection in modulated laser fields [16] and periodically driven Josephson junctions [17].

The present notes give an extended version of a shorter set of lecture notes on "dynamical localization, dissipation and noise" which appear in [34].

2 Dissipative quantum dynamics

a. Model systems

The interaction of a system with Hamiltonian $H_S(t)$ interacting with the environment with Hamiltonian H_B is described by

$$H = H_S(t) + H_B + H_{int}. \tag{2.1}$$

The environment is modelled by its normal modes, i.e. uncoupled harmonic oscillators

$$H_B = \sum_{k,\lambda} \hbar\omega_k b_{k\lambda}^+ b_{k\lambda} \tag{2.2}$$

with the mode operators $b_{k\lambda}, b^{+}_{k\lambda}$ satisfying Bose commutation relations

$$
\begin{aligned}
[b_{k\lambda},\ b_{k'\lambda'}] &= 0 = [b^{+}_{k\lambda},\ b^{+}_{k'\lambda'}] \\
[b_{k\lambda},\ b^{+}_{k'\lambda'}] &= \delta_{kk'}\delta_{\lambda\lambda'}
\end{aligned}
\tag{2.3}
$$

where k, λ are a mode-index and a polarization index respectively. Among the many different models one may construct we shall consider two examples:

1. Dipole coupling model for the kicked rotor.
   Here we choose $H_S(t)$ as in (1.4) and

   $$H_{int} = \sum_k g_k \sum_\lambda (b^{+}_{k\lambda} + b_{k\lambda}) f_\lambda(q) \tag{2.4}$$

   where $\lambda = 1, 2$ and

   $$f_1(q) = \cos q,\ f_2(q) = \sin q. \tag{2.5}$$

   The kicked rotor considered here has angular momentum p around the z-axis and a dipole-moment $\vec{d}$ transverse to the z-axis with x, y-components equal to $e\cos q$, $e\sin q$, respectively, where e is a charge. These components couple with equal coupling constant g_k to independent linearly polarized modes of the environment with polarization vectors $\vec{e}_\lambda$. We note that we may write H_{int} as a scalar product $H_{int} = \vec{A}\cdot\vec{d}$ with $\vec{A} = \frac{1}{e}\sum_{k,\lambda} g_k \vec{e}_\lambda (b^{+}_{k\lambda} + b_{k\lambda})$. Therefore H is invariant under rotations and conserves the total angular momentum around the z-axis

   $$
   \begin{aligned}
   J_{\text{tot}} = p + \hbar \sum_k \Big[&\frac{1}{\sqrt{2}}(b^{+}_{k1} + i b^{+}_{k2}) \frac{1}{\sqrt{2}}(b_{k1} - i b_{k2}) \\
   &- \frac{1}{\sqrt{2}}(b^{+}_{k1} - i b^{+}_{k2}) \frac{1}{\sqrt{2}}(b_{k1} + i b_{k2}) \Big] = \text{const.}
   \end{aligned}
   \tag{2.6}
   $$

2. Quantum optical model
   We choose $H_S(t)$ as in (1.12) and

   $$H_{int} = \sum_{k,\lambda} g_{k\lambda} \left(a^{+} e^{i\omega_0 t} b_{k\lambda} + a e^{-i\omega_0 t} b^{+}_{k\lambda} \right) \tag{2.7}$$

   with the same properties of the mode operators $b_{k\lambda}$ as before. The ansatz (2.7) incorporates the rotating wave approximation based on the fact that ω_0 is a large optical frequency, so that only modes $b_{k\lambda}$ with frequencies ω_k near ω_0 will couple effectively, and the influence of terms oscillating with frequencies $\omega_0 + \omega_k$ will be extremely small on average. Therefore such terms are omitted in eq. (2.7).

   The goal is now to eliminate the modes of the environment to end up with a reduced dynamical description of the dissipative subsystem of interest. Methods to achieve that goal are described next.

b. Wigner-Weisskopf theory and quantum measurements

The simplest and earliest example of the treatment of the interaction of a microscopic quantum system with the environment is the Wigner-Weisskopf theory [20]. Here the system is an atom, in the simplest case just a two-level atom consisting of a ground state $|0\rangle$ and an excited state $|1\rangle$. The environment is the vacuum state of the electromagnetic field to which the atom is coupled. In fact, it is important to realize, that, in principle, even the vacuum state is an environment, and may change the dynamics of a quantum system into a dissipative and noisy dynamics. The state of an atom may in principle, and nowadays even in practice, be monitored by observing the emitted electromagnetic quanta. Of course, there is a back-reaction on the dynamics of the atom at each moment when an emitted photon is observed, and, even more remarkably, also when it is observed that no photon is emitted. The Wigner-Weisskopf theory also predicts this back-reaction and therefore gives a surprisingly complete account of the quantum dynamics of a continuously measured quantum system. For a recent detailed development of this theory see [35-37].

In order to exemplify the application of the Wigner-Weisskopf theory and to clarify the relation between dissipative and continuously measured quantum dynamics, we shall discuss a periodically driven two-level atom continuously monitored in time by a photon counter. A detailed theory of resonance fluorescence along these lines has been given in [37]. This system is still too simple to display chaos or dynamical localization. (For a treatment of quantum chaos in two-level systems coupled to a single quantized boson mode see [38].) However, it serves well to bring out in a simple case the main ingredients of any quantum theory of open systems, which are shared by the more sophisticated approaches we describe later.

We start by considering the undriven atom, i.e. in (2.1) we take

$$H_s = \hbar\omega_0 |1\rangle\langle 1| \tag{2.8}$$

and we use (2.2), (2.7) with

$$a^+ \to |1\rangle\langle 0|, \quad a = |0\rangle\langle 1|. \tag{2.9}$$

The energy of the state $|0\rangle$ has been put equal to 0. The initial state at $t = 0$ we take as $|\psi(0)\rangle = |1\rangle|\text{vac}\rangle$ where $|\text{vac}\rangle$ is the electromagnetic vacuum state. The wave-function for $t > 0$ can be written as

$$\begin{aligned} |\psi(t)\rangle = {} & C_1(t)e^{-i\omega_0 t}|1\rangle|\text{vac}\rangle \\ & + \sum_{k\lambda} C_{k\lambda}(t)e^{-i\omega_{k\lambda} t} b^+_{k\lambda}|0\rangle|\text{vac}\rangle \end{aligned} \tag{2.10}$$

and the Schrödinger equation reduces to

$$\dot{C}_1 = -\frac{i}{\hbar}\sum_{k\lambda} g_{k\lambda} e^{-i(\omega_{k\lambda}-\omega_0)t} C_{k\lambda}$$
$$\dot{C}_{k\lambda} = -\frac{i}{\hbar} g^*_{k\lambda} e^{i(\omega_{k\lambda}-\omega_0)t} C_1 \tag{2.11}$$

The second equation may be formally integrated and inserted in the first equation

$$\dot{C}_1 = -\frac{1}{\hbar^2}\int_0^t dt' \sum_{k\lambda} |g_{k\lambda}|^2 e^{-i(\omega_{k\lambda}-\omega_0)(t-t')} C_1(t'). \tag{2.12}$$

If $|g_{k\lambda}|^2$ is slowly varying as a function of k, λ over a certain bandwidth γ of frequancies $\omega_{k\lambda}$ around ω_0, where γ is the inverse of the typical time-scale on which $C_1(t)$ varies and will be specified later, then the prefactor of $C_1(t')$ in the integral of eq. (2.13) decays on a time-scale much shorter than γ^{-1}. Therefore $C_1(t')$ can be replaced by $C_1(t)$ and be pulled out in front of the integral. The remaining integral over t' is independent of t and can be done in an approximation where the lower border is extended back to $t' = -\infty$. The result is

$$\dot{C}_1(t) = -(\gamma + i\delta)C_1(t)$$
$$\gamma + i\delta = \frac{1}{\hbar^2}\sum_{k\lambda} |g_{k\lambda}|^2 \left\{\pi\delta(\omega_{k\lambda} - \omega_0) - iP\frac{1}{\omega_{k\lambda} - \omega_0}\right\}. \tag{2.13}$$

From a description in terms of the full wave-function (2.10) we now turn to a reduced description in terms of a statistical operator of the atom alone. We define

$$\varrho_S^{(n)} = Tr_B P^{(n)} |\psi(t)\rangle\langle\psi(t)| \tag{2.14}$$

where $P^{(n)}$ is the projection operator ($P^{(n)^2} = P^{(n)}$) on the n-photon subspace. Thus, (2.14) defines the statistical operator of the atom conditional on the number of emitted photons being n. For the present case n can only take the values 0 or 1. If $n = 0$, we have $P^{(0)}|\psi(t)\rangle\langle\psi(t)| = |C_1|^2|1\rangle\langle 1|$ and we have the equation of motion

$$\dot{\varrho}_S^{(0)} = -\frac{i}{\hbar}\left[H_{\text{eff}}, \varrho_S^{(0)}\right] \tag{2.15}$$

with $H_{\text{eff}} = \hbar(\omega_0 + \delta - i\gamma)|1\rangle\langle 1|$. We note, that even when no photons are emitted, the original dynamics described by H_s is modified and now described by the non-hermitian Hamiltonian H_{eff}. On the other hand under these conditions the pure state $|\psi(0)\rangle = |1\rangle$ remains a pure (but decaying) state $C_1(t)|1\rangle$.

Turning now to $n = 1$ we have

$$\varrho_S^{(1)} = (1 - |C_1(t)|^2)|0\rangle\langle 0| = |0\rangle\langle 0| - |0\rangle\langle 1|\varrho_S^{(0)}|1\rangle\langle 0| \tag{2.16}$$

and

$$\frac{d\varrho_S^{(1)}}{dt} = 2\gamma|0\rangle\langle 1|\varrho_S^{(0)}|1\rangle\langle 0| \tag{2.17}$$

which is an inhomogeneous equation and describes the random emission event.

In order to generalize this description to a periodically driven two-level atom [35], we have to include a driving term in H_{eff}

$$H_{\text{eff}} = \hbar(\omega_0 + \delta - i\gamma)|1\rangle\langle 1| - (\vec{\mu}\cdot\vec{E}e^{-i\omega t}|1\rangle\langle 0| + h.c.) \tag{2.18}$$

and allow for an arbitrary number of photons $n \geq 0$ in the emitted field. The master equations (2.15) (2.17) then generalize to [35,36,37]

$$\frac{d}{dt}\varrho_S^{(n)} = -\frac{i}{\hbar}\left[H_{\text{eff}}, \varrho_S^{(n)}\right] + J\varrho_S^{(n-1)} \tag{2.19}$$

with

$$J\varrho_S^{(n-1)} = 2\gamma|0\rangle\langle 1|\varrho_S^{(n-1)}|1\rangle\langle 0|(1-\delta_{n,0}) \tag{2.20}$$

To describe the evolution of the driven and observed atom this set of equations must be solved for all n starting with $n = 0$. Let us denote the evolution under Heff without the emission of a photon as

$$\varrho_S^{(0)}(t) = S_{t,0}\,\varrho_S^{(0)}. \tag{2.21}$$

Then $P^{(0)}(t) = \text{Tr}_S\,\varrho_S^{(0)}(0) = \text{Tr}_S S_{t,0}\,\varrho_S^{(0)}$ is the probability that no photon is emitted within time t. The evolution for $n = 1$ is then obtained as

$$\varrho_S^{(1)}(t) = \int_0^t dt_1\; S_{t,t_1} J S_{t_1,0}\,\varrho_S^{(0)} \tag{2.22}$$

and $P^{(1)}(t)$ is the probability that one photon is emitted within time t. Iterating this procedure one finds the probability that n photons are emitted within time t as

$$P^{(n)}(t) = \int_0^t dt_n \ldots \int_0^{t_2} dt_1 \text{Tr}_S\left\{S_{t,t_n} J S_{t_n,t_{n-1}} \ldots J S_{t_1,0}\,\varrho_S^{(0)}\right\}. \tag{2.23}$$

Indeed the expression in curly brackets gives the time-evolved statistical operator of the atom conditional on the observation that photons have been emitted in (t_1, dt_1), $(t_2, dt_2), \ldots, (t_n, dt_n)$ within the time t. A theory of "quantum jumps" in atoms has been given in this description in [37]. In the present lectures we shall not investigate the detailed conditional dynamics provided by eq. (2.23). In fact, such a study has not yet been carried out for classically chaotic quantum systems. It would be very interesting and might offer a new perspective on the fundamental question of the meaning of chaos in quantum theory. In the absence of such more detailed results we shall consider here only the unconditional dynamics obtained from eq. (2.23) by summing over the photon number n

$$\varrho_S(t) = \sum_n \varrho_S^{(n)}(t) \tag{2.24}$$

$\varrho_S(t)$ describes the dissipative dynamics of the atom, i.e. the subsystem S. In the case of a measurement the dynamics described by $\varrho_S(t)$ corresponds to the situation, that the measurement device has been put into place, but the results of the measurement are discarded. It also describes the ensemble which one obtains in the limit if the measurement is frequently repeated under identical conditions. The influence of continuous measurements on dynamical localization has been studied within this restricted framework in [39].

Having thus clarified the intimate relation between dissipative dynamics and "measured dynamics" in quantum theory we shall not dwell on the measurement-aspect of our results, in the following.

c. Quantum Langevin equation

The reduced dynamics in this description is very easy to derive, and the central quantities describing dissipation appear very directly. On the other hand, except in cases where H_S describes free particles or linear systems or can at least be approximated by such systems, no methods of solution, not even numerical ones, are known. For a discussion of the quantum Langevin equation see [22]. We shall consider the dipole coupling model by this method.

The Heisenberg equations of motion for p, q, $b_{k\lambda}$ are easy to write down

$$\begin{aligned}
\ddot{q} &= -K \sin q \delta^{(1)}(t) - \sum_{k\lambda} g_k f'_\lambda(q)(b^+_{k\lambda} + b_{k\lambda}) \\
\dot{b}_{k\lambda} &= -i\omega_k b_{k\lambda} - \frac{i}{\hbar} g_{k\lambda} f_\lambda(q).
\end{aligned} \tag{2.25}$$

The last equation may be formally solved

$$b_{k\lambda}(t) = -\frac{ig_k}{\hbar} \int_{-\infty}^t d\tau \; e^{-i\omega_k(t-\tau)+\varepsilon\tau} f_\lambda(q(\tau)) + b_{k\lambda}(-\infty) e^{-i\omega_k t} \tag{2.26}$$

where $\varepsilon \to +0$ is required for convergence, and the result can be inserted in the first equation carefully preserving the ordering of the operators. After a partial integration and taking $\varepsilon \to 0$ we obtain,

$$\begin{aligned}
\ddot{q}(t) = &- K \sin q(t) \delta^{(1)}(t) \\
&- \int_{-\infty}^t d\tau \; \gamma(t-\tau) \sum_\lambda f'_\lambda(q(t)) \frac{d}{d\tau} f_\lambda(q(\tau)) \\
&+ \sum_\lambda f'_\lambda(q(t)) \xi_\lambda(t)
\end{aligned} \tag{2.27}$$

with the damping kernel

$$\gamma(t) = \frac{2}{\hbar}\sum_k \frac{g_k^2}{\omega_k}\cos\omega_k t \tag{2.28}$$

and the Langevin operators

$$\xi_\lambda(t) = -\sum_k g_k\Big(b_{k\lambda}(-\infty)e^{-i\omega_k t} + h.c.\Big). \tag{2.29}$$

From the Heisenberg equation of motion for $f_\lambda(q)$ we obtain

$$\frac{d}{dt}f_\lambda(q) = f_\lambda'(q)\dot{q} + \frac{\hbar}{2i}f_\lambda''(q). \tag{2.30}$$

The operators ξ_λ defined by (2.12) satisfy the fundamental commutation relation

$$[\xi_\lambda(t),\ \xi_\lambda(t')] = i\hbar\delta_{\lambda\lambda'}\dot{\gamma}(t-t'). \tag{2.31}$$

For a thermal environment with inverse temperature β the fluctuation dissipation relation also fixes their anti-commutator

$$\begin{aligned} S_{\lambda\lambda'}(t,t') &= \frac{1}{2}\Big\langle\Big\{\xi_\lambda(t),\ \xi_{\lambda'}(t')\Big\}_+\Big\rangle \\ &= \delta_{\lambda\lambda'}\int_0^\infty \frac{d\omega}{2\pi}\hat{\gamma}(\omega)\hbar\omega\coth\frac{1}{2}\beta\hbar\omega\cos\omega(t-t') \end{aligned} \tag{2.32}$$

where $\hat{\gamma}(\omega)$ is the Fourier transform of $\gamma(t)$. (2.10) is difficult to handle, but it serves to bring out in the most direct manner the fundamental physical quantities $\gamma(t)$ and $S_{\lambda\lambda'}(t,t')$. We point out that the operator orderings in the second and third term of (2.10) are strictly correlated with each other.

Let us now use this formulation for a discussion of the Markov approximation for a thermal environment in the present model. It requires two limits:

1. Constant friction ("Ohmic dissipation"):

   $$\hat{\gamma}(\omega) = 2\gamma, \quad \gamma(t) = 2\gamma\delta(t). \tag{2.33}$$

   In this limit we simply obtain, $-\gamma\dot{q}(t)$ for the second term on the right hand side of (2.10), using (2.13) and the explicit form of the $f_\lambda(q)$.

2. High temperature:
   The Markov approximation also requires that $S_{\lambda\lambda'}(t,t')$ reduces to a δ-function of $t-t'$. This is the case, according to the expression (2.15), only for time-scales $|t-t'| >> \frac{\hbar\beta}{2\pi}$ where (2.15) may be replaced by

   $$S_{\lambda\lambda'}(t-t') \to \frac{2\gamma}{\beta}\delta(t-t'). \tag{2.34}$$

Thus in order to validate the Markov approximation we must have sufficiently high temperature. Non-Markovian effects will ***always*** appear for sufficiently low temperature. This result is typical for systems in low-temperature physics like Josephson junctions. Indeed, in the example mentioned in (1.11) the variable q of the present model would be identified with the Josephson phase. A very different result is obtained in quantum optical examples which we consider next.

d. Master equation

In our discussion of the master equation [24] we shall treat the quantum optical model of section 2.

During the kicks dissipation is completely negligible. Therefore we only have to consider the time-interval between two kicks. We shall derive the equation of motion satisfied by the statistical operator ϱ_S of the subsystem S in that time interval. In the interaction representation with respect to H_B the von Neumann equation of the total system reads

$$i\hbar\dot{\varrho}(t) = [\tilde{H}(t),\ \varrho(t)]. \tag{2.35}$$

It can be formally integrated, assuming that $\varrho(t)$ at $t \to -\infty$ i.e. long before the time interval in question, factorizes: $\varrho(-\infty) = \varrho_S(-\infty) \otimes \varrho_B$ where $\varrho_B = Z^{-1}\exp(-\beta H_B)$ is the canonical operator. We obtain from (2.35)

$$\begin{aligned} \varrho(t) - \varrho_S(-\infty) \otimes \varrho_B + i\int_{-\infty}^{t} dt'\ \frac{\chi^{(3)}}{2}[(a^+a)^2,\ \varrho(t')] \\ = -\frac{i}{\hbar}\int_{-\infty}^{t} dt'\ [\tilde{H}_{int}(t'),\ \varrho(t')]. \end{aligned} \tag{2.36}$$

The perturbation series is generated by iterating this equation, taking the right-hand side equal to zero in zeroth order. The zeroth order solution takes the form $\varrho(t) = \varrho_S(t) \otimes \varrho_B$. We stop after the second order and take the trace over the Hilbert space of the environment in order to obtain $\varrho_S(t) = Tr_B\varrho(t)$. We obtain

$$\begin{aligned} &\varrho_S(t) - \varrho_S(-\infty) + i\int_{-\infty}^{t} dt'\ \frac{\chi^{(3)}}{2}[(a^+a)^2,\ \varrho_S(t')] \\ &-\frac{\chi^{(3)}}{2\hbar}\int_{-\infty}^{t} dt' \int_{-\infty}^{t'} dt''\ \mathrm{Tr}_B\Big[(a^+a)^2,\ [\tilde{H}_{\mathrm{int}}(t''),\ \varrho_S(t'') \otimes \varrho_B]\Big] \\ &= -\frac{i}{\hbar}\int_{-\infty}^{t} dt'\ Tr_B[\tilde{H}_{int}(t'),\ \varrho_S(t') \otimes \varrho_B] \\ &-\frac{1}{\hbar^2}\int_{-\infty}^{t} dt' \int_{-\infty}^{t'} dt''\ Tr_B\Big[\tilde{H}_{int}(t'),\ [\tilde{H}_{\mathrm{int}}(t''),\ \varrho_S(t'') \otimes \varrho_B]\Big]. \end{aligned} \tag{2.37}$$

Then the trace can be carried out using $Tr_B \varrho_B b_{k\lambda} = 0$. It gives rise to the functions (with $Tr \varrho_B b^+_{k\lambda} b_{k\lambda} = \bar{n}_{k\lambda}$)

$$
\begin{aligned}
J_\pm(t' - t'') &= \sum_{k\lambda} \frac{g^2_{k\lambda}}{\hbar^2} e^{\pm i(\omega_0 - \omega_k)(t' - t'')} \\
S_\pm(t' - t'') &= \sum_{k\lambda} \frac{g^2_{k\lambda}}{\hbar^2} \bar{n}_{k\lambda} e^{\pm i(\omega_0 - \omega_k)(t' - t'')}
\end{aligned} \tag{2.38}
$$

for $t' > t''$, replacing (2.28) and (2.32) in the present example, respectively. We now wish to pass to the Markov approximation, which here consists in approximating

$$
\begin{aligned}
\int_{-\infty}^{t'} dt''\ \varrho_S(t'') J_\pm(t' - t'') &\simeq \Big(\kappa(\omega_0) \mp i\Delta(\omega_0)\Big) \varrho_S(t') \\
\int_{-\infty}^{t'} dt''\ \varrho_S(t'') S_\pm(t' - t'') &\simeq \Big(\kappa(\omega_0)\bar{n}(\omega_0) \mp i\Delta_s(\omega_0)\Big) \varrho_S(t')
\end{aligned} \tag{2.39}
$$

where

$$
\begin{aligned}
\kappa(\omega) - i\Delta(\omega) &= \int_{-\infty}^{0} dt' \sum_{k\lambda} \frac{g^2_{k\lambda}}{\hbar^2} e^{-i(\omega - \omega_k)t'} \\
\kappa(\omega)\bar{n}(\omega) - i\Delta_s(\omega) &= \int_{-\infty}^{0} dt' \sum_{k\lambda} \frac{g^2_{k\lambda}}{\hbar^2} \bar{n}_{k\lambda} e^{-i(\omega - \omega_k)t'} .
\end{aligned} \tag{2.40}
$$

The imaginary parts of these expressions are given by principal part integrals

$$
\begin{aligned}
\Delta(\omega) &= P \int_{-\infty}^{+\infty} \frac{d\omega'}{\pi} \frac{\kappa(\omega')}{\omega - \omega'} \\
\Delta_s(\omega) &= P \int_{-\infty}^{+\infty} \frac{d\omega'}{\pi} \frac{\kappa(\omega')\bar{n}(\omega')}{\omega - \omega'}
\end{aligned} \tag{2.41}
$$

and give rise to small frequency shifts which we shall neglect (or assume to be absorbed in the Hamiltonian H_S). The real parts of (2.40) define damping constants

$$
\begin{aligned}
\kappa(\omega) &= \frac{\pi}{\hbar^2} \sum_{k\lambda} g^2_{k\lambda} \delta(\omega_k - \omega) \\
\kappa(\omega)\bar{n}(\omega) &= \frac{\pi}{\hbar^2} \sum_{k\lambda} g^2_{k\lambda} \bar{n}_{k\lambda} \delta(\omega_k - \omega).
\end{aligned} \tag{2.42}
$$

Clearly the transition from (2.38) to (2.39) is justified if the frequency dependence of $\kappa(\omega)$ and $\kappa(\omega)\bar{n}(\omega)$ is negligible in a frequency interval around ω_0 which may be tiny compared to ω_0, but must be sufficiently large so that its inverse is much smaller than the dynamical time-scale of interest, which is here given by a kicking period. Thus in the present example, typical for quantum optics, the foundation of the Markov approximation is completely different from the first example. The temperature β^{-1}, which only enters in $\bar{n}(\omega_0) = (\exp \beta\hbar\omega_0 - 1)^{-1}$, may be taken to be effectively zero ($\beta\hbar\omega_0 << 1$) as long as $\beta^{-1} << \hbar\kappa$, without invalidating the Markov approximation, yielding $\bar{n}(\omega_0) \simeq 0$.

Returning now to (2.37) with the approximation (2.39) and taking the time derivative we obtain

$$\begin{aligned}\dot{\varrho}_S = & -i\frac{\chi^{(3)}}{2}[(a^+a)^2, \varrho_S] + \\ & + \kappa(1+\bar{n})([a, \varrho_S a^+] + [a\varrho_S, a^+]) \\ & + \kappa\bar{n}([a^+, \varrho_S a] + [a^+\varrho_S, a]).\end{aligned} \tag{2.43}$$

which is the desired master equation. By a formal trick eq.(2.43) may also be turned into a master equation for the kicked rotor. To this purpose we rewrite it in the l-representation

$$a^+a|l>= l|l> \quad l \geq 0, \text{ integer} \tag{2.44}$$

and note that in this representation the unperturbed energy

$$\frac{\chi^{(3)}}{2}(a^+a)^2|l>= \frac{\chi^{(3)}}{2}l^2|l> \tag{2.45}$$

is the same as for the rotor, apart from a trivial scale factor, and apart from the fact that the integer l is restricted to $l \geq 0$, in the present case. Formally extending the master equation to negative l also (and removing the scale factor), we obtain for the rotor with $\langle l|\varrho_S|l' >= \varrho_{l,l'}$

$$\begin{aligned}\dot{\varrho}_{l,l'} = & -\frac{i\hbar}{2}(l^2 - l'^2)\varrho_{l,l'} \\ & +2\kappa\Big(\sqrt{(|l|+1)(|l'|+1)}\varrho_{l+l/|l|,l'+l'/|l'|} - \frac{(|l|+|l'|)}{2}\varrho_{l,l'}\Big)\end{aligned} \tag{2.46}$$

where we used the assumption $\bar{n} \simeq 0$, for simplicity.

In [28] this master equation was derived directly from a formal microscopic model. It is used in section 3 to analyze the influence of dissipation on dynamical localization in the kicked-rotor-model. We note, however, that the units of p, q (and consequently also $\hbar$) used in the present lectures differ by factors 2π (and for $\hbar$ by $(2\pi)^2$) from the units used in [28].

e. Influence functional method

In this most powerful method for the treatment of dissipative quantum systems the real-time path-integral representation of the time-dependent density matrix of the total system is used as a starting point and the trivial Gaussian integration over the harmonic oscillators of the environment is then performed explicitely and exactly. For a detailed discussion of the influence functional method see [26]. Non-Markovian effects can be treated by this method by evaluating the remaining non-trivial path integrals over the influence functional in the WKB approximation.

Here we wish to discuss again the Markovian limit only, see e.g. [27]. In that limit the influence functional technique can be used to give an alternative derivation of the master equation, in which the formal perturbative expansion given in eqs. (2.19), (2.20) above is completely avoided. In fact, it turns out that within the Markov approximation the result obtained in the second order of the perturbative expansion is exact. Even more: it would be incorrect or at least inconsistent to continue that expansion to higher orders without at the same time abandoning the Markov approximation. These claims are substantiated by the following derivation of the master equation. Readers who feel content with the derivation already given may skip this section and proceed to chapter 3.

We consider again the example furnished by the Hamiltonian (2.1) where H_S, H_B and H_{int} are given by eqs. (1.12), (2.2), and (2.7), respectively. It will be useful to employ the representation in the overcomplete non-orthogonal basis of coherent states defined by

$$\begin{aligned} a|\alpha\rangle &= \alpha|\alpha\rangle \\ b_{k\lambda}|\{\beta\} &= \beta_{k\lambda}|\{\beta\}\rangle \end{aligned} \tag{2.47}$$

with the properties

$$\begin{aligned} \langle\alpha|\alpha'\rangle &= e^{\alpha^*\alpha' - \frac{|\alpha|^2}{2} - \frac{|\alpha'|^2}{2}} \\ \int \frac{d^2\alpha}{\pi}|\alpha\rangle\langle\alpha| &= 1 \end{aligned} \tag{2.48}$$

and similar corresponding relations for the states $|\{\beta\}\rangle$. The thermal equilibrium state of the environment in this representation has the form

$$\varrho_B = \prod_{k\lambda}\int \frac{d^2\beta_{k\lambda}}{\pi\bar{n}_{k\lambda}} e^{-|\beta_{k\lambda}|^2/\bar{n}_{k\lambda}}|\beta_{k\lambda}\rangle\langle\beta_{k\lambda}|. \tag{2.49}$$

We wish to evaluate the expression defined by

$$\varrho_S(\alpha,\alpha',t) = \langle\alpha|\mathrm{Tr}_B\varrho(t)|\alpha'\rangle \tag{2.50}$$

where

$$\varrho(t) = U(t,t_0)\varrho(t_0)U^+(t,t_0)$$

$$\varrho(t_0) = \varrho_S(t_0) \otimes \varrho_B \tag{2.51}$$

$$U(t,t_0) = \left(e^{-\frac{i}{\hbar}\int_{t_0}^{t} dt \left\{ H_S(a^+,a) + \sum_{k\lambda} g_{k\lambda}\left(a^+ b_{k\lambda} e^{i(\omega_0-\omega_{k\lambda})\tau} + h.c.\right)\right\}} \right)_+$$

Time ordering is denoted by $(\ldots)_+$. Eventually we may let $t_0 \to -\infty$. In order to make optimal use of the coherent state representation we assume $\varrho_S(t_0)$ to be given in normal order as a function of a^+, a (i.e. all a are ordered to the right of the a^+. Then

$$\langle \bar{\alpha}|\varrho(t_0)|\bar{\alpha}'\rangle = \varrho_S(\bar{\alpha}^*, \bar{\alpha}' t_0) \tag{2.52}$$

is a function of $\bar{\alpha}^*$, $\bar{\alpha}'$ only, not of $\bar{\alpha}$ and $\bar{\alpha}^*$. Inserting eqs. (2.34) into (2.33), using the completeness relation of the $|\alpha\rangle$ to the left and right of $\varrho(t_0)$ and evaluating the trace over B in the basis of the $|\{\beta\}\rangle$ we can put eq. (2.33) into the form

$$\varrho_S(\alpha,\alpha',t) = \int \frac{d^2\bar{\alpha}}{\pi}\frac{d^2\bar{\alpha}'}{\pi} J(\alpha,\alpha',t;\bar{\alpha},\bar{\alpha}',t_0)\varrho_S(\bar{\alpha},\bar{\alpha}',t_0) \tag{2.53}$$

with the influence functional

$$J(\alpha,\alpha',t;\bar{\alpha},\bar{\alpha}',t_0) = \prod_{k\lambda}\int \frac{d^2\beta_{k\lambda}}{\pi}\frac{d^2\bar{\beta}_{k\lambda}}{\pi\bar{n}_{k\lambda}} G_{tt_0}(\alpha,\bar{\alpha};\beta,\bar{\beta})G^*_{tt_0}(\alpha',\bar{\alpha}';\beta,\bar{\beta})e^{-\frac{|\bar{\beta}_{k\lambda}|^2}{\bar{n}_{k\lambda}}} \tag{2.54}$$

where

$$G_{tt_0}(\alpha,\bar{\alpha};\beta,\bar{\beta}) = \langle\alpha|\langle\{\beta\}|U(t,t_0)|\{\bar{\beta}\}\rangle|\bar{\alpha}\rangle. \tag{2.55}$$

The expression (2.55) will now be reduced to a path integral by discretizing the time integral in the definition $U(t,t_0)$ (eq. (2.51)), factorizing the exponential function into individual time-steps, which is allowed for sufficiently fine step size ε, and inserting the completeness relation for $|\alpha\rangle$ and $|\{\bar{\beta}\}\rangle$ after each time-step. Denoting the amplitudes after time-step j by α_j and β_j and suppressing the indices k,λ on the β_j for clarity, we can write for the resulting multiple integral

$$G(\alpha,\bar{\alpha};\beta\bar{\beta}) = \int \mathcal{D}\mu[\alpha]\mathcal{D}^2\left[\frac{\beta}{\pi}\right] e^{-\frac{i\varepsilon}{\hbar}\sum_{j=1}^{N}\left(H_S(\alpha_j^*,\alpha_{j-1}) + H_{\text{int}}(\alpha_j^*,\alpha_{j-1},\beta_j^*,\beta_{j-1})\right)}$$
$$e^{\frac{1}{2}\sum_{j=1}^{N}\sum_{k\lambda}[\beta_{j-1}(\beta_j^*-\beta_{j-1}^*) - \beta_j^*(\beta_j-\beta_{j-1})]} \tag{2.56}$$

with

$$\mathcal{D}\mu[\alpha] = \prod_{i=1}^{N-1}\left(\frac{d^2\alpha_i}{\pi}\right) e^{\frac{1}{2}\sum_{j=1}^{N}[\alpha_{j-1}(\alpha_j^*-\alpha_{j-1}^*) - \alpha_j^*(\alpha_j-\alpha_{j-1})]}$$
$$\mathcal{D}^2[\frac{\beta}{\pi}] = \prod_{i=1}^{N-1}\left(\frac{d^2\beta_i}{\pi}\right) \tag{2.57}$$

and the "boundary" condition and $\beta_N^* = \beta^*$, $\beta_0 = \bar{\beta}$. In order to arrive at the expression (2.56) we made use of the fact that H_S and H_{int} are in normally ordered form, i.e. for example $\langle \alpha_j | H_S | \alpha_{j-1} \rangle$ is a function of α_j^* and α_{j-1} only, (and not of α_j or α_{j-1}^*). Taking formally $\varepsilon = t_j - t_{j-1} \to 0$ and using the boundary conditions for β_N^* and β_0 we may rearrange

$$\begin{aligned}
&\frac{1}{2}\sum_{j=1}^{N}\left[\beta_{j-1}(\beta_j^* - \beta_{j-1}^*) - \beta_j^*(\beta_j - \beta_{j-1})\right] \\
&= -\frac{1}{2}|\beta|^2 - \frac{1}{2}|\beta_0|^2 + \frac{1}{2}\beta_N^*\beta_{N-1} + \frac{1}{2}\beta_0\beta_1^* \\
&\quad + \frac{1}{2}\int_{t_0}^{t} dt(\beta(\tau)\dot{\beta}^*(\tau) - \beta^*(\tau)\dot{\beta}(\tau)).
\end{aligned} \tag{2.58}$$

Now we observe that the exponential in the integrand of eq.(2.56) is quadratic in the β, therefore the path integral over β is Gaussian and can be done by extremizing the exponent subject to the "boundary" condition. The extremizing path satisfies

$$\begin{aligned}
\dot{\beta}_{k\lambda} &= -\frac{i}{\hbar} g_{k\lambda}\alpha(\tau)e^{-i(\omega_0 - \omega_{k\lambda})\tau} \\
\dot{\beta}_{k\lambda}^* &= \frac{i}{\hbar} g_{k\lambda}\alpha^*(\tau)e^{i(\omega_0 - \omega_{k\lambda})\tau}
\end{aligned} \tag{2.59}$$

with

$$\beta_{k\lambda}(t_0) = \bar{\beta}_{k\lambda}\,, \quad \beta_{k\lambda}^*(t_0) = \beta_{k\lambda}^*.$$

The solutions are given by

$$\begin{aligned}
\beta_{k\lambda}(\tau) &= \bar{\beta}_{k\lambda} - \frac{i}{\hbar}\int_{t_0}^{\tau} g_{k\lambda}\alpha(\tau')e^{-i(\omega_0-\omega_{k\lambda})\tau'}d\tau' \\
\beta_{k\lambda}^*(\tau) &= \beta_{k\lambda}^* - \frac{i}{\hbar}\int_{\tau}^{t} g_{k\lambda}\alpha^*(\tau')e^{i(\omega_0-\omega_{k\lambda})\tau'}d\tau'.
\end{aligned} \tag{2.60}$$

Inserting the extremizing paths we obtain after some rearrangament

$$\begin{aligned}
&G_{tt_0}(\alpha, \bar{\alpha}; \beta, \bar{\beta}) = \\
&e^{-\frac{i}{\hbar}\int_{t_0}^{t} d\tau H_S(\alpha^*, \alpha) - \int_{t_0}^{t} d\tau \int_{t_0}^{\tau} d\tau' J_+(\tau - \tau')\alpha^*(\tau)\alpha(\tau')} \\
&\cdot e^{-\frac{1}{2}(|\beta_{k\lambda}|^2 + |\bar{\beta}_{k\lambda}|^2 - 2\beta_{k\lambda}^*\bar{\beta}_{k\lambda})} \\
&\cdot e^{-\frac{i}{\hbar}\sum_{k\lambda} g_{k\lambda}\int_{t_0}^{\tau} d\tau \alpha^*(\tau)\bar{\beta}_{k\lambda}e^{i(\omega_0 - \omega_{k\lambda})\tau}} \\
&\cdot e^{-\frac{i}{\hbar}\sum_{k\lambda} g_{k\lambda}\int_{t_0}^{t} d\tau \alpha(\tau)\beta_{k\lambda}^* e^{-i(\omega_0 - \omega_{k\lambda})\tau}}
\end{aligned} \tag{2.61}$$

The function J_+ was defined in eq. (2.38). We are careless with normalization constants, because the result can always be normalized at the end. This expression, and its complex conjugate are now used in eq. (2.54) where the remaining Gaussian integrals over $\bar{\beta}_{k\lambda}$ and $\beta_{k\lambda}$ can now also be performed. In fact, it turns out to be most convenient to perform only the $\beta_{k\lambda}$-integral, and to leave the integral over $\bar{\beta}_{k\lambda}$, i.e. the average over the thermal noise of the reservoir, for a later step, defining for the time being the complex Gaussian noise

$$\xi(\tau) = \sum_{k\lambda} \frac{g_{k\lambda}}{\hbar} \bar{\beta}_{k\lambda} e^{-i(\omega_0 - \omega_{k\lambda})\tau} \tag{2.62}$$

with the correlation functions

$$\langle \xi(\tau) \rangle = 0; \quad \langle \xi(\tau)\xi(\tau') \rangle = S_-(\tau - \tau') \tag{2.63}$$

where S_- was defined in (2.38).

Then we obtain the following exact expression

$$J(\alpha, \bar{\alpha}, t;\ \alpha', \bar{\alpha}', t_0) = \int \mathcal{D}\mu[\alpha]\mathcal{D}\mu[\alpha'] \, e^{-\frac{i}{\hbar}\int_{t_0}^{t} d\tau H_S(\alpha^*, \alpha)} \, e^{+\frac{i}{\hbar}\int_{t_0}^{t} d\tau H_S(\alpha'^*, \alpha')} R[\alpha, \alpha'] \tag{2.64}$$

where

$$\begin{aligned} R[\alpha, \alpha'] = {} & e^{\int_{t_0}^{t} d\tau \int_{t_0}^{\tau} d\tau' J_-(\tau - \tau')(\alpha(\tau) - \alpha'(\tau))\alpha'^*(\tau')} \\ & \cdot e^{-\int_{t_0}^{t} d\tau \int_{t_0}^{\tau} d\tau' J_+(\tau - \tau')(\alpha^*(\tau) - \alpha'^*(\tau))\alpha'(\tau')} \\ & \cdot \left\langle e^{-i\int_{t_0}^{t} d\tau [\xi^*(\tau)(\alpha(\tau) - \alpha'(\tau)) + \xi(\tau)(\alpha^*(\tau) - \alpha'^*(\tau))]} \right\rangle_\xi \end{aligned} \tag{2.65}$$

Here $\langle \ldots \rangle_\xi$ denotes the average over ξ. At this point we introduce the Markov approximation and use according to eq. (2.39) $J_\pm(\tau - \tau') \approx 2(\kappa \mp i\Delta)\delta(\tau - \tau')$. Then the double time integrals in eq. (2.64) reduce to single time integrals and eq. (2.64) can be written in the form

$$\begin{aligned} R[\alpha, \alpha'] = {} & \exp \int_{t_0}^{t} d\tau \left[2\kappa\alpha\alpha'^* - (\kappa - i\Delta)|\alpha|^2 - (\kappa + 1\Delta)|\alpha'|^2 \right] \\ & \left\langle \exp\left\{ -i \int_{t_0}^{t} d\tau [\xi^*(\alpha - \alpha') + \xi(\alpha^* - \alpha'^*)] \right\} \right\rangle_\xi \end{aligned} \tag{2.66}$$

This finishes our evaluation of the influence functional $J(\alpha, \bar{\alpha}, t;\ \alpha', \bar{\alpha}', t_0)$. Apart from the remaining average over the classical Gaussian noise ξ it is given by the path integral over the exponential of a single time integral, which as a preparation for the next step we rewrite in the form

$$J(\alpha,\bar{\alpha},t;\,\alpha',\bar{\alpha}',t_0)=\int\mathcal{D}\mu[\alpha]\mathcal{D}\mu[\alpha']\langle e^{-\frac{i}{\hbar}\varepsilon\sum_{j=1}^{N}H_{\xi j}}\rangle. \tag{2.67}$$

Here we returned to discretized time using eq. (2.58) and defined

$$\begin{aligned} H_{\xi j}=&\,H_S(\alpha_j^*,\alpha_{j-1})-H_S(\alpha_{j-1}'^*,\alpha_j')+\\ &+i\hbar(2\kappa\alpha_{j-1}\alpha_{j-1}'^*)-(\kappa-i\Delta)\alpha_j^*\alpha_{j-1}-(\kappa+i\Delta)\alpha_{j-1}'^*\alpha_j'\\ &+\hbar\xi^*(\tau_j)(\alpha_{j-1}-\alpha_j')+\hbar\xi(\tau_j)(\alpha_j^*-\alpha_{j-1}'^*) \end{aligned} \tag{2.68}$$

We emphasize that $H_{\xi j}$ depends on α_j^*, α_{j-1}, $\alpha_{j-1}'^*$, α_j' and <u>not</u> on their complex conjugates. This is a consequence of the normally ordered form of H_S+H_{int}. We can now evaluate the time derivative for fixed ξ

$$i\hbar\langle\alpha_j|\dot{\varrho}_{S\xi}(t)|\alpha_j'\rangle=i\hbar\dot{\varrho}_S(\alpha_j^*,\alpha_j')\mid_{\xi\text{ fixed}}=i\hbar\dot{\varrho}_{S\xi}(\alpha_j^*,\alpha_j') \tag{2.69}$$

assuming again that $\dot{\varrho}_S$ is normally ordered and obtain

$$i\hbar\dot{\varrho}_{S\xi}(\alpha_j^*,\alpha_j',t)=\int\left(\frac{d^2\alpha_{j-1}}{\pi}\right)\int\left(\frac{d^2\alpha_{j-1}'}{\pi}\right)H_{\xi j}\varrho_{S\xi}(\alpha_{j-1}^*,\alpha_{j-1}'). \tag{2.70}$$

This equivalent to the operator equation

$$i\hbar\dot{\varrho}_{S\xi}(t)=\hat{H}_\xi(t)\varrho_{S\xi}(t) \tag{2.71}$$

where the operator $\hat{H}_\xi\varrho_{S\xi}$ is obtained from the c-number expression $H_{\xi j}\times\varrho_{S\xi}(\alpha_{j-1}^*,\alpha_{j-1}')$ by the following replacement and ordering rules:

$$\begin{aligned} \varrho_{S\xi}(\alpha_{j-1}^*,\alpha_{j-1}') &\;\rightarrow\; \varrho_{S\xi}(a^+,a)\quad\text{normally ordered}\\ \alpha_j^* &\;\rightarrow\; a^+\quad\text{leftmost position}\\ \alpha_{j-1} &\;\rightarrow\; a\quad\text{left of and next to }\varrho_S\\ \alpha_j' &\;\rightarrow\; a\quad\text{rightmost position}\\ \alpha_{j-1}' &\;\rightarrow\; a^+\quad\text{right of and next to }\varrho_S. \end{aligned} \tag{2.72}$$

These rules are easily verified by taking the matrix elements of eq. (2.71) with coherent states.

Thus we obtain

$$\begin{aligned} i\hbar\dot{\varrho}_{S\xi} = & [H_S(a^+,a),\varrho_{S\xi}] \\ & +i\hbar(2\kappa a\varrho_{S\xi}a^+ - (\kappa - i\Delta)a^+a\varrho_{S\xi} - (\kappa + i\Delta)\varrho_{S\xi}a^+a) \\ & +\hbar\xi^*(t)(a\varrho_{S\xi} - \varrho_{S\xi}a) + \hbar\xi(t)(a^+\varrho_{S\xi} - \varrho_{S\xi}a^+). \end{aligned} \quad (2.73)$$

What we really wish to calculate is $\langle\dot{\varrho}_{S\xi}\rangle_\xi = \dot{\varrho}_S$. The average over ξ can be performed in eq. (2.73) using the time derivative of the relation

$$\left\langle e^{-i\int_{t_0}^t d\tau(\xi^*(\tau)\Lambda + \xi(\tau)\Lambda^+)}\right\rangle_\xi = \\ = e^{-\frac{1}{2}(\Lambda\Lambda^+\int_{t_0}^t d\tau\, d\tau' S_-(\tau-\tau') + h.c.)} \quad (2.74)$$

which holds for any not explicitely time-dependent operator Λ and its adjoint Λ^+. In the present case $\Lambda(\ldots) = [a,(\ldots)]$, $\Lambda^+(\ldots) = [a^+,(\ldots)]$ are mutually adjoint with respect to the scalar product $(\varrho_1,\varrho_2) = \mathrm{Tr}\varrho_1^+\varrho_2$, and $\Lambda\Lambda^+ = \Lambda^+\Lambda$. We take $S_-(\tau-\tau')$ again in the Markovian limit (2.39) and obtain for (2.74) the expression

$$\exp\left\{t\cdot 2\kappa\bar{n}[a,[(\ldots),\ a^+]]\right\}. \quad (2.75)$$

Using the time-derivative of eq. (2.74), (2.75) in the ξ-average of (2.74) we obtain the master equation in its final form

$$\begin{aligned} i\hbar\dot{\varrho}_S = & [H_S(a^+,a),\varrho_S] - \hbar\Delta[a^+a,\varrho_S] \\ & + i\hbar\kappa([a,\varrho_S,a^+] + [a\varrho_S,a^+]) \\ & + 2i\hbar\kappa\bar{n}[a,[\varrho_S,a^+]]. \end{aligned} \quad (2.76)$$

If we absorb the frequency shift Δ in the system Hamiltonian H_S this coincides with the master equation derived in the preceding section in a less clean but apparently simpler manner. This concludes our derivation via the influence functional method. We conclude that the only approximation which is really necessary for the validity of the master equation for the present model is the Markov approximation (2.39)

3 Dynamical localization in the dissipative kicked-rotor model

a. Quantum map

We solve the master equation (2.46) between the two kicks at the times $t = n$ and $t = n+1$. Then the kick at time $t = n+1$ is applied. The solution takes the form [28]

$$\varrho_{l',m'}(n+1) = \sum_{l,m} G(l'm'|lm)\varrho_{l,m}(n) \tag{3.1}$$

with the following explicit expression for the kernel G (with $\lambda = \exp(-2\kappa)$)

$$\begin{aligned} G(l'm'|lm) = \lambda^{(|l|+|m|)/2}\Big\{ & G_c(l'm'|lm) \\ & + \sum_{j\geq 1} \Theta_{l\cdot m} \binom{|l|}{j}^{1/2} \binom{|m|}{j}^{1/2} \left(\frac{\lambda^{-1} - e^{-i(|l|-|m|)}}{1 + i(|l|-|m|)/|\ln\lambda|} \right)^j \cdot \\ & \cdot\, G_c(l'm'|l - jl/|l|,\ m - jm/|m|) \Big\} \end{aligned} \tag{3.2}$$

where $\Theta_{l\cdot m} = \frac{1}{2} + lm/2|lm|$ and G_c is the unitary kernel describing the conservative map

$$\begin{aligned} G_c(l'm'|lm) &= <l'|U|l> (<m'|U|m>)^* \\ U &= e^{\frac{i}{\hbar}K\cos q} e^{-\frac{i}{\hbar}p^2/2}. \end{aligned} \tag{3.3}$$

The sum in (3.2) is taken over the number j of quanta of size $\hbar$ of the angular momentum p absorbed by the environment in between two subsequent kicks. Thus for fixed j quantum coherence is fully preserved but between two terms of different j there is no quantum coherence left (the sum over j is over probabilities, not probability amplitudes).

b. Semi-classical limit, quantum noise

The semi-classical limit of the quantum map is most easily discussed by introducing the Wigner phase-space distribution

$$W_l(q) = Tr\left(\varrho \sum_m \int_{-1/2}^{1/2} dx\, e^{i[m(q-\hat{q}) + x(l - \frac{\hat{p}}{\hbar})]}\right) \tag{3.4}$$

and then taking the saddle point approximation for $\hbar \to 0$ in the resulting kernel G. The result to leading (Gaussian) order, for

$$(K^2\hbar)^{1/3} << (1-\lambda)|p_n| \tag{3.5}$$

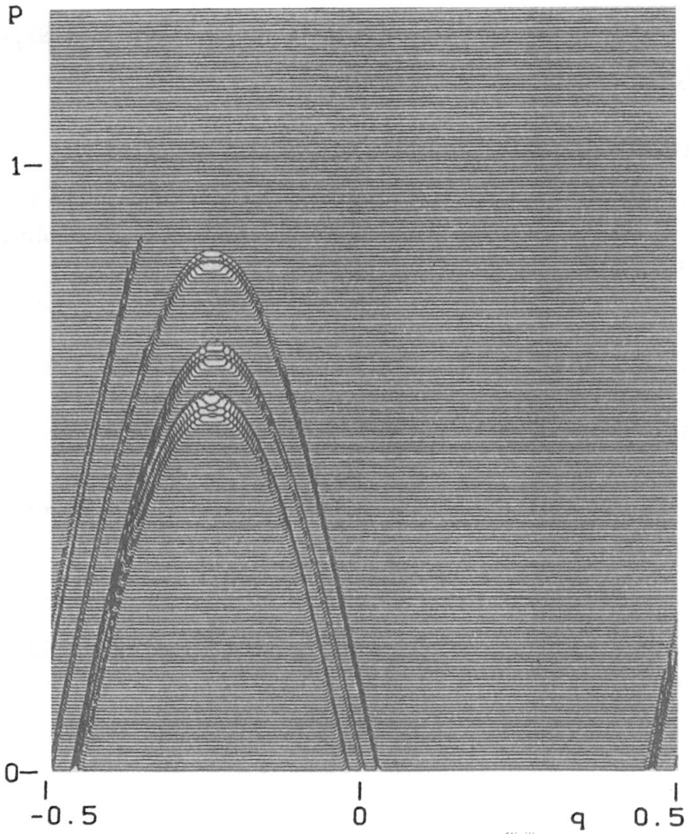

Fig. 3 Strange attractor of the damped kicked rotor for $K = 5$, $\lambda = 0.3$ with its invariant measure in the phase plane. Here and in the following p and q are given in units of 2π (from [28]).

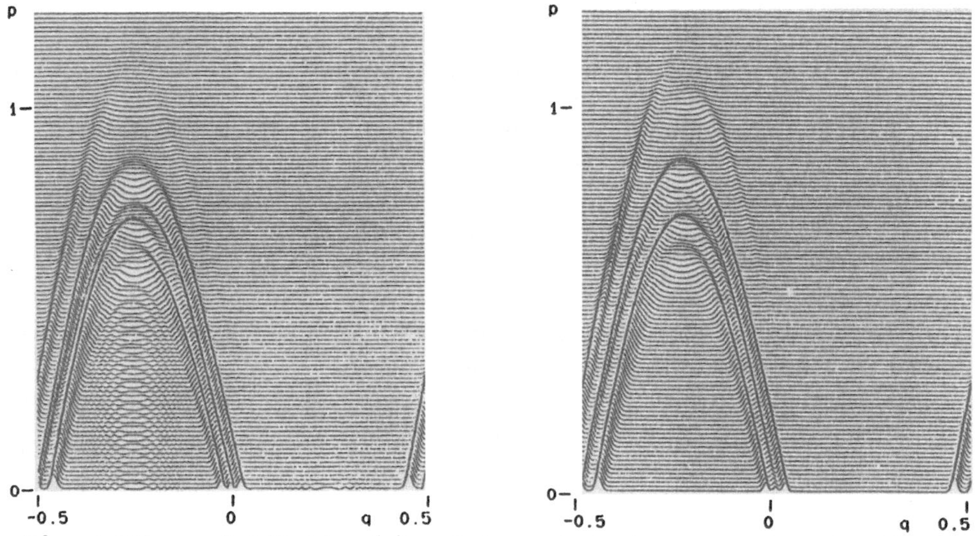

Fig. 4 Wigner distribution (a) and its semi-classical approximation (b) in the steady state for the case of fig. 3 and $\hbar/2\pi = 0.01$ (from [28]).

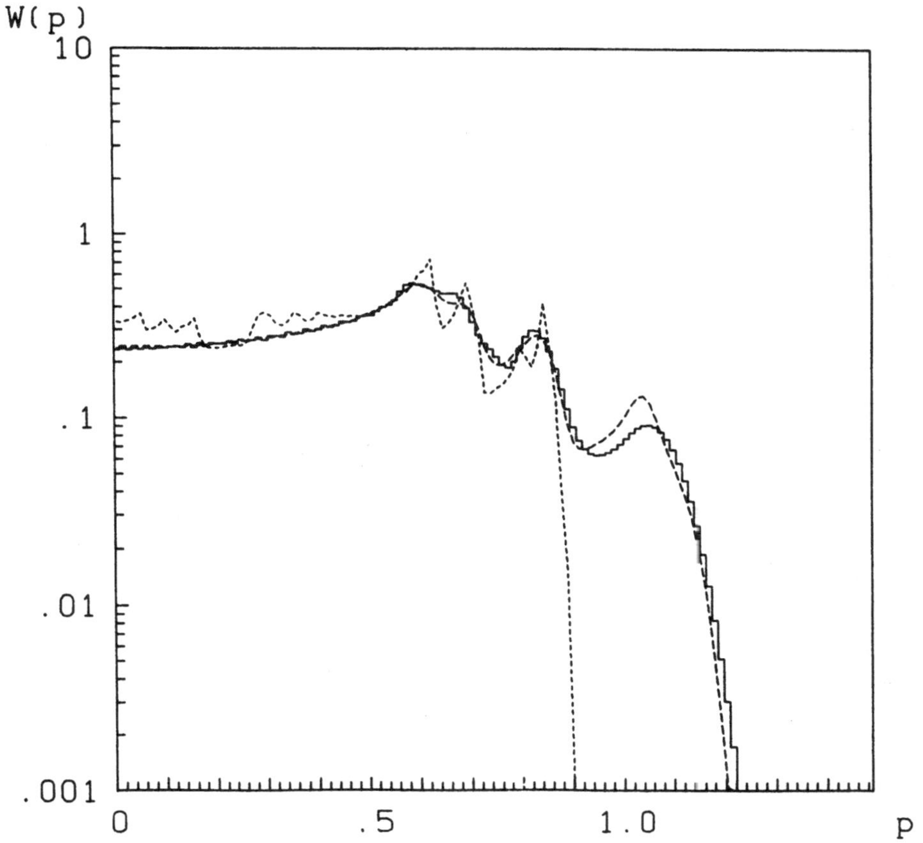

Fig. 5 Angular momentum distribution for the case of fig. 4; quantum (full line), semi-classical (dashed line), classical (dotted line) (from [28]).

is mathematically equivalent to a c-number stochastic map [40]

$$\begin{aligned} p_{n+1} &= \lambda p_n - K \sin q_{n+1} + \eta_n \\ q_{n+1} &= \left(q_n + \frac{1-\lambda}{|\ln \lambda|} p_n + \zeta_n \right) (\mathrm{mod}\ 2\pi) \end{aligned} \tag{3.6}$$

with the classical Gaussian noise defined by the correlation functions [28]

$$\begin{aligned} < \eta_n > &= 0 = < \zeta_n > \\ < \eta_n \eta_n' > &= \hbar \delta_{nn'} |p_n| \lambda (1-\lambda) \\ < \eta_n \zeta_n' > &= \hbar \delta_{nn'} |p_n| \lambda \frac{|\ln \lambda| - (1-\lambda)}{|\ln \lambda|} \\ < \zeta_n \zeta_n' > &= \hbar \delta_{nn'} \left(\frac{1-\lambda}{4\lambda |p_n|} + \frac{\lambda}{1-\lambda} \left(1 - \frac{(1-\lambda)}{|\ln \lambda|} \right)^2 |p_n| \right). \end{aligned} \tag{3.7}$$

The deterministic part of (3.6), after some rescaling, is known as the Zaslavski map.

The semi-classical reduction to a classical stochastic process with noise intensity $\sim \hbar$ of a quantum system with sufficiently strong dissipation (and no thermal noise, because we deliberately took $\bar{n} = 0$) is the theoretical foundation of the notion "quantum noise" frequently used in this context. Taking thermal noise into account ($\bar{n} \neq 0$) the form (3.6) subsists, but the noise intensities in (3.7) receive additional terms proportional to $\bar{n}$.

In fig. 3 we show the strange attractor with its invariant measure for the classical Zaslavski map with $K = 5$, $\lambda = 0.3$. In fig. 4a the Wigner distribution obtained for the steady state of the quantum map is shown for the same case, with $(\hbar/2\pi) = 10^{-2}$ and is compared in fig. 4b with the approximation obtained from the classical stochastic map (3.6) (3.7). In fig. 5 the probability distribution of the angular momentum obtained by integrating the phase-space distribution over q is shown for the classical, the semi-classical and the quantum case for the same set of parameters. These figures illustrate the fate of a classical strange attractor in quantum theory. The sharp classical phase-space measure is smeared out by quantum noise, caused by the random quantum emission into the environment responsible for dissipation. Further smearing would result from thermal noise, which we have deliberately suppressed assuming $T \approx 0$.

These results are thus dominated by quantum noise. Quantum coherence effects, and in particular dynamical localization are clearly absent. In order to see the latter effects we must turn to the case of weak dissipation, where (3.5) is no longer satisfied.

c. Dynamical localization and weak dissipation

For sufficiently weak dissipation the (localized) quasi-energy states $|\kappa>$ of the non-dissipative map form a useful basis in which the effects of dissipation can be discussed by perturbation theory. The quasi-energy states are, of course, not known explicitely, but a reasonable approximation can be obtained by assuming them to be of the form, in the $|l>$-representation

$$< l|\kappa > \simeq (2/\xi)^{1/2} \exp\left(-\frac{|l-l_\kappa|}{\xi}\right) \exp(i\phi_\kappa(l)) \tag{3.8}$$

with a random phase $\phi_\kappa(l)$.

Due to dissipation there occur transitions between the quasi-energy states which therefore decay exponentially

$$|\kappa_n > \simeq e^{(-i\omega_\kappa - \frac{\gamma_\kappa}{2})n} |\kappa > . \tag{3.9}$$

The transition rate (number of transitions per kicking period) between quasi-energy states from $|\kappa>$ to $|\kappa'>$ can be obtained directly from the master equation

$$R_{\kappa\kappa'} = \sum_{l',m',l,m} < \kappa'|l' >< m'|\kappa' > G(l'm'|lm) < l|\kappa >< \kappa|m > . \tag{3.10}$$

Expanding to first order in $(1-\lambda)$ we find

$$\begin{aligned} R_{\kappa\kappa'} - \delta_{\kappa\kappa'} = |\ln\lambda| \Big\{ & - \Big|\sum_l < \kappa'|l > |l| < l|\kappa > \Big|^2 \\ & + \Big|\sum_{l>0} \sqrt{|l|} < \kappa'|l-1 >< l|\kappa > \Big|^2 \\ & + \Big|\sum_{l<0} \sqrt{|l|} < \kappa'|l+1 >< l|\kappa > \Big|^2 \Big\}. \end{aligned} \tag{3.11}$$

The decay rate γ_κ of the state κ follows from

$$\gamma_\kappa = \sum_{\kappa' \neq \kappa} R_{\kappa\kappa'} \tag{3.12}$$

and the normalization condition

$$\sum_{\kappa'} R_{\kappa\kappa'} = 1 \tag{3.13}$$

as

$$\gamma_\kappa = 1 - R_{\kappa\kappa}. \tag{3.14}$$

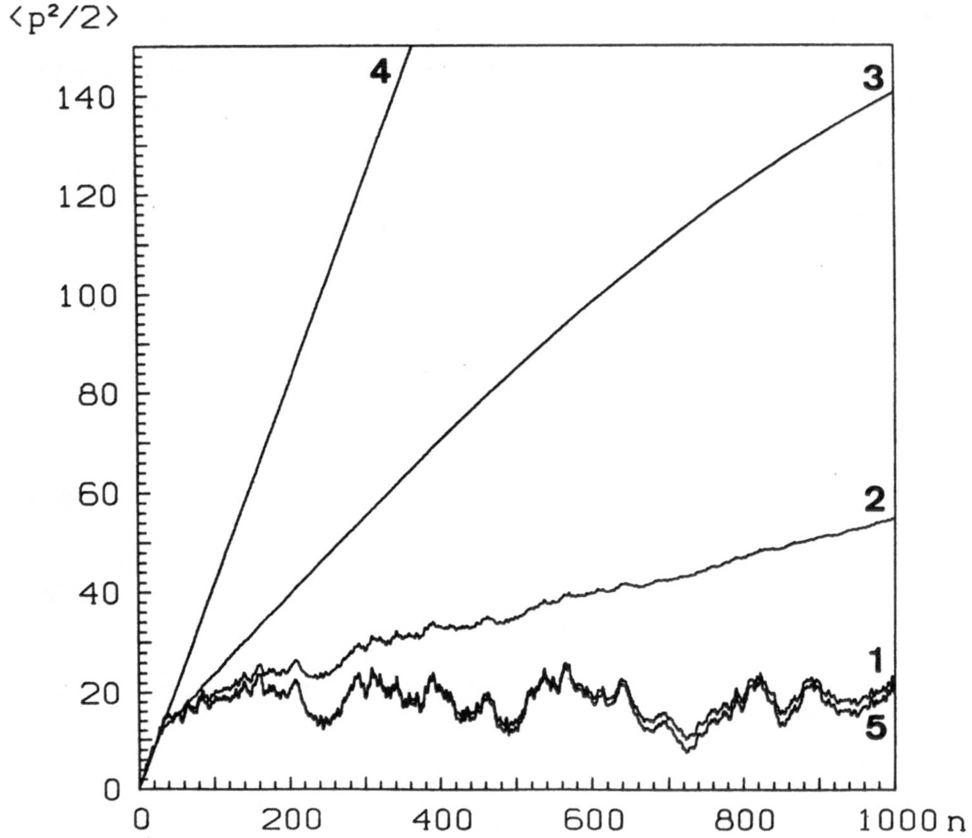

Fig. 6 Mean kinetic energy of the kicked rotor as a function of time n (in kicking periods) for $K = 10$, $\hbar/2\pi = 0.075 \cdot (1 + \sqrt{5})$; $\lambda = 1$ (curve 5); $1 - \lambda = 5 \cdot 10^{-6}$ (curve 1); $1 - \lambda = 10^{-4}$ (curve 2); $1 - \lambda = 10^{-3}$ (curve 3); $\hbar = 0$, $\lambda = 1$ (curve 4) (from [28]).

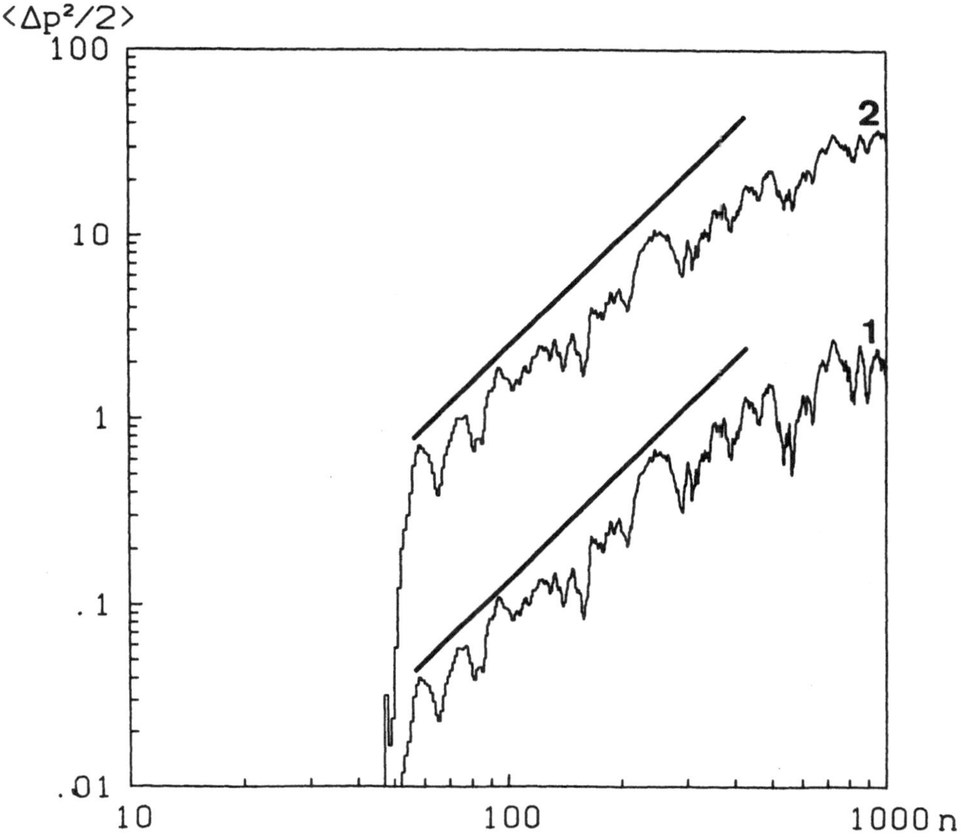

Fig. 7 Doubly logarithmic plot of curves 1, 2 of fig. 6 (from [28]).

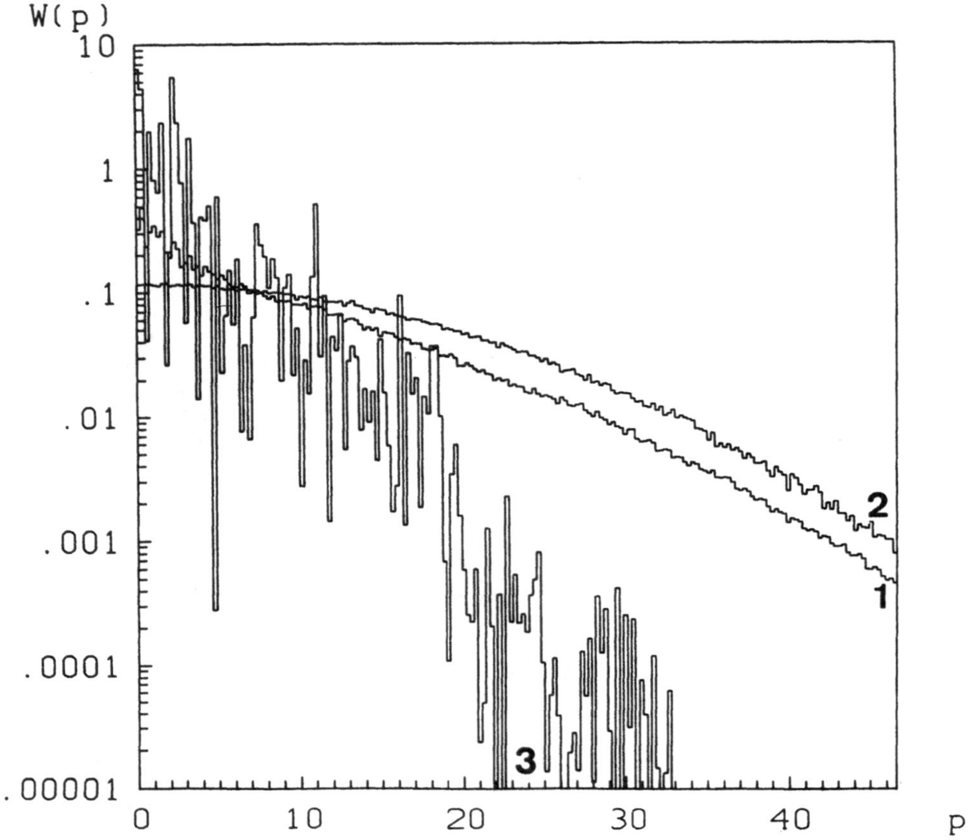

Fig. 8 Probability distribution of the angular momentum after 10^3 kicks for $\hbar/2\pi = 0.075(1+\sqrt{5})$, $K = 10$, $1-\lambda = 10^{-4}$ (curve 1) compared with the classical result (curve 2) and the quantum result for $\lambda = 1$ (curve 3) (from [28]).

The same result can also be obtained directly from perturbation theory. Using the randomness of the phases $\phi_\kappa(l)$ in (3.8) in order to evaluate (3.11) approximately we obtain

$$\gamma_\kappa \simeq |l_\kappa|(1-\lambda)(1+0(1-\lambda,\ \xi^{-1}))$$
$$R_{\kappa,\kappa'} \simeq \frac{|l_\kappa|+|l_{\kappa'}|}{\xi} e^{-2|l_\kappa - l_{\kappa'}|/\xi}(1-\lambda)(1+0(1-\lambda,\ \xi^{-1})). \quad (3.15)$$

Choosing the initial state $|l=0>$ we may in the prefactors of these expressions replace $|l_\kappa|$, $|l'_\kappa|$ by ξ. We can define "weak dissipation" more precisely by the condition

$$\gamma_\kappa \cdot n^* = \gamma_\kappa \cdot \xi < 1 \quad (3.16)$$

where $n^* = \xi$ is the characteristic time (in kicking periods) which the system needs to establish dynamical localization. The relevant quantum numbers κ are determined by the initial state. From the transition rate $R_{\kappa\kappa'}$ a dissipation induced diffusion process of the angular momentum $l = p/\hbar$ on time-scales $n > n_c = \gamma_\kappa$ follows with the diffusion constant

$$\bar{D}_\kappa = \sum_{\kappa'}(l_\kappa - l'_\kappa)^2 R_{\kappa\kappa'} \simeq 2(1-\lambda)|l_\kappa|\xi^2. \quad (3.17)$$

In fig. 6 the result of a numerical simulation [28] is shown (for $K = 10$, $\hbar/2\pi = 0.15(\frac{1+\sqrt{5}}{2})$, $1-\lambda = 0,\ 5.10^{-6},\ 10^{-4},\ 10^{-3}$) and compared with the corresponding classical case without dissipation.

Dynamical localization for $n > n^*$ in the dissipationless case $\lambda = 1$ can be clearly seen, and its replacement, for times $n > n_c$, by a slow dissipation-induced diffusion process whose rate is proportional to $(1-\lambda)$ as predicted by (3.17).

At the beginning the dissipation-induced increase of $< p^2 >$ grows quadratically in n, as can be seen in the doubly-logarithmic plot of a part of the same data in fig. 7.

This quadratic increase can be understood by the following scaling argument, which extends a well-known argument by Chirikov et al. [9]: For the sake of this argument we first consider the case without dissipation. As is well-known the mean-square $< \Delta p^2 > = < (p_n - p_0)^2 >$ then satisfies the scaling law

$$< \Delta p^2 > = \hbar^2 \xi^2 f(\frac{n}{n^*}),\ n^* = \xi \quad (3.18)$$

with a scaling function $f(x)$ with the properties

$$f(x) = \begin{cases} 1 & x >> 1 \\ x/\alpha & x << 1 \end{cases} \quad (3.19)$$

The saturation of $f(x)$ for $x >> 1$ describes localization. Its physical reason can be seen in the gradual resolution, as a function of time, of the spacings between the discrete quasi-energy levels. Let $q(x)$, with $x = n/\xi$, be the fraction of quasi-energy level pairs which are spectrally resolved after the time n (in kicking periods) has elapsed. Due to the resolution of these quasi-energy level pairs the diffusion rate of p is reduced from D (the classical value) to $D(1-q(x))$, with $D = \hbar^2\xi/\alpha$, i.e.

$$\frac{\partial < \Delta p^2 >}{\partial n} = D(1-q(x)) = \hbar^2 \xi f'(x) \tag{3.20}$$

where the second equality follows from (3.18). It relates the scaling function $f(x)$ to $q(x)$. (By the uncertainty principle $q(x)$ also gives the quasi-energy level pair distribution function, $q(n/\xi)$ = probability for an ordered pair of quasi-energy levels to be separated by less than the distance $2\pi/n$)).

Now we return to the case with dissipation, where dissipation-induced transitions between resolved pairs of quasi-energy levels give rise to an excess diffusion.

We consider the fraction $dq(x)$ of level pairs resolved in the time interval $(x\xi, (x+dx)\xi)$. Its contribution to the excess diffusion for $n > x\xi$ is

$$\delta < \Delta p^2 >= 2\hbar^2(1-\lambda)|l_\kappa|\xi^2 dq(x)(n - x\xi). \tag{3.21}$$

where we used (3.17). Thus, the total excess diffusion up to time n is

$$\delta < \Delta p^2 >= 2\hbar^2(1-\lambda)|l_\kappa|\xi^2 \int_0^{q(\frac{n}{\xi})} dq(x)(n - x\xi) \tag{3.22}$$

By partial integration and the use of (3.20) we obtain

$$\delta < \Delta p^2 >= 2\hbar^2(1-\lambda)|l_\kappa|\xi^2 \Big(\frac{n}{\xi} - \alpha f(\frac{n}{\xi})\Big). \tag{3.23}$$

It can be seen that the right hand side is proportional to n^2 in the domain where $f(x) = x/\alpha + O(x^2)$, while for $n >> \xi$ the linear diffusive increase of $\delta < \Delta p^2 >$ takes over, which can also be seen in fig. 7.

Finally, we point out the dependence of the dissipation-induced diffusion rate on the center of localization l_κ of the quasi-energy states. For $|l_\kappa| \to 0$ the diffusion rate $\bar{D}_\kappa$ and the decay rate γ_γ become arbitrarily small. The effect on the distribution of angular momentum can be seen in fig. 8. Curve 1 is obtained for $n = 10^3$ time steps, starting with the initial state $|l = 0>$, with the parameters $\hbar/2\pi = 0.15(\frac{1\pm\sqrt{5}}{2})$, $K = 10$, $1-\lambda = 10^{-4}$. For comparison the corresponding classical result (curve 2) and the quantum result for $\lambda = 1$ (curve 3) are also shown. It can be seen that for $p << \hbar\xi$ localization persists, i.e. the quasi-energy states in that domain have not yet decayed. As the time n increases the size of this domain decreases inversely proportional to n.

For times long compared to the classical relaxation time $n >> (1-\lambda)^{-1}$ the quantum system approaches a unique steady state. It is an important feature of dissipative systems that for $\hbar \to 0$ the quantum steady-state approaches the classical steady state (e.g. in a description using the Wigner phase-space distribution). Thus (barring exceptional cases) in dissipative systems the limits $t \to \infty$ and $\hbar \to 0$ commute. If the dissipation becomes smaller and smaller, one has to wait longer and longer before the steady state is reached. For vanishing dissipation a steady state is never reached and the limits $\hbar \to 0$ and $t \to \infty$ no longer commute.

4 Dynamically localized electromagnetic field in a high-Q cavity

The results of the preceding section can be applied, with some suitable changes, to the quantum optical system described by the Hamiltonian (1.12) and (2.1), (2.2), (2.7).

Let us first estimate the diffusion in the classical limit of the dissipationless system described by (1.12). Its Heisenberg equation of motions take the form

$$\dot{a} = -\frac{1}{2}\chi^{(3)}a^+a^2 - \frac{i\Omega_0}{2}e^{i(\omega_0-\omega)t}\delta^{(1)}(t) \tag{4.1}$$

and correspondingly for a^+. In the classical limit a may be replaced by the expectation value in a coherent state $\alpha = \langle\alpha|a|\alpha\rangle$ and we obtain the classical equation of motion

$$\dot{\alpha} = -\frac{1}{2}\chi^{(3)}|\alpha|^2\alpha - \frac{i\Omega_0}{2}e^{i(\omega_0-\omega)t}\delta^{(1)}(t) \tag{4.2}$$

which can be integrated over the kicking period, e.g. $t = n+\varepsilon$ to $t = n+1+\varepsilon$ with the result

$$\alpha_{n+1} = \alpha_n e^{-\frac{i}{\hbar}\chi^{(3)}|\alpha_n|^2} - \frac{i\Omega_0}{2}e^{i(\omega_0-\omega)(n+1)} \tag{4.3}$$

It follows that

$$|\alpha_{n+1}|^2 - |\alpha_n|^2 = -\frac{i\Omega_0}{2}\left(\alpha_n^* e^{\frac{i}{2}\chi^{(3)}|\alpha_n|^2 + i(\omega_0-\omega)(n+1)} - c.c.\right) + \frac{\Omega_0^2}{4}. \tag{4.4}$$

If the phase-factor on the right-hand side is assumed to be random we obtain the simple diffusion law

$$\langle|\alpha_{n+1}|^2\rangle - \langle|\alpha_n|^2\rangle = \frac{\Omega_0^2}{4} \tag{4.5}$$

i.e. the complex amplitude spreads itself diffusively in the complex plane. Its probability distribution P satisfies the diffusion equation

$$\frac{\partial P}{\partial n} = \frac{\Omega_0^2}{4} \frac{\partial^2 P}{\partial \alpha^* \partial \alpha}. \tag{4.6}$$

For the distribution of the action $\mathrm{I} = \hbar|\alpha|^2$ the diffusion equation

$$\frac{\partial P(\mathrm{I}, n)}{\partial n} = \frac{\partial}{\partial \mathrm{I}} \mathcal{D}(\mathrm{I}) \frac{\partial P(\mathrm{I}, n)}{\partial \mathrm{I}} \tag{4.7}$$

follows with

$$\mathcal{D}(\mathrm{I}) = \hbar \frac{\Omega_0^2}{2} \mathrm{I}. \tag{4.8}$$

This is eq. (1.13). Thus, the diffusion rate of the intensity increases linearly with the intensity. This intensity-dependence of the diffusion constant is the main difference with the kicked rotor model.

Now we turn to the quantum model without dissipation. Remarkably, an exact solution of the Heisenberg equations for one kicking period can be given in the form

$$a_{n+1} = e^{-\frac{i}{2}\chi^{(3)} a_n^+ a_n} a_n - \frac{i\Omega_0}{2} e^{i(\omega_0 - \omega)(n+1)} \tag{4.9}$$

and we obtain, in place of eq. (4.4)

$$a_{n+1}^+ a_{n+1} - a_n^+ a_n = -\frac{i\Omega_0}{2} \left(a_n^+ e^{\frac{i}{2}\chi^{(3)} a_n^+ a_n + i(\omega_0 - \omega)(n+1)} - h.c. \right) + \frac{\Omega_0^2}{4}. \tag{4.10}$$

Because of the quantization of $a_n^+ a_n$ the phase-factor on the right hand side preserves phase-coherence and classical diffusion gives way to quantum localization. The estimate of the number n^* of kicks after which this happens, proceeds along standard lines and yields $n^* \approx \xi(\mathrm{I})$ where ξ is the localization length, i.e. the number of quasi-energy states contained in a state with fixed quantum number $a^+ a = \mathrm{I}/\hbar$. Consistency requires that $\hbar^2 \xi^2(\mathrm{I}) = \alpha \mathcal{D}(\mathrm{I}) n^*$, i.e. we obtain (1.13)

$$\xi(\mathrm{I}) = \alpha \mathcal{D}(\mathrm{I})/\hbar^2 = \alpha \frac{\Omega_0^2}{2} \mathrm{I}/\hbar. \tag{4.11}$$

Thus, the localization length increases linearly with I. For a numerical example we refer to [19] where the localized wave-packet obtained from the $\mathrm{I} = 20\hbar$ state after 10^3 kicks for $\Omega_0 = 10^{-2}$, $\hbar = 1$ was calculated. We note, however, that in this numerical example the localization length is much smaller than 1, in which case localization is a "trivial" consequence of the localization of the basis (2.44) and can be understood by perturbation theory.

Now we turn to the influence of dissipation on this system. In the present model dissipation is due to losses of the cavity, e.g. due to radiation leaking from the cavity. In particular, such a leakage of radiation is inevitable, if a measurement of the mode-intensity is to be performed. This gives a clear physical picture of the relation between the problem of a quantum measurement and dissipation. We may use the master equation (2.43) in which we shall again neglect thermal noise, for simplicity. In the photon-number representation (2.44) it takes the form, without the kicking term,

$$\dot{\varrho}_{l,l'} = -i\frac{\chi^{(3)}}{2}(l^2 - l'^2)\varrho_{l,l'} + 2\kappa\left(\sqrt{(l+1)(l'+1)}\varrho_{l+1,l'+1} - \frac{l+l'}{2}\varrho_{l,l}\right). \quad (4.12)$$

Its solution over one kicking period including one kick takes the form (3.1) with

$$\begin{aligned} G(l'm'|lm) &= \lambda^{(l+m)/2}\Big\{G_c(l'm'|lm) \\ &\quad + \sum_{j\geq 1}\binom{l}{j}^{1/2}\binom{m}{j}^{1/2}\left(\frac{\lambda^{-1} - e^{-1\chi^{(3)}(l-m)}}{1 - i\chi^{(3)}(l-m)/|\ln\lambda|}\right)^j \\ &\quad \cdot\; G_c(l'm'|l-j, m-j)\Big\}e^{i(\omega_0-\omega)(l-m)} \qquad (4.13) \\ G_c(l'm'|lm) &= \langle l'|U|l\rangle(\langle m'|U|m\rangle)^* \\ U &= \exp\left[-i\frac{\Omega_0}{2}(a^+ + a)\right]\exp\left(-i\frac{\chi^{(3)}}{2}(a^+a)^2\right). \end{aligned}$$

The estimates (3.15) for the decay-rates γ_k of quasi-energy states and the transition rates $R_{\kappa\kappa'}$ between them still apply. The condition for weak dissipation $\gamma_\kappa\xi < 1$ now implies

$$\frac{\alpha}{2}(1-\lambda)\Omega_0^2 \mathrm{I}^2/\hbar^3 < 1. \quad (4.14)$$

Thus, for sufficiently large actions I this conditions will always be violated, i.e. localization will be completely destroyed by dissipation in that regime. On the other hand, for sufficiently small actions the condition (4.14) can be satisfied if $(1-\lambda)\Omega_0^2$ is also sufficiently small. In that regime dynamical localization is maintained over a characteristic time $\gamma_\kappa^{-1} \simeq \hbar/(\mathrm{I}_\kappa(1-\lambda))$. The diffusion constant $\bar{D}(\mathrm{I})$ of the dissipation-induced diffusion in the localization regime is given by (4.8) which yields the functional dependence $\bar{D}(\mathrm{I}) \sim \mathrm{I}^3$. However, the strong increase of $\bar{D}(\mathrm{I})$ with the third power of I is at least partially balanced by an increase of the localization length proportional to I with the net result that the rate of decay of localization increases only proportional to I.

5 Rydberg atoms in a noisy wave-guide

a. Basic effects and ideas for an experiment

In the preceeding sections the influence of the weak coupling to the environment on a quantum system exhibiting dynamical localization was studied for the example of the kicked rotor and the kicked nonlinear cavity oscillator. We summarize the main result, the distinction of four different dynamical regimes:

1. Very short times: $n << n^* = \xi$:
   Here the quantum system mimicks the classical behavior of the system i.e. chaotic diffusion, because by the uncertainty principle the discreteness of the quasi-energy levels is not yet resolved. There is no influence of the environment yet as the coupling is very weak.

2. Intermediate times $\xi \lesssim n \lesssim n_c = \gamma_\kappa^{-1}$:
   Here dynamical localization occurs. The environment has only a very small influence due to the weak coupling.

3. Long times $n_c \lesssim n \lesssim n_d = (1-\lambda)^{-1}$:
   Environment-induced diffusion occurs with a rate proportional to the square of the coupling constant.

4. Very long times $n_d < n$:
   The steady state is reached (if the environment absorbs energy from the system, i.e. if there is dissipation).

In the present section we discuss how these results are compared with experiment [32,33].

A sufficiently good realization of the kicked rotor is furnished by a microwave-driven Rydberg atom. In that case the interaction with the environment is realized by the coupling to the electromagnetic field, either its thermal and vacuum fluctuations, or artificial and experimentally controlled noise. In order to have the latter possibility the interaction region is realized in a wave-guide, rather than a microwave-cavity, as in other experiments. In order to have variable interaction times, which are the same for all atoms, a pulsed microwave field is used and an atomic beam with negligible dispersion during the interaction time. The experiment is run in a region of parameter space where ionization by the microwave field is negligible and the probability distribution over the Rydberg states after the interaction with the microwave field is measured. The actual experiment has been carried out with Rubidium-atoms for which a detailed theory of dynamical localization has not yet been worked out. However, a very reasonable qualitative comparison can be made with a simple 1-dimensional theory for hydrogen atoms, to which we turn next.

b. Theory

As the theory closely follows the lines of sections 2, 3 we can be brief. The Hamiltonian

$$H(t) = H_S(t) + H_{int} + H_B \tag{5.1}$$

consists of the parts $H_S(t)$ given by (1.6), $H_{int} = \hbar \sum_k x(g_k b_k + g_k^* b_k^+)$ analogous to (2.4) and H_B as in (2.2) (where k, λ is combined to a single index k for brevity). The derivation of the master equation proceeds as in section 2c, with the important modification that now we use the representation provided by the Floquet basis $|\varphi_\alpha(t)>$

$$U_S(t)|\varphi_\alpha(0)> = e^{-i\mu_\alpha t}|\varphi_\alpha(t)> \tag{5.2}$$

where $U_S(t)$ is the time-evolution operator generated by $H_S(t)$, which includes the microwave field. The interaction Hamiltonian, in interaction representation with respect to $H_S(t) + H_B$ can then be written as

$$\tilde{H}_{int}(t) = \hbar \sum_i g_i \sum_{\alpha\beta} \left\{ |\varphi_\alpha(0)><\varphi_\beta(0)| b_i X_{\alpha\beta}^{(i)}(t) + h.c. \right\} \tag{5.3}$$

with

$$X_{\alpha\beta}^{(i)}(t) = \sum_k \frac{1 + sgn\Omega_{\alpha\beta}(k)}{2} e^{i(\Omega_{\alpha\beta}(k) - \omega_i)t} \bar{X}_{\alpha\beta}(k) \tag{5.4}$$

where we used the observations

$$\begin{aligned} \Omega_{\alpha\beta}(k) &= \mu_\alpha - \mu_\beta + k\omega \\ \bar{X}_{\alpha\beta}(k) &= \frac{\omega}{2\pi} \int_0^{2\pi/\omega} dt\ e^{-ik\omega t} < \varphi_\alpha(t)|x|\Phi_\beta(t)> \end{aligned} \tag{5.5}$$

where ω is the frequency of the coherent microwave field. In (5.3) we introduced a rotating wave approximation similar to (2.7), i.e. we neglected terms rotating with the frequencies $\Omega_{\alpha\beta}(k) + \omega_i$.

The elimination of the reservoir now proceeds as in section IId, including the Markov assumption. In the present case the latter amounts to assuming that

$$\begin{aligned} \gamma_{\alpha\beta}^{(\lambda)}(k) = 2\pi \varrho_\lambda |g(\Omega_{\alpha\beta}(k) + \omega)|^2 \\ |\bar{X}_{\alpha\beta}(k)|^2 \Theta(\Omega_{\alpha\beta}(k)) \end{aligned} \tag{5.6}$$

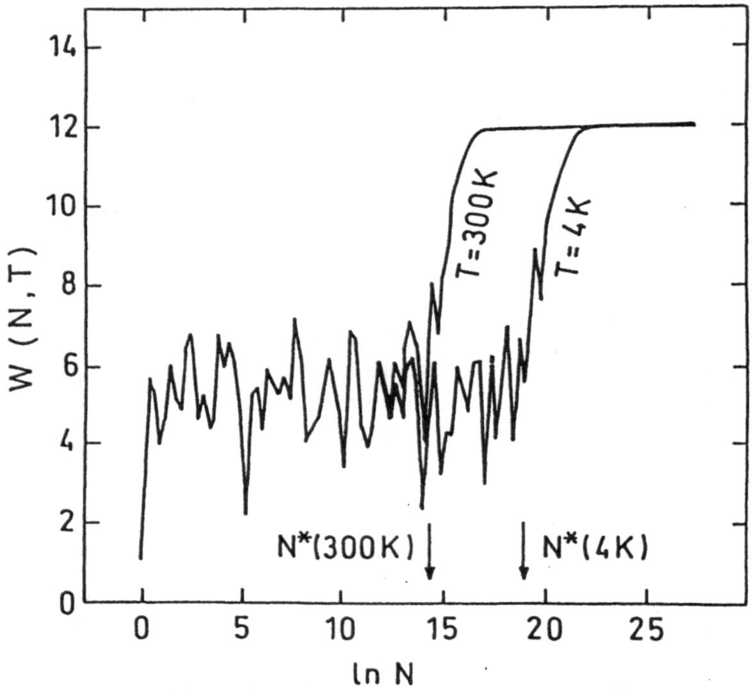

Fig. 9 Shannon-width of the P_n-function as a function of time for two different temperatures (from [33]).

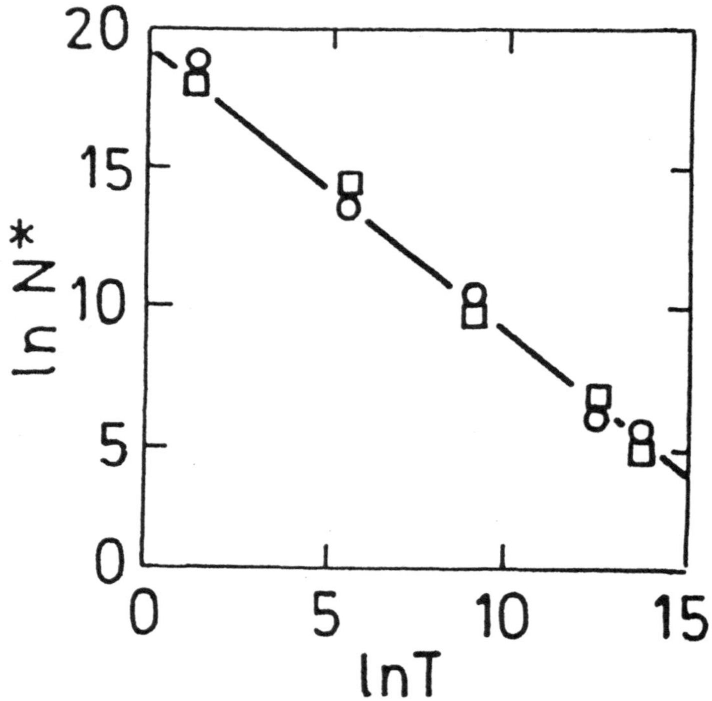

Fig. 10 The relation between temperature and break time N^* for two different field strengths (squares: $F = 4V/cm$; circles: $F = 8V/cm$). The slope of the straight line reflects the relation $N^* \cdot T = \text{const}$ (from [33]).

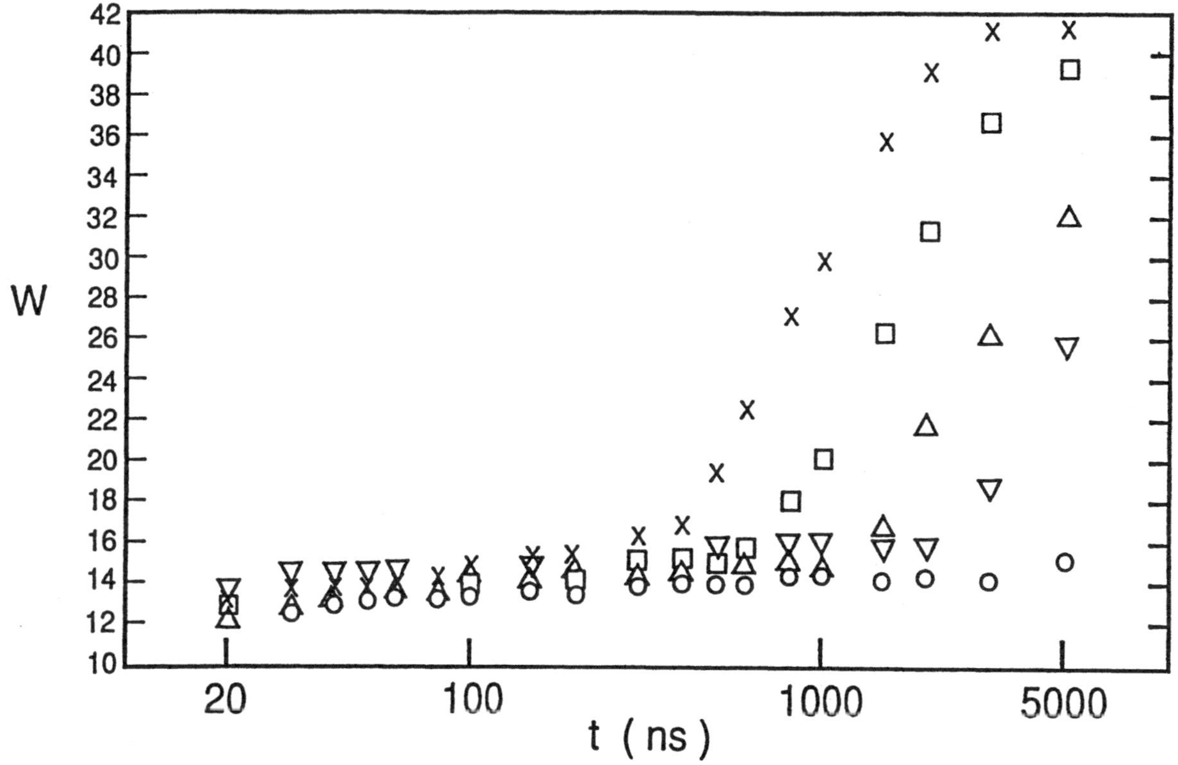

Fig. 11 Shannon width of the P_n distribution extracted from the measurements as a function of time. Coherent microwave power 6.3 μW. Circles: coherent microwave only; additional noise power: crosses 12.6μW; squares 6.3 μW; pyramids 3.2 μW; triangles 1.6 μW (from [33]).

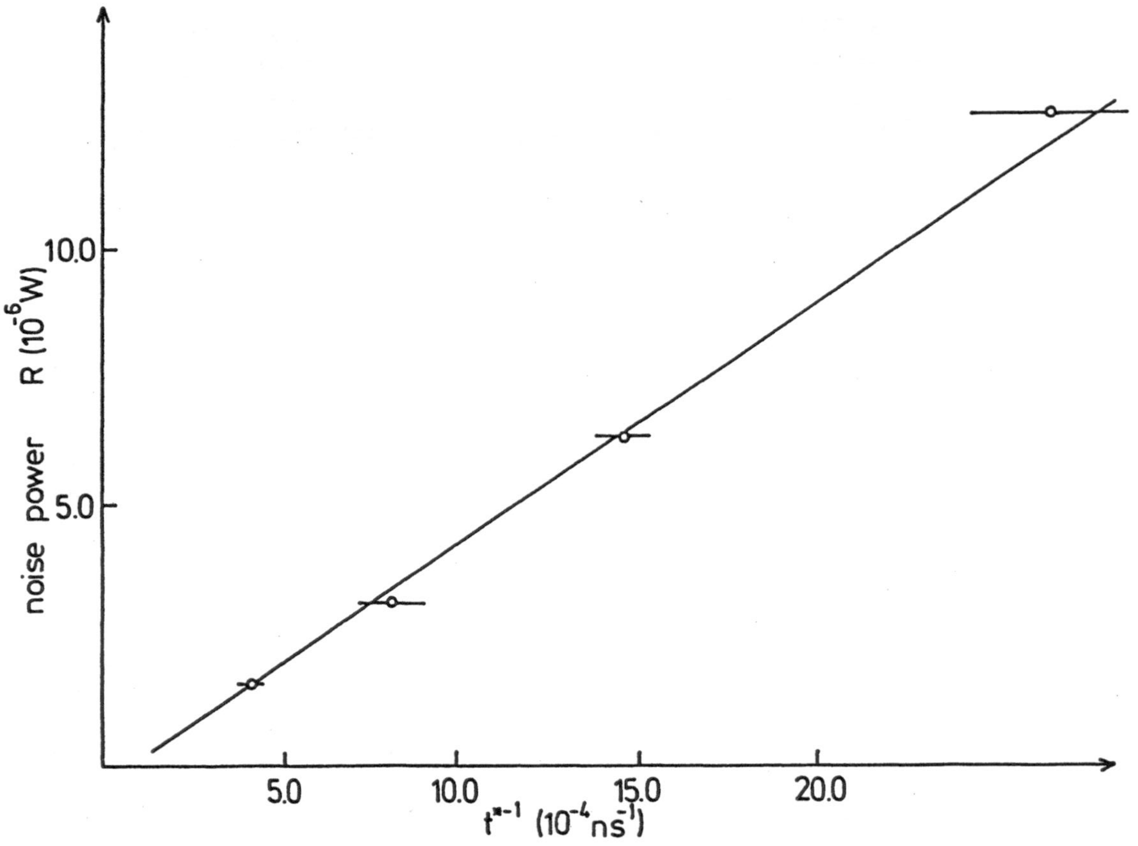

Fig. 12 Relation between noise power and break time t^* extracted from fig. 11 (from [33]).

is independent of ω in a frequency domain $\Delta\omega$ large compared to $\gamma_{\alpha\beta}^{(\lambda)}(k)$. Here $\varrho_\lambda(\omega)$ is the density of states of the modes of the wave-guide (TM and TE) distinguished by the index λ. $\Theta(x)$ is the step function. The final master equation in the Floquet basis and the interaction picture then reads

$$\begin{aligned} \dot{\tilde{\varrho}}_{\alpha\alpha} &= \sum_\mu (M_{\mu\alpha}\tilde{\varrho}_{\mu\mu} - M_{\alpha\mu}\tilde{\varrho}_{\alpha\alpha}) \\ \dot{\tilde{\varrho}}_{\alpha\beta} &= -\frac{1}{2}\tilde{\varrho}_{\alpha\beta}\sum_\mu (M_{\alpha\mu} + M_{\beta\mu}) \end{aligned} \tag{5.7}$$

where

$$M_{\alpha\beta} = \sum_{k,\lambda}\left\{\gamma_{\alpha\beta}^{(\lambda)}(k) + \bar{n}^{(\lambda)}(|\Omega_{\alpha\beta}(k)|)\cdot(\gamma_{\alpha\beta}^{(\lambda)}(k) + \gamma_{\beta\alpha}^{(\lambda)}(k))\right\} \tag{5.8}$$

with $\bar{n}^{(\lambda)}(\omega) = (e^{\beta\hbar\omega} - 1)^{-1}$ for a thermal environment or

$$\bar{n}^{(\lambda)}(\omega) = \frac{2\pi}{\hbar\omega}R_\lambda(\omega) \tag{5.9}$$

for artificial noise with the spectral density $R_\lambda(\omega)$ of the noise power in the mode λ. The master equation (4.7) is easily solved

$$\begin{aligned} \varrho_{\alpha\beta}(t) &= e^{-i(\mu_\alpha-\mu_\beta)t-\frac{t}{2}\sum_\gamma(M_{\alpha\gamma}+M_{\beta\gamma})}\varrho_{\alpha\beta}(0) \\ \varrho_{\alpha\alpha}(t) &= \sum_\beta (e^{\Lambda t})_{\alpha\beta}\varrho_{\beta\beta}(0) \end{aligned} \tag{5.10}$$

with

$$\Lambda_{\alpha\beta} = -M_{\beta\alpha} + \delta_{\alpha\beta}\sum_\mu M_{\alpha\mu}. \tag{5.11}$$

The localization of the distribution over the Rydberg states is quantified by the Shannon width W

$$\begin{aligned} W &= e^S \\ S &= -\sum_n P_n \ln P_n \end{aligned} \tag{5.12}$$

where P_n is the occupation probability of the Rydberg state with principal quantum number n.

A numerical computation using a basis of 12 atomic states ($n = 69\ldots80$) has been performed [32,33] for a rectangular wave-guide ($\simeq 1 \times 2cm$), with $\varepsilon = 1.56 \times 10^{-9}$ a.u. $\omega = 1.6 \times 10^{-6}$ a.u., and initial state $n_0 = 71$. The result for $W(NT)$ as a function of $\ln N$ (N = number of microwave periods, T = temperature of the environment) is shown in fig. 9 for two different temperatures. The 4 dynamical regions mentioned in section a and here labelled by 1 to 4 can be clearly seen in this figure. The coherence time $n_c = \gamma_\kappa^{-1}$, $\gamma_\kappa = \frac{1}{2}\sum_\gamma M_{\kappa\gamma}$ is here denoted by N^*. The numerical dependence of N^* on T can be extracted and is shown in fig. 10 for $\varepsilon = 4V/cm$ (squares) and $\varepsilon = 8V/cm$ (circles) and compared with the theoretical prediction $N^*T =$ const.

c. Experiment

An experiment [33] has been performed with *Rb* atoms in a coherent microwave whose frequency could be varied in the range from 8 to 18 GHz with a pulse length between $20ns$ and $5\mu s$. Artificial noise was used to realize the coupling to the environment. The details of the experiment, and the method to extract the Shannon width W from the data are described in [33]. In fig. 11 the Shannon width W is plotted as a function of the interaction time for various values of the noise power. Again the four dynamical regimes can be distinguished and are labelled 1 to 4. In particular a localized regime with $\xi \simeq W \simeq 14$ is followed by a noise-induced spreading of the wave-packet. The time t^* after which the localization breaks up can be extracted from the/data (fig. 12) and scales inversely proportional to the noise power, as predicted by theory. The main qualitative results of the experiment are therefore in agreement with the numerical results of the theoretical model based on 1-dimensional hydrogen atoms, and moreover with the qualitative results obtained for the kicked rotor model. In particular, nontrivial dynamical localization with $\xi \simeq W >> 1$ is found in the experimental data of fig. 11, also observed is the environment-induced break-up after a time proportional to the square of the coupling strength or to the noise intensity.

A more detailed quantitative analysis of the experimental data would, of course, be desirable. It would have to include the detailed level structure of *Rb*, possibly also the effects of the continuum and of the core-electrons, and effects of the pulse-shape. This remains a task for the future.

6 Dynamical localization in the periodically driven pendulum

The pendulum is one of the simplest nonlinear dynamical systems. It is completely integrable. That property is immediately lost, however, when an additional periodic driving is put on. Therefore the periodically driven pendulum has often been used as a typical example of a non-integrable dynamical system. As we shall show, this system, when quantized, shows dynamical localization under suitable conditions. Accessible physical realizations of quantized pendula exist, i.e. experimental tests of the teoretical predictions can be proposed.

In the present chapter we shall discuss first the appearance of chaos in the classical driven pendulum, then analyze dynamical localization in the quantized system and finally turn to two proposals for an experimental check. The content of this part of the lectures is based on ref. [41], [16], [17].

a. Classical pendulum

The undriven pendulum is described by the Hamiltonian

$$H = \frac{P^2}{2} - k\cos\phi \qquad (6.1)$$

where we haven chosen convenient units. The phase-space of the pendulum is either a cylinder $-\pi < \phi \leq \pi$, $-\infty < p < +\infty$ or the whole plane $-\infty < \phi < +\infty$, $-\infty < p < +\infty$, depending on whether we are able, or not, experimentally to keep track of the number of revolutions of the pendulum in anti-clockwise (positive) or clockwise (negative) direction. Using the cylindrical phase-space we have the familiar "resonance" with an elliptic (stable) fixed point at $\phi = 0 = P$ surrounded by the bounded orbits describing oscillations, and the hyperbolic (unstable) fixed point at $\phi = \pm\pi$, $P = 0$ whose stable and unstable manifold coincide and form a smooth separatrix. Outside the separatrix there are the unbounded, running solutions, in which the pendulum rotates in either direction. Now we put on a periodic external field or torque of frequency $\omega = 1$. The resulting Hamiltonian we give in three equivalent forms, related by canonical transformations. All three forms are frequently used in practice.

$$H_1 = \frac{p^2}{2} - k\cos\varphi - \lambda\varphi\sin t. \qquad (6.2)$$

This form can only be used if the real plane is adopted as a phase space. Via the canonical transformation

$$\begin{aligned} \phi &= \varphi + \lambda\sin t \\ P &= p + \lambda\cos t \\ H_2 &= H_1 + \lambda\varphi\sin t + \lambda P\cos t - \frac{\lambda^2}{2}\cos^2 t \end{aligned} \qquad (6.3)$$

we arrive at the second form

$$H_2 = \frac{P^2}{2} - k\cos(\phi - \lambda\sin t). \qquad (6.4)$$

In this form the Hamiltonian H_2 is 2π-periodic in ϕ and the cylindrical phase space may be used. We note that the canonical transformation (6.4) cannot be defined on a cylindrical phase space. A further canonical transformation

$$\begin{aligned} \varphi &= \phi - \lambda\sin t \\ P &= P \\ H_3 &= H_2 - \lambda P\cos t + \frac{\lambda^2}{2}\cos^2 t \end{aligned} \qquad (6.5)$$

brings us to the third form of the Hamiltonian

$$H_3 = \frac{(P - \lambda \cos t)^2}{2} - k \cos \varphi \tag{6.6}$$

which is also 2π-periodic in φ. The transformation (6.6) is well-defined on the cylindrical phase space. Alternatively, we may go from H_1 to H_3 via the canonical transformation

$$\begin{aligned} \varphi &= \varphi \\ P &= p + \lambda \cos t \\ H_3 &= H_1 + \lambda \varphi \sin t \end{aligned} \tag{6.7}$$

which cannot be restricted to a cylindrical phase space.

Now we discuss the response of the pendulum to the external driving force. It is useful to consider the stroboscopic map of the P, ϕ plane into itself induced by the Hamiltonian dynamics over one period of the external force. Here we find the typical scenario for the onset of chaos: for small but finite driving amplitude λ the hyperbolic fixed point survives, but its stable and unstable manifolds no longer coincide. Instead they intersect each other inifinitely often in homoclinic points forming a "homoclinic tangle" with horse-shoes and chaos in a narrow chaotic separatrix layer, replacing the original smooth separatrix. In addition all oscillatory orbits whose periods are rational multiples of the driving period are destroyed and replaced by an alternating chain of elliptic (stable) and hyperbolic (unstable) periodic orbits. The stable and unstable manifolds of the hyperbolic periodic orbits again intersect and form a heteroclinic tangle, in secondary separatrix layers around the "elliptic islands" i.e. in the domain of stability of the elliptic periodic orbits. This structure is repeated infinitely often within each elliptic island on an ever smaller, rapidly decreasing scale. When the amplitude λ has sufficiently large regular right-running ($P > 0$) and left-running ($P < 0$) solutions still exist but all oscillatory orbits, even those whose periods are the "most irrational" multiples of the driving period (in a sense which can be made precise by the continued fraction decomposition), are destroyed and the chaotic domain percolates in the whole domain between the left-running ($P < 0$) and right-running ($P > 0$) regular solution. This is the domain of large-scale chaos, which has a sharp threshold at some finite value $\lambda = \lambda_c(k)$.

We can also enter into the chaotic domain from the opposite direction, namely by increasing the parameter k starting from very small values, keeping the parameter λ fixed. The initial regular motion is then that of a free, periodically driven rotor. Increasing the parameter k from zero, all rotating orbits whose periods are rationally related to the driving period are destroyed and replaced by resonances. The threshold to global chaos is now passed at some critical value $k = k_c(\lambda)$ when the "primary" resonances at integer multiples of the driving frequency overlap.

We now give a more quantitative discussion following [41]. The standard manner to determine the chaos threshold is to expand the Hamiltonian into sums of stationary resonance terms.

Thus,

$$k\cos(\phi - \lambda\sin t) = \sum_{m=\infty}^{+\infty} kJ_m(\lambda)\cos(\phi - mt) \tag{6.8}$$

where the $J_m(\lambda)$ are the Bessel functions. The terms on the right hand side describe the primary resonances at

$$P = \dot{\phi} = m \tag{6.9}$$

of mutual distance $\delta = 1$ and width

$$\Delta P_m = 4\sqrt{k|J_m(\lambda)|}. \tag{6.10}$$

For $|m| > \lambda$ the widths of the resonances decay exponentially, for $|m| < \lambda$ and $\lambda >> 1$ the estimate $|J_m(\lambda)| \lesssim (\pi\lambda)^{1/2}$ can be used. Thus the analysis predicts infinitely many resonances (6.11), of which only those with $|m| \lesssim \lambda$ have a sizeable width. The onset of chaos can now be determined from "resonance overlap" by the "2/3-rule" [41]

$$S = \frac{\Delta P_m}{\delta} > 0.63 \tag{6.11}$$

which yields the condition

$$k > \frac{\sqrt{\pi\lambda}}{40}. \tag{6.12}$$

Another useful way [41] to visualize the effect of the driving field on the undriven resonance around $\dot{\phi} = P = 0$ is to determine the moving center of the driven resonance from the condition

$$\varphi = \frac{d}{dt}(\phi - \lambda\sin t) = 0 \tag{6.13}$$

which yields

$$\dot{\phi} = P = P_R(\tau) = \lambda\cos t. \tag{6.14}$$

Thus, the resonance is no longer a stationary fixed region around $P = 0$, but it is a moving region around a time-dependent value of P oscillating at the driving frequency with the amplitude λ. The width of the moving resonance remains practically time-independent and is given by $\Delta P = 4\sqrt{k}$. Thus, practically every point in the domain $|P| < \lambda$ is crossed by the resonance twice during each period. It is useful to distinguish the limiting cases of fast crossing, $\lambda >> k$, where the resonance moves much faster than the system points it crosses, and the opposite limit of slow crossing, $\lambda >> k$. We shall here consider only the case of the fast crossing.

As long as a system point ϕ, P with $|P| < \lambda$ is outside the moving resonance it behaves effectively like a free rotor. When the resonance crossing occurs this rotor has some effectively random phase $\varphi_r = \phi_r - \lambda \sin t$, which cannot change much during the fast crossing. (At the time of crossing $\dot{\varphi}_r = 0$, by definition). However, the action P changes during each crossing by approximately [41]

$$\Delta P \simeq -\sqrt{2\pi}\frac{k}{\sqrt{\lambda}}\sin(\varphi_r \pm \pi/4). \tag{6.15}$$

As there are two crossings per period $T = 2\pi$ with approximately uncorrelated values of φ_r this gives a diffusion constant

$$\mathcal{D} \simeq \frac{\langle \Delta P \rangle}{T/2} = \frac{k^2}{\lambda}. \tag{6.16}$$

Thus, there will be a diffusive spreading of P over the domain $|P| < \lambda$ and we predict for the steady state

$$\langle \Delta P^2 \rangle \simeq \frac{1}{\sqrt{3}}\lambda. \tag{6.17}$$

b. Quantized system

When quantizing the pendulum a decision has to be made whether to represent the states in the Hilbert space of 2π-periodic functions of ϕ or on the Hilbert space of square integrable functions over the real line $-\infty < \phi < \infty$. In the first case we have

$$\begin{aligned} &-\pi < \phi \leq \pi \\ &P|l\rangle = \hbar l|l\rangle \quad l \text{ integer}, -\infty < l < +\infty \end{aligned} \tag{6.18}$$

in the second

$$-\infty < \phi < +\infty$$

$$P|l + x\rangle = \hbar(l + \kappa)|l + \kappa\rangle \quad l \text{ integer}, -\infty < l < +\infty \tag{6.19}$$

$$\kappa \text{ real}, \quad 0 \leq \kappa < 1.$$

Thus, let us adopt the second possibility for the moment. Then the fractional part of the eigenvalues $l+x$ of P, the quasi-momentum x, remains conserved,

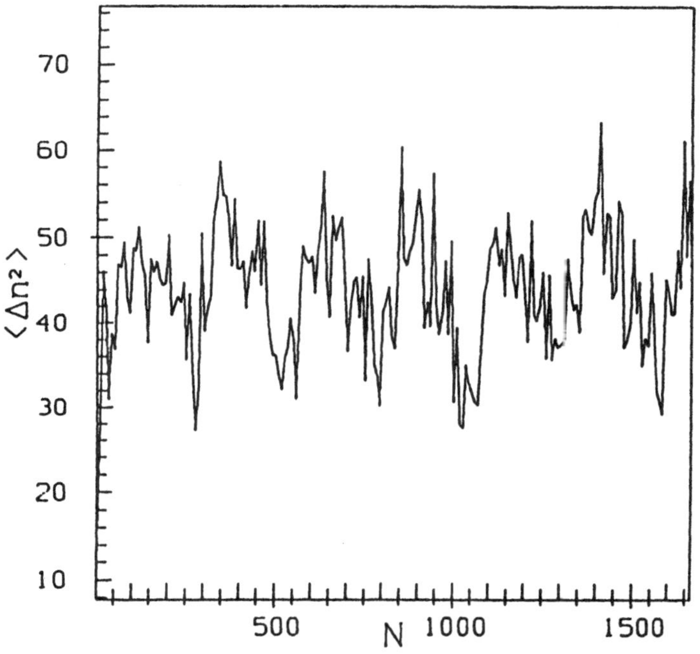

Fig. 13 Mean square of the number of occupied levels of the cosine potential vs. the number N of cycles of the external field for $\lambda = 85.0$, $k = 15.0$, and $\hbar = 1.58$ (from [17]).

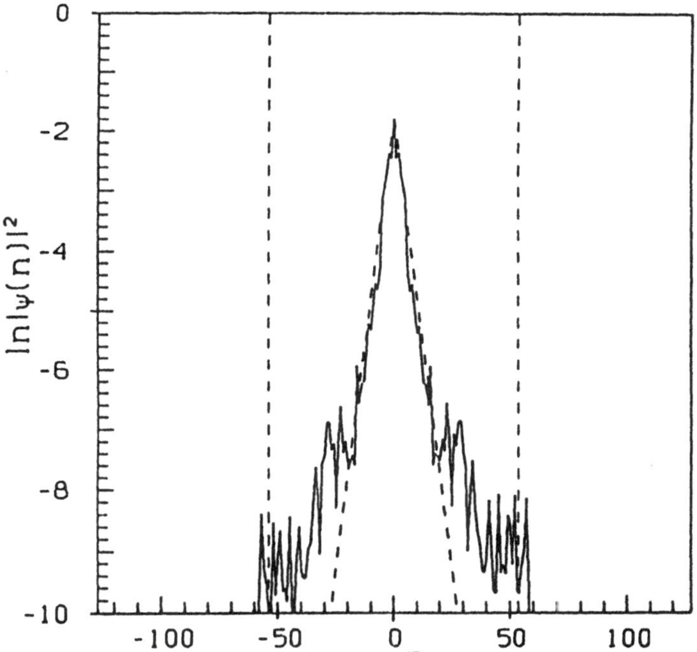

Fig. 14 Logarithm of the time-averaged occupation probability corresponding to Fig. 13. Dashed lines give the border $|n| = \lambda/\hbar$ of the classical chaotic domain and the exponential falloff with the localization length l_D (from [17]).

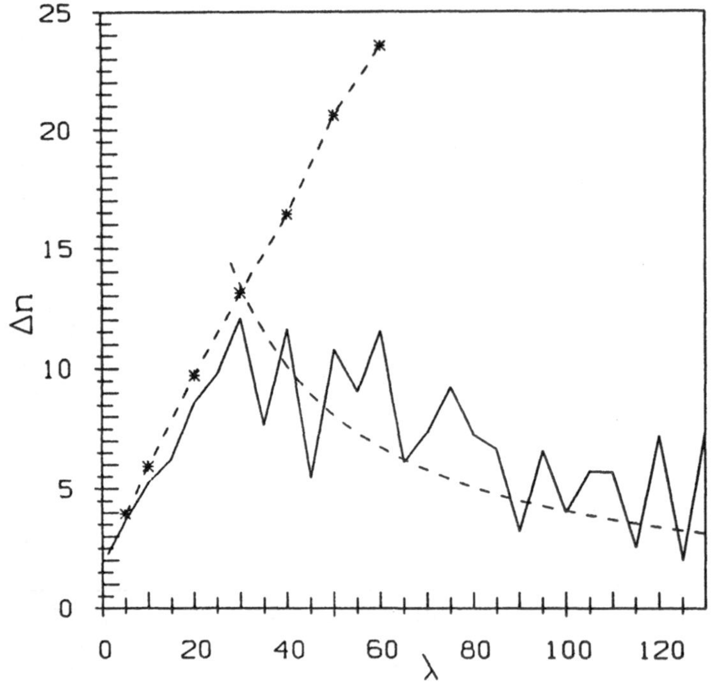

Fig. 15 Root mean square of the number of occupied levels versus the normalized amplitude λ of the driving field for the same values of the parameters k, $\hbar$ as in Fig. 13. Classical results, indicated by $*$'s, are joined by a dashed line. Another dashed line gives the analytical result for the quantum regime (from [17]).

due to the 2π-periodicity of the Hamiltonians H_2 and H_3 in ϕ or φ. Indeed, by the Bloch theorem the solution of the Schrödinger equation

$$i\hbar\dot{\psi}(\varphi,t)=\left\{\frac{1}{2}\Big(-i\hbar\frac{\partial}{\partial\varphi}-\lambda\cos t\Big)^2-k\cos\varphi\right\}\psi(\varphi,t) \tag{6.20}$$

can be represented as

$$\psi(\varphi,t)=\int_0^1 d\kappa\; e^{i\kappa\varphi}u_\kappa(\varphi,t) \tag{6.21}$$

with $u_\kappa(\varphi,t)=u_\kappa(\varphi+2\pi,t)$ 2π-periodic and satisfying

$$i\hbar u_\kappa(\varphi,t)=\left\{\frac{1}{2}\Big(-i\hbar\frac{\partial}{\partial\varphi}+\hbar\kappa-\lambda\cos t\Big)^2-k\cos\varphi\right\}u_\kappa(\varphi,t) \tag{6.22}$$

which describes the quantum dynamics in the Hilbert space of 2π-periodic functions, however, at the cost of modifying the kinetic energy (in a way which breaks the parity symmetry $\varphi\to-\varphi$, $t\to t+\pi$ but leaves the time reversal symmetry $i\to-i$, $t\to-t$, $\varphi\to-\varphi$ unbroken). Thus, the system behaves like an ensemble of independent pendula with a given distribution of κ-values. The initial condition is fixed by

$$u_\kappa(\varphi,0)=\sum_{m=-\infty}^{+\infty} e^{-i\kappa(\varphi+2\pi m)}\psi(\varphi+2\pi m,0) \tag{6.23}$$

and expectation values of 2π-periodic observables Ω can be calculated as ensemble averages

$$\langle\psi(t)|\Omega|\psi(t)\rangle=\mathrm{Tr}\varrho(t)\Omega \tag{6.24}$$

with

$$\varrho(t)=\int_0^1 d\kappa\;|u_\kappa(t)\rangle\langle u_\kappa(t)|. \tag{6.25}$$

Classically, the shift of P by $\hbar\kappa$ has no significant influence on the diffusion of P. Quantum mechanically, the diffusion will be replaced by localization, the average localization length being given by

$$\xi=\alpha\mathcal{D}/\hbar^2=\alpha k^2/(\lambda\hbar^2) \tag{6.26}$$

with $\alpha\simeq0.5$. As the κ-term does not break time-reversal symmetry there is no reason to expect a dependence of α on κ, i.e. all pendula of the ensemble will have the same average localization length (apart from fluctuations), which we could also have calculated using the first quantization scheme (6.18). The results of a numerical study of dynamical localization in the pendulum (using the space of 2π-periodic functions) is shown in figs. 13,

14, 15, where we present the time-evolution of $\langle \Delta l^2 \rangle$, the time-averaged occupation probability of the $|l\rangle$ states, and the root mean square of the number of excited $|l\rangle$ states as a function of the driving amplitude λ. These figures clearly show the localization effect. Fig. 15 shows the cross-over from the nonlocalized domain for small λ, where the localization length ξ is larger than the classically accessible chaotic domain $|P|/\hbar < \lambda/\hbar$, to the domain where the fluctuations of P are limited by localization.

c. Coupling to the environment

The coupling of the pendulum to the environment can be of two types: (i) The coupling preserves the 2π-periodicity of H_2 or H_3 in ϕ or φ. Then there is no way to distinguish between ϕ and $\phi + 2\pi$ and the quasi-momentum κ remains conserved (or the first quantization scheme (6.18) is used). (ii) The coupling destroys the 2π-periodicity of H_2 or H_3 in ϕ or φ. Then the second quantization scheme (6.19) has to be used and the quasi-momentum κ is no longer preserved. This effect occurs e.g. for quasi-particle tunnelling in Josephson junctions. We shall here only discuss the first possibility. For a discussion of the effects of different couplings of noise to the kicked rotor in the two different quantization schemes see [30]. For the coupling to the environment we use the Hamiltonian H_B of (2.2) and minimal coupling in the form

$$H_S + H_{\text{int}} = \frac{1}{2}\left(P - \sum_{k\lambda} \frac{g_{k\lambda}}{\omega_k}(b_{k\lambda}^{+} + b_{k\lambda})\right)^2 - k\cos(\phi - \lambda \sin t). \tag{6.27}$$

We employ the Langevin method of section 2c to eliminate the environmental degrees of freedom resulting in the generalized Heisenberg-Langevin equation of motion

$$\ddot{\phi} + \int_{-\infty}^{t} d\tau\, \gamma(t-\tau)\dot{\phi} + k\sin(\phi - \lambda\sin t) = \xi(t) \tag{6.28}$$

with

$$\begin{aligned}
&\gamma(t) = \sum_{k\lambda} \frac{g_{k\lambda}^2}{2\omega_k}\cos\omega_k t \\
&[\xi(t), \xi(t')] = i\hbar\dot{\gamma}(t-t') \\
&\frac{1}{2}\langle \xi(t)\xi(t') + \xi(t')\xi(t)\rangle = \int_0^\infty \frac{d\omega}{2\pi}\hat{\gamma}(\omega)\hbar\omega \coth\frac{\beta\hbar\omega}{2}\cos\omega(t-t').
\end{aligned} \tag{6.29}$$

Constant friction ("Ohmic dissipation") results if (2.33) is satisfied, i.e. if

$$\gamma = 2\pi\mathcal{D}(\omega)\frac{g^2(\omega)}{\omega} \tag{6.30}$$

is independent of frequency, where $\mathcal{D}(\omega)$ is the spectral density of states of the oscillators and $g(\omega_k) = g_{k\lambda}$ is the frequency dependent coupling constant.

Due to the presence of the dissipation the quasi-energy states acquire a finite life-time. For "Ohmic dissipation" it is given by

$$n_c = (\gamma\xi)^{-1} \tag{6.31}$$

where the factor ξ applies for a state localized near $P = 0$ whose typical P-scale is given by $\hbar\xi$. Thus, dynamical localization and the dissipation-induced subsequent diffusion may be observed in the present system if

$$n_c = (\gamma\xi)^{-1} < 2\pi n^* = 2\pi\xi \tag{6.32}$$

which is the "weak dissipation" condition (3.16) for the present case.

d. Experimental realization by the deflection of an atomic beam in a modulated standing light wave

We shall consider the deflection of atomic De Broglie waves by a standing light wave, the dual process of the more familiar diffraction of a light wave by a matter grating. We shall argue that this process is a quantum mechanical realization of the pendulum. In order to realize the periodically driven pendulum, in which dynamical localization occurs, the optical path-length of the light-wave to and from the mirror must be periodically modulated, e.g. by oscillations of the mirror position, or by mudulating the refractive index. We now describe the system following ref. [16].

We consider a beam of two-level atoms (states $|g\rangle$, $|e\rangle$ with energy difference $\hbar\omega_0$, dipole momentum d), which are initially in their ground state $|g\rangle$. They move in the z-direction with kinetic energy E_0 and then pass through a single classical standing-wave light field $\vec{\mathcal{E}}(x,t) = \vec{e}_y(\mathcal{E}_0 \cos k_L x \; e^{-i\omega_L t} + c.c.)$. We assume that the x-coordinate of the mirror, reflecting the incoming travelling light wave - and thereby determining the position of the nodes of the standing light wave - oscillates around its average with $\Delta L \sin\omega t$, so that the nodes are harmonically oscillating in the same way, i.e. $\vec{\mathcal{E}}(x,t)$ passes into $\vec{\mathcal{E}}(x - \Delta L \sin\omega t, t)$. This can be achieved e.g. by appropriately driving a piezoelectrical crystal. The dipole and rotating wave approximation then yield the Hamiltonian

$$H = \frac{p^2}{2M} + \hbar\omega_0|e\rangle\langle e| - (d\mathcal{E}_0 \cos(k_L(x - \Delta L \sin\omega t))e^{-i\omega_L t}\sigma_+ + h.c.) \tag{6.33}$$

where p is the center-of-mass momentum of the atoms (with mass M) and $\sigma_\pm$ are Pauli spin-operators. In a reference frame moving with $v = \sqrt{2E_0/M}$ in the z-direction there remains only the transverse atomic center-of-mass motion in the x-direction and we can represent the atomic state as $\psi_g(x,t)|g\rangle + \psi_e(x,t)e^{-i\omega_L t}|e\rangle$ with equations of motion

$$i\hbar\frac{\partial\psi_g}{\partial t} = -\frac{\hbar^2}{2M}\frac{\partial^2\psi_g}{\partial x^2} - \frac{\hbar\Omega}{2}\cos(k_L(x-\Delta L\sin\omega t))\psi_e \tag{6.34}$$

$$i\hbar\frac{\partial\psi_e}{\partial t} = -\frac{\hbar^2}{2M}\frac{\partial^2\psi_e}{\partial x^2} + \hbar\delta_L\psi_e - \frac{\hbar\Omega}{2}\cos(k_L(x-\Delta L\sin\omega t))\psi_g \tag{6.35}$$

Here we neglect spontanous emission from the upper atomic level, which is justified for sufficiently high detuning $\delta_L = \omega_0 - \omega_L$; $\Omega/2 = d\mathcal{E}_0/\hbar$ is the Rabi frequency. Adiabatic elimination of the excited state amplitude with the assumption that the detuning δ_L is large compared to the Rabi frequency Ω and the excited state kinetic energy term leads to

$$i\hbar\frac{\partial\psi_g}{\partial t} = -\frac{\hbar^2}{2M}\frac{\partial^2\psi_g}{\partial x^2} - \frac{\hbar\Omega_{\text{eff}}}{4}\cos^2(k_L(x-\Delta L\sin\omega t))\psi_g \tag{6.36}$$

where $\Omega_{\text{eff}} = \Omega^2/\delta_L$ is the effective Rabi frequency. So the dynamic of the atoms in the ground state - with an energy shift of $\hbar\Omega_{\text{eff}}/8$ - is governed by the Hamiltonian

$$H = \frac{p_x^2}{2M} - \frac{\hbar\Omega_{\text{eff}}}{8}\cos(2k_L(x-\Delta L\sin\omega t)) \tag{6.37}$$

Since the probability to find an atom in the excited state is negligible in our case, the properties of the atoms are completely determined by the ground state amplitude. After a rescaling $t' = \omega t$, $\varphi = 2k_L x$, $P = [2k_L/(M\omega)]p_x$, $H' = [4k_L^2/(M\omega^2)]H$ we get (with omission of the dashes) the dimensionless Hamiltonian (6.4) with the (classical) parameters $k = \varepsilon_r\Omega_{\text{eff}}/\omega^2$ and $\lambda = 2k_L\Delta L$, where $\varepsilon_r = \hbar k_L^2/(2M)$ is the recoilshift. The quantized system in addition contains the rescaled Planck constant $\hbar\!\!\!k = 8\varepsilon_r/\omega$ via the commutator $[P,\varphi] = -i\hbar\!\!\!k$. Thus, the analysis for the preceding subsections can be applied directly. For our discussion of dynamical localization below we shall require a large chaotic domain $\lambda \gg 1$, i.e. $\Delta L \gg 1/(2k_L)$. In the classical domain, by applying (6.17), we predict momentum fluctuations with root mean square $\sqrt{\langle p_x^2\rangle} \simeq M\omega\Delta L/\sqrt{3}$ due to diffusion of P_x.

In the quantum system an initial state, localized near $n = 0$ and given by a linear superposition of about ξ Floquet states, first spreads by classical diffusion and then develops into an exponentially localized distribution $|\psi| \sim \exp(-|n|/\xi_D)$ with a localization length $\xi_D \simeq 2\xi$. Thus fluctuations of the transverse momentum are quantum mechanically reduced to $\sqrt{\langle p^2\rangle} \simeq \hbar\!\!\!k\ \xi_D/\sqrt{2}$ or $\sqrt{\langle p_x^2\rangle} \simeq [\sqrt{2}\pi\hbar\Omega_{\text{eff}}{}^2/(64\omega^2)]/\Delta L$. For fixed external frequency, they decrease inversely proportional to ΔL, contrary to the classical case where they increase proportionally to ΔL. This finding provides us with a clear signature of the effect which should be observable, if the classical restriction of the fluctuations by the width of the chaotic domain ($\simeq \lambda/\sqrt{3}$) is larger than the quantum restriction due to dynamical localization ($\simeq \hbar\!\!\!k\, l_D/\sqrt{2}$), i.e. $\Delta L > [\sqrt{\sqrt{6}\pi\hbar/(64M\omega^3)}]\Omega_{\text{eff}}$. The experimental conditions under which the effect should be observable can now be summarized. E.g. for Ytterbium atoms, optically pumped to a two-state system (atomic frequency $\omega_0/(2\pi) \simeq 5.40 \cdot 10^{14}$Hz) and passing orthogonally a modulated standing light wave with detuning $\delta_L/(2\pi) \simeq 4.0$GHz (wavenumber $k_L \simeq 1.13 \cdot 10^7\text{m}^{-1}$), driving frequency of the mirror $\omega/(2\pi) \simeq 125$kHz, and Rabi frequency $\Omega/(2\pi) \simeq 140$MHz, we have $k \simeq 1.2$, $\hbar\!\!\!k \simeq 0.24$, and $\lambda \simeq 2.26 \cdot 10^7\text{m}^{-1}\Delta L$. In order to observe the classical to quantum cross-over displayed in fig 15 the amplitude of the mirror oscillations should then be varied in the range $0.1 \ldots 5.0\mu$m. The cross-over occurs at about 0.3μm. The localization needs about $l_D/2$ periods of the mirror oscillations to establish itself (cf. fig. 13). The interaction time of the atoms with the standing-wave light field, t_{int}, therefore has to be large compared to $l_D\pi/\omega$. In this example $l_D \simeq 23$ at the classical-quantum cross-over, which amounts to $t_{int} \gg 90\mu$s.

The predicted effect rests entirely on coherence, therefore, as we have discussed, dissipation and noise have to be kept sufficiently low. In the present example, the special form of the coupling to the environment discussed in the preceding section does not apply. Instead there could be noise in the driving laser field, whose linewidth must be small compared to the inverse interaction time. In addition spontaneous decay of the upper atomic level must be suppressed, which is achieved by detuning sufficiently far from resonance. With the above given parameters and a spontaneous decay rate of $\gamma/(2\pi) \simeq 183$kHz, the number of spontaneous decays of an atom during an interaction time $t_{int} \simeq 300\mu$s (which is attained for example by Ytterbium atoms of kinetic energy $E_0 \simeq 1.4 \cdot 10^{-23}$J passing an interaction region of about 3mm) is $N = (\Omega/2)^2\gamma t_{int}/((2\delta_L)^2 + \gamma^2) \simeq 0.03$ and therefore negligible. The measurement of the transverse atomic momentum can be realized without back action on the system.

The experiment has so far not been performed. If it could be done, it would be a beautiful achievement, both from the theoretical and the experimental point of view.

e. Dynamical localization in Josephson junctions

Another realization of the quantum mechanical pendulum is furnished by a Josephson junction, under appropriate conditions (see e.g. [42]). A periodically driven pendulum is obtained by impressing an external periodically varying current I_{ex} through the junction

$$I_{ex} = I_J \sin\varphi + c\dot{v}. \tag{6.38}$$

Here φ is the phase-difference of the order parameter of the superconductors on both sides of the junction, C the capacitance of the junction, V the voltage difference, $I_J \sin\varphi$ the current due to tunnelling of Cooper pairs. Josephson relation $\varphi = \frac{2e}{\hbar}V$ (e = elementary charge) closes eq. (6.38). With the choice

$$I_{ex}(t) = I_0 \sin\Omega t \tag{6.39}$$

and the definitions

$$\begin{gathered}
\phi = \varphi + \lambda \sin\Omega t \\
P = \frac{2e}{\hbar}V + \lambda\Omega\cos\Omega t \\
\lambda = \frac{2e}{\hbar C\Omega^2} I_0 \ , \ k = \frac{2e}{\hbar C\Omega^2} I_J
\end{gathered} \tag{6.40}$$

and rescaling the time $\Omega t \to t$, the Hamiltonian (6.4) can again be applied. When quantizing the system canonically, the dimensionless parameter

$$\not{k} = \frac{(2e)^2}{\hbar C\Omega} \tag{6.41}$$

plays the role of Planck's constant, in the present units. The eigenvalues of the operator P are given by $\not{k}\, l$. As P is related to the voltage V by eq. (6.41), and the total charge Q on the capacitance of the junction is $Q = CV$, it can be seen that in an eigenstate of P we have in physical units

$$Q = 2el - (I_0/\Omega)\cos\Omega t. \tag{6.42}$$

Thus, the quantization of P quantizes the number of Cooper pairs on the capacitance, while the total charge Q on the capacitance is not quantized.

If all dissipative effects are negligible, we expect again dynamical localization to appear in this system[17]. E.g. for a capacitance of $C = 10^{-13}F$, the Josephson current $I_J = 10^{-7}$ Amp, a driving frequency $\Omega/2\pi = 10^{10}$ Hz we have $k = 0.8$, $\lambda = 0.8 I_0/I_J$, $\not{k} = 0.16$. If one varies I_0 in the range of 0.1. to 10μ Amp one would expect to see the classical to quantum cross-over at $I_0 \simeq 0.7\mu$ AMp at a localization length of $\xi_D \simeq 28$. However, a consideration of the coupling to the environment is essential in this system. First of all, there is a coupling of the Cooper pairs to the unpaired quasi-particle states, leading to quasi-particle tunnelling, i.e. a finite shunt resistance. In order to avoid this unwanted effect it is necessary to freeze out the excitation of quasi-particles by making the temperature much lower than the energy gap of the superconductor. A different source of dissipation comes from imperfections of the current source, and the impedance of the leads. A finite dissipation rate is also due to the necessity to extract energy from the junction at a non-zero rate in order to measure the voltage fluctuations V. This effect could be minimized by taking advantage of the fact that current-driven junctions in series oscillate independently of each other (see e.g. [42]), unless they are coupled in special way, e.g. by being placed in a common cavity or wave-guide. Thus, an array of N independent junctions in series would allow us to reduce the required rate of extracting energy from each individual junction by a factor of $1/N$. Again, the epxeriment proposed here has not yet been done. There remains therefore the challenge to increase the short list of systems (so far consisting of Rydberg atoms in microwave fields only) for which dynamical localization is observed in the laboratory.

7 Acknowledgement:

This work was supported by the Deutsche Forschungsgemeinschaft through the Sonderforschungsbereich 237 "Unordnung und große Fluktuationen". I wish to thank the organization committee of the 1992 Chris Engelbrecht Summer School in Theoretical Physics, in particular Professor Dieter Heiss, for the invitation to present these lectures and for their hospitality. I would also like to thank all participants of this stimulating school for pertinent questions and remarks, which greatly improved my understanding of the subject.

References

[1] P. W. Anderson, *Rev. Mod. Phys.* **50**:191 (1978); P. A. Lee, T. V. Ramakrishnan, *Rev. Mod. Phys.* **57**:287 (1985).

[2] A. J. Lichtenberg, M. A. Lieberman, "Regular and Stochastic Motion", (Berlin, Springer, 1983).

[3] G. Casati, B. V. Chirikov, D. L. Shepelyansky, I. Guarneri, *Phys. Rep.* **154**: 77 (1987).

[4] B. V. Chirikov, *Phys. Rep.* **52**:263 (1979).

[5] J. M. Greene, *J. Math. Phys.* **20**:1183 (1981).

[6] A. B. Rechester, R. B. White, *Phys. Rev. Lett.* **44**:1586 (1980); A. B. Rechester, M. N. Rosenbluth; R. B. White, *Phys. Rev.* **A23**:2664 (1981).

[7] T. Dittrich, R. Graham, *Naturwissenschaften* **76**:401 (1989).

[8] G. Casati, B. V. Chirikov, F. M. Izrailev, J. Ford, in Stochastic Behavior in Classical and Quantum Hamiltonian Systems, *Lecture Notes in Physics* **Vol. 93**, ed. G. Casati and J. Ford (Springer, Berlin 1979).

[9] B. V. Chririkov, F. M. Izrailev, D. L. Shepelyansky, *Sov. Sci. Rev.* **C2**:209 (1981).

[10] T. Hogg, B. A. Huberman, *Phys. Rev. Lett.* **48**:711 (1982); *Phys. Rev.* **A28**:22 (1983).

[11] S. Fishman, D. R. Grempel, R. E. Prange, *Phys. Rev. Lett.* **49**:509 (1982); D. R. Grempel, R. E. Prange, S. Fishman, *Phys. Rev.* **A29**:1639 (1984).

[12] D. L. Shepelyansky, *Physica* **8D**:208 (1983).

[13] G. Casati, I. Guarneri, D. L. Shepelyansky, *IEEE Journal of Quant. Electr.* **QE24**: 1420 (1988).

[14] R. Graham, M. Höhnerbach, *Phys. Rev. Lett.* **64**: 637 (1990) *Phys. Rev.* **A43**: 3966 (1991) *ibid* **A45** (1992).

[15] P. A. Dando, D. Richards, *J. Phys.* **B23**: 3179 (1990).

[16] R. Graham, M. Schlautmann, P. Zoller, *Phys. Rev.* **A45**:R19 (1992)

[17] R. Graham, M. Schlautmann, D. L. Shepelyansky, *Phys. Rev. Lett.* **67**: 255 (1991).

[18] R. Graham, J. Keymer, *Phys. Rev.* **A44**:6281 (1991).

[19] J. R. Kuklinski, *Phys. Rev. Lett.* **64**: 2507 (1990).

[20] V. Weisskopf, E. Wigner, *Z. Phys.* **63**:54 (1930).

[21] G. W. Ford, M. Kac, P. Mazur, *J. Math. Phys.* **6**: 504 (1965).

[22] G. W. Ford, J. T. Lewis, R. F. O'Connell, *Phys. Rev.* **A37**: 4419 (1988).

[23] R. K. Wangsness, F. Bloch, *Phys. Rev.* **89**: 728 (1953).

[24] C. W. Gardiner, "Quantum Noise" (Berlin, Springer (1992).

[25] R. P. Feynman, F. L. Vernon, *Ann. Phys. (N.Y.)* **24**: 118 (1963).

[26] H. Grabert, P. Schramm, G.-L. Ingold, *Phys. Rep.* **168**: 115 (1988).

[27] A. H. Castro Neto, A. O. Caldeira, Phys. Rev. **A42**:6884 (1990).

[28] T. Dittrich, R. Graham, *Z. Phys.* **B62**: 515 (1986); *Europhys. Lett.* **4**: 263 (1987); *ibid* **7**: 287 (1988); *Ann. Phys. (N.Y.)* **200**: 363 (1990).

[29] S. Adachi, M. Toda, K. Ikeda, *Phys. Rev. Lett.* **61**: 655/(1988).

[30] D. Cohen, *Phys. Rev.* **A43**: 639 (1991); *ibid* **A44**: 2292 (1991).

[31] E. Ott, T. M. Antonsen Jr., J. D. Hanson, *Phys. Rev. Lett.* **23**: 2187 (1984).

[32] R. Blümel, R. Graham, L. Sirko, U. Smilansky, H. Walther, K. Yamada, *Phys. Rev. Lett.* **62**: 341 (1989).

[33] R. Blümel, A. Buchleitner, R. Graham, L. Sirko, U. Smilansky, H. Walther, *Phys. Rev.* **A44**: 4521 (1990).

[34] R. Graham, in Proceedings of the International School of Physics "Enrico Fermi" 1991, "Quantum Chaos", ed. G. Casati, I. Guarneri, U. Smilansky, (North Holland, Amsterdam 1992).

[35] B. R. Mollow, *Phys. Rev.* **A12**:1919 (1975).

[36] M. D. Srinivas, E. B. Davies, *Opt. Acta* **28**:981 (1981).

[37] P. Zoller, M. Marte, D. F. Walls, *Phys. Rev.* **A35**:198 (1987).

[38] R. Graham, M. Höhnerbach, *Phys. Lett.* **101A**:61 (1984); *Acta Phys. Austr.* **50**:45 (1984); *Z. Physik* **B57**:233 (1984); in "Quantum Measurement and Chaos", ed. E. R. Pike and S. Sarkar, (Plenum, New York 1987).

[39] T. Dittrich, R. Graham, *Phys. Rev.* **A42**:4647 (1990).

[40] R. Graham, *Europhys. Lett.* **3**:259 (1987).

[41] B. V. Chirikov, D, L. Shepelyansky, *Sov. Phys. Tech. Phys.* **27**:156 (1982).

[42] K. K. Likharev, "Dynamics of Josephson Junctions and Circuits" (Gordon and Breanch, New York 1986).